Disinfection of Wastewater and Water for Reuse

Geo. Clifford White
Consulting Engineer

Van Nostrand Reinhold Environmental Engineering Series

VNR VAN NOSTRAND REINHOLD COMPANY
NEW YORK   CINCINNATI   ATLANTA   DALLAS   SAN FRANCISCO
LONDON   TORONTO   MELBOURNE

TD
429
.W47

Van Nostrand Reinhold Company Regional Offices:
New York Cincinnati Atlanta Dallas San Francisco

Van Nostrand Reinhold Company International Offices:
London Toronto Melbourne

Published by Van Nostrand Reinhold Company
135 West 50th Street, New York, N.Y. 10020

Published simultaneously in Canada by Van Nostrand Reinhold Ltd.

15 14 13 12 11 10 9 8 7 6 5 4 3 2 1

Library of Congress Cataloging in Publication Data

White, Geo. Clifford.
  Disinfection of wastewater and water for reuse.

  (Van Nostrand Reinhold environmental engineering
series)
  Includes index.
  1. Water reuse. 2. Sewage—Purification.
3. Water—Purification. I. Title.
TD429.W47   628'.32   78-18457
ISBN 0-442-29405-0

For my wife Charlotte

Van Nostrand Reinhold Environmental Engineering Series

THE VAN NOSTRAND REINHOLD ENVIRONMENTAL ENGINEERING SERIES is dedicated to the presentation of current and vital information relative to the engineering aspects of controlling man's physical environment. Systems and subsystems available to exercise control of both the indoor and outdoor environment continue to become more sophisticated and to involve a number of engineering disciplines. The aim of the series is to provide books which, though often concerned with the life cycle—design, installation, and operation and maintenance—of a specific system or subsystem, are complementary when viewed in their relationship to the total environment.

The Van Nostrand Reinhold Environmental Engineering Series includes books concerned with the engineering of mechanical systems designed (1) to control the environment within structures, including those in which manufacturing processes are carried out, and (2) to control the exterior environment through control of waste products expelled by inhabitants of structures and from manufacturing processes. The series includes books on heating, air conditioning and ventilation, control of air and water pollution, control of the acoustic environment, sanitary engineering and waste disposal, illumination, and piping systems for transporting media of all kinds.

Van Nostrand Reinhold Environmental Engineering Series

ADVANCED WASTEWATER TREATMENT, by Russell L. Culp and Gordon L. Culp

ARCHITECTURAL INTERIOR SYSTEMS—Lighting, Air Conditioning, Acoustics, John E. Flynn and Arthur W. Segil

SOLID WASTE MANAGEMENT, by D. Joseph Hagerty, Joseph L. Pavoni and John E. Heer, Jr.

THERMAL INSULATION, by John F. Malloy

AIR POLLUTION AND INDUSTRY, edited by Richard D. Ross

INDUSTRIAL WASTE DISPOSAL, edited by Richard D. Ross

MICROBIAL CONTAMINATION CONTROL FACILITIES, by Robert S. Rurkle and G. Briggs Phillips

SOUND, NOISE, AND VIBRATION CONTROL (Second Edition), by Lyle F. Yerges

NEW CONCEPTS IN WATER PURIFICATION, by Gordon L. Culp and Russell L. Culp

HANDBOOK OF SOLID WASTE DISPOSAL: MATERIALS AND ENERGY RECOVERY, by Joseph L. Pavoni, John E. Heer, Jr., and D. Joseph Hagerty

ENVIRONMENTAL ASSESSMENTS AND STATEMENTS, by John E. Heer, Jr. and D. Joseph Hagerty

ENVIRONMENTAL IMPACT ANALYSIS: A New Dimension in Decision Making, by R. K. Jain, L. V. Urban and G. S. Stacey

CONTROL SYSTEMS FOR HEATING, VENTILATING, AND AIR CONDITIONING (Second Edition), by Roger W. Haines

WATER QUALITY MANAGEMENT PLANNING, edited by Joseph L. Pavoni

HANDBOOK OF ADVANCED WASTEWATER TREATMENT (Second Edition), by Russell L. Culp, George Mack Wesner and Gordon L. Culp

Preface

This book has been prepared in light of the tremendous amount of interest generated over the last five years in the field of wastewater disinfection. It has prompted me to update Chapter 7 of the *Handbook of Chlorination.*

As a special consultant, I am required to keep abreast of current techniques in the field of disinfection. This includes potable water and wastewater. It also includes the responsibility for knowing the capabilities of disinfecting processes, whether chlorination-dechlorination, chlorine dioxide, bromine, bromine chloride, ozone, ultraviolet radiation, gamma radiation, iodine, ultrasonics, or on-site generation of disinfecting compounds. This book represents the latest information I have been able to gather on the various approaches to wastewater disinfection.

To do this has been a formidable task. Since 1972, my library of some 35 three-ring binders has grown to over 120. These binders contain papers on the subject of disinfection, personal notes and reports, correspondence, and manufacturers' literature covering the handling and use of various disinfectants.

Reading, collating, and filing this material is a tedious but rewarding task. My efforts to keep pace with what is happening have provided me and my wife, Charlotte, with some fascinating trips to foreign lands and new friends around the world.

When I finished the manuscript for *Handbook of Chlorination* in 1971, the idea of a widespread need for dechlorination was not taken seriously. Now, however, it is quite common to refer to the chlorination process in wastewater disinfection as the "chlorination-dechlorination method." I have used this terminology in this book and have likewise included the latest current practices and design of this system.

Another important factor that influenced the contents of this text was the recent discovery of the formation of objectionable compounds in waters and wastewaters treated with chlorine compounds. This aroused the profession to search for "alternatives" to chlorination. Therefore, I have made a strenuous and honest effort to give the reader a fair and practical evaluation of other methods of treatment.

In addition, I have addressed the possible alternatives in the chlorination system where safety in handling is of prime consideration. In Chapter 5, "Hypo-chlorination," the reader will discover this subject updated to include current practices using imported hypochlorite and on-site generation.

The reader will also find the latest methods, particularly those of foreign origin, on the use of chlorine dioxide and bromine. Of equal importance is the updating in the current practices of the use of ozone in wastewater treatment.

I hope of course that the material in this book will prove to be of use to designers, operators, and students, and to be a helpful guide for the regulatory agencies. Since it is a physical impossibility for me to serve all the consultants and operators who have questions on this subject, it is my fervent hope that this text may be an acceptable substitute for my personal services.

I am most appreciative of all the continuing help I have received from my many colleagues—without whom I could not have completed this book. I hope this work will please them. I would like to acknowledge these people here—the list is long—but I would be sure to omit someone, which would be a grave offense. At any rate, you all know who you are—*my many thanks to all of you.*

Geo. Clifford White
San Francisco

Contents

1

Concepts of wastewater disinfection

HISTORICAL BACKGROUND

The history of wastewater disinfection is essentially the history of chlorination, and the use of chlorine has generally been associated with the search for means to control disease in man.

Chlorine is never found in its free state in nature. It appears in nature in the form of chlorides (Cl). It was discovered by the Swedish chemist Scheele in 1774; who, however, was unaware of its elemental nature. Davy, in 1810 definitely proved it to be an element and named it chlorine after the Greek word for green.

Chlorine was first used as a disinfectant about 1800 by de Morveau, in France, and by Cruikshank, in England.[1] The earliest recorded practice of sewage chlorination on a large scale was in 1854 when the Royal Sewage Commission used chloride of lime to deodorize London sewage. In 1879 William Soper of England used chlorinated lime to treat the feces of typhoid patients before disposal into a sewer. The first use of chlorine on a plant scale for disinfection was made at the wastewater treatment plant in Hamburg, Germany in 1893.[1] This application was the result of a disastrous waterborne typhoid epidemic there. The first recorded use for this purpose in the United States was at Brewster, New York in 1894.[1]

Brewster was a small village in the Croton drainage area of the New York City water supply. This installation is noteworthy because of the method of chlorine production. Chlorine was made at the site by electrolytic decomposition of a brine solution. The resulting chlorine solution was then applied to the Brewster sewage plant discharge. At the time this particular method of generating chlorine was known as the Woolf process. This system operated successfully for 17 yr. until it was destroyed by fire in 1911.

The following are milestone dates:

1879—Chlorinated lime was used for disinfection of sewage effluents in England and France.
1893—First known use of chlorine for the destruction of pathogenic bacteria in sewage effluents at Brewster, New York, and Hamburg, Germany.
1906—Studies by Phelps and Carpenter at MIT confirmed the disinfecting power of chlorinated lime reported by British and German investigators.
1909—Commercial development of liquid chlorine.
1913—Commercial development of first practical gas chlorinator.
1922—Development of first flow paced automatic gas chlorinator.
1928—High-test hypochlorites were developed.
1958—Twenty-two hundred wastewater plants serving 38 million people were equipped with chlorination facilities.
1961—First installation of a chlorine residual controlled disinfection system for wastewater effluent at Napa, California.

From the 1920's onward, liquid-gas chlorine has been the principal wastewater disinfectant.

1970—Wastewater disinfection by chlorination becomes predictable and controllable.

In 1970, a few cities where the treatment plants were located in highly congested areas, disinfection facilities were changed over to hypochlorite to avoid potential hazards of stored chlorine gas. Some of these have gone back to chlorine gas because of the significant monetary savings. By 1972 on-site chlorine gas generating equipment, similar to the Woolf process of 1893, began to appear on the market.

In 1975 chlorine was first used on a plant scale for nitrogen removal from wastewater.

Active interest in wastewater disinfection began in the United States about 1945. Up to that time the primary use of chlorine in sewage disposal systems was for odor control, hydrogen sulfide destruction, and prevention of septicity. Most of the sewage-treatment plants practicing disinfection during that time belonged to the United States Armed Forces. It was military policy during World War II that sewage effluents at all Army bases in the United States had to be chlorinated. Today as a result of the 1970 Federal Water Pollution Control Act, almost all wastewater-treatment plants are subjected to some disinfection requirement.

Chlorine also plays an important role in the treatment of cyanide wastes which are highly toxic. Cyanide wastes must be treated before being discharged to either a sewage collection system or a receiving water. When they are discharging to a sewer the cyanides need only be reduced to cyanates, but when going to a receiving water the cyanides must be completely destroyed to elemental carbon and nitrogen. The new Federal regulations for industrial discharges makes treatment of cyanide wastes imperative. The state of the art for on-site cyanide-waste treatment has developed to a point where packaged systems are readily available for the individual discharger.

Chlorine still plays an active role in controlling plant odors, and for preventing septicity wherever it may occur in the various unit processes.

IMPORTANCE OF DISINFECTION

Today the emphasis is on the disinfection of all wastewater effluents in the United States. *Sewage disinfection* is defined as the process of destroying pathogenic microorganisms in the wastewater stream by physical or chemical means. This is best accomplished by the use of chemical agents such as aqueous solutions of chlorine, chlorine dioxide, hypochlorite, bromine, bromine chloride, ozone, or combinations of these chemicals. Other means which have been used in special conditions and with varying degrees of success are ultra violet light radiation, gamma radiation, sonics, heat, and silver ions. Present disinfection practices depend almost exclusively on chlorine compounds.

From the viewpoint of health, the disinfection process is the most important stage of wastewater treatment. The objectives of wastewater disinfection are: to prevent the spread of disease, to protect potable water supplies, bathing beaches, receiving waters used for boating and water contact sports, and shellfish growing areas.

Public Health Agency Perspective

The public health agency is deeply committed to the theory of multiple barriers or multiple points of control between sewage discharge and water supply intake. These barriers or points of control include wastewater treatment, land confinement, dilution, time, distance, and potable water treatment. Any type of treatment is fallible, so reliance on natural barriers should be maintained as long as possible. Where the natural barriers are eroding, increased emphasis and reliability must be placed onto the artificial barriers of treatment processes. The factors which operate against and deteriorate the effectiveness of natural barriers are: increased population, increased mobility of population, increased recreation, more leisure time, increased sewage discharge, and increased water use. Inequities of rainfall distribution combined with increased water consumption produces increased water recycling of wastewater. This leads to an overall decrease in dilution, time, and distance factors between sewage discharges and potable water intakes.

All of these factors support the efforts of regulatory agencies to improve the quality of wastewater effluents prior to their discharge into the environment. For many areas of use this is the only protection available. Therefore, disinfection is the last remaining barrier against the transmission of waterborne diseases.

Contemporary Practices in the USA

Over the last 25–30 yrs. the various regulatory agencies, local, state, and now the EPA (Environmental Protection Agency) have been seeking a set of guide lines for proof of disinfection in various receiving water situations.

The most persistent and aggressive pursuit of wastewater disinfection requirements has been by the California State Department of Public Health, United States. The adequacy of disinfection is evaluated in compliance with a prescribed MPN (most probable number) of coliform organisms as determined by Standard Methods[2] for the "confirmed" test procedure. For example, in ocean and saline bay waters used for recreation, 80 percent of the receiving water samples must fall within a coliform MPN of 1000/100 ml. This is approximately equivalent to a median MPN of 230/100 ml.* For other situations, more restrictive median or average coliform concentrations may govern. The discharger is required to disinfect to the degree necessary to maintain a suitable quality in the receiving waters. Often the discharger is given the option of demonstrating compliance by meeting the receiving water quality in the effluent itself. This eliminates the costly process of monitoring several sampling stations in the receiving waters. In most situations the discharger usually is required to apply the disinfection requirements only to the effluent quality. This practice has evolved to a point in the State of California where practically all cases are based on the MPN total coliform in the plant effluent.

The evolution of these requirements is of particular interest because it relates largely to geographical considerations.

Other states of the union for various reasons have not been as concerned with the protection of receiving waters as has California. The predominant factor which led to the adoption of the aforementioned coliform requirements revolves

---

*The MPN of 230/100 ml is used when the statistical procedure utilizes a 5-tube dilution. For a 3-tube dilution the MPN would be 240/100 ml.

around the extensive development and use of the California coastline and coastal waters for recreational and shellfish growing purposes. The California coastline is some 720 miles in length. Along this coastline are some of the most beautiful bathing and water sports beaches in the world.

Evolution of Disinfection Requirements

It was the fouling of one of the most beautiful expanses of resort beach areas in California that captured the attention and energies of the California State Department of Public Health (see Fig. 1-1.). Beginning about 1920, the Bureau of Sanitary Engineering a division of this agency under the able direction of C. G. Gillespie, entered into a twenty-yr. battle with the City of Los Angeles to clean up their Hyperion outfall discharge which was solely responsible for the fouling of a ten-mile stretch of this magnificent beach. Obviously Los Angeles was the huge stumbling block in the overall California antipollution program envisioned by Mr. Gillespie and his colleagues.

In order to convince the City of Los Angeles that the Hyperion outfall discharge (170 mgd) was polluting this beach area (which currently attracts about one million people on summer weekends) the Bureau of Sanitary Engineering, California State Department of Health, made a year-long study of ten miles of beach in Santa Monica Bay in 1941-1942.[3]

The report of this investigation dated June 26, 1943 led to the immediate quarantine of this ten-mile stretch of beach, on the grounds that both the beach and surf waters were polluted with sewage and were therefore dangerous to one's health. After a length of time considered sufficient to correct this hazardous condition of gross pollution, the California State Board of Public Health took the City of Los Angeles to court on a suit based on pollution as determined by the "coliform count" in the bathing waters. Other factors were considered but it was the coliform count that became the most persuasive piece of evidence.

In their law suit against the city, the State Board of Health maintained that 1000 *Escherichia coli*/100 ml as a limiting standard in the surf waters assured safety for the bathers. The State of California won the suit because the judge hearing the case held that the coliform standards used by the Board of Public Health were reasonable after it was demonstrated that there was clear evidence of pollution and physical nuisance in areas where these bacterial limits were exceeded.

This historic case established a statistical coliform concentration baseline which defines the difference between pollution and pollution-free recreational contact waters in the open surf. This has been accepted by the Sanitary engineering profession in the United States as a landmark achievement.

Fig. 1-1   Los Angeles County Beach, California, 1974. (*Los Angeles Times* Photo.)

Rationale for Coliform Concentration Requirements

The most comprehensive study of inland recreational waters by Stevenson[4] in 1950-1953 concluded that an MPN coliform concentration of 2300/100 ml may be a threshold quality associated with an increase in the incidence of disease. For those who may have an interest in the comparison between fecal coliform and total coliform, the use of data developed some years later on the fecal coliform content of the same waters and by the use of ratios, a geometric mean fecal coliform content of 400/100 ml was determined to be equivalent to the 2300/100 ml total coliform number.

In theory two different standards could be set for a potable water and for a wastewater discharge. One would be a standard of assured safety where there is no health concern. The other standard would be the threshold of unsafe water (2300/100 ml) and might be used as an indication that quarantine action or abatement action should be undertaken. The California State Department of Health has recommended discharge requirements which, based on experience and judgment, will result in a water quality where there is no health concern. In other words their quality requirements reflect an inherent factor of public health safety, whether it be potable water, wastewater discharging to variable use receiving waters, or water reuse situations. These concepts are described as follows:

1. The limit for surf waters is about 500 times the pollution allowed by the United States Public Health Service standards for potable water (2.2 MPN/100 ml). The 2.2 standard has been universally accepted by water and health experts for many years. Comparing relative ingestion of potable water versus seawater in the course of surf bathing (2-3 cc/swim), the figures seem compatible.

2. There is no scientific evidence to indicate that water within this standard causes ill health.

3. The level of indicator organisms is seldom reached or exceeded in saline waters where the cause is not obviously recent waste contamination.

4. A less severe standard might show "approved" areas to lie within visible areas of grease and detritus of waste origin and would therefore seem to be lacking in common sense and decency.

The application of this standard for disinfection first appeared as a required chlorine residual after a specified contact time which would produce the desired quality in the receiving waters. At this time it was thought that disinfection would meet these standards if a 0.5 to 0.75 mg/liter orthotolidine residual at the end of 30 min. contact time was accomplished. Contact chambers were built to give a theoretical 30 min. detention time at average flow based on the volume of the chamber. Effective mixing and contact chamber short circuiting was never considered. During this period a great many chlorination facilities went into operation (1947-1957), and it became obvious that a residual-contact period requirement often produced effluents of quite different bacterial quality at different plants. After many years of testing and surveillance of these installations, the Bureau of Sanitary Engineering of the California State Department of Health concluded that it was practical, feasible, and superior to prescribe a coliform count directly to the plant effluent, rather than a chlorine residual value as evidence of disinfection.

Current Coliform Requirement in California

The numbers presently in effect are: 80 percent of samples less than 1000/100 ml for coastal bathing waters (equivalent to a median of 230/100 ml); median of 70/100 ml for shellfish growing areas, and a median of 23/100 ml for confined waters used for bathing or other water contact sports assuming the dilution is at least 100 to 1. The requirement for discharge into ephemeral streams or other areas where public exposure to effluents receiving little dilution is for an essentially coliform-free effluent, that is, a median MPN not greater than 2.2/100 ml. There is a subtle implication of the necessity for good operation and adequate treatment to achieve the 23/100 ml requirement. This has been found to be a meetable standard, but it requires a properly operated treatment plant with an effective disinfection system to consistently achieve it. The severe effluent standard of 2.2/100 ml implies the necessity for some type of advanced treatment prior to disinfection to reliably meet this level of disinfection effectiveness, and thereby suggests some virus removal capability for the system beyond that which normally occurs. These are not alternative requirements; only implications of what might be needed to meet the requirement.

There may be cases where the quality of the plant effluent bears no relation to the receiving water quality. For example, a receiving water might have a consistent coliform concentration as high as 2300/100 ml or even greater. If the State Department of Health had decided that disinfection of a wastewater discharging to such a receiving water was necessary, the discharger would not be allowed to simply meet the water quality of the receiving water (2300/100 ml or greater) because this would not be evidence of effluent disinfection.

Administration of Requirements

Administratively, this is how the control system presently operates in California:

1. The discharger applies to the appropriate Regional Water Quality Control Board for permission to discharge wastewater at a given location.
2. The Board then notifies all interested agencies for recommendations.
3. The State Department of Public Health submits its recommendation for the disinfection requirement to the Regional Board.

4. The Board then holds a public hearing to discuss and establish the requirements with the discharger.

The Regional Water Quality Control Board can issue cease and desist action on a discharger for violation of any portion of the requirements including the requirements on disinfection. Further, the Board can and has placed a ban on further connections to the discharger's collection system if the requirements are not met and impose a heavy fine for each day of violation.

Total Coliforms Versus Fecal Coliforms as a Standard

The fecal coliform determination is the latest in a long history of selective tests to separate the strains of coliform bacteria found in wastewater. It has been largely adopted by the United States Environmental Protection Agency for various disinfection requirements: shellfish areas, recreational waters, etc.

Coliforms from the intestines of dogs and cats are mainly *E. coli*. The fecal coliforms from humans and livestock account for about 97 percent of the total. Fish do not have permanent coliform flora in their intestines. The presence of coliforms in fish is evidence of pollution in the water of their habitat. Fecal coliforms are not abundant in soil (as is *E. coli*) since they die off rapidly when deposited in the soil.[5]

Currently, there are no means for distinguishing between fecal coliforms of man and those of other warm blooded animals.[6] Their presence in significant numbers is, however, indicative of fresh pollution. All fecal coliforms in a stream may be accepted as being of fecal origin, whereas an unknown portion of the total coliform bacteria may be of other origins. Determination of fecal coliforms provides a superior indicator of fecal pollution. There is no argument on this point. The question remains: Is the fecal coliform concept a valid parameter for disinfection? The differential fecal coliform test demonstrates the ability to enumerate coliform bacteria originating from fecal sources while suppressing those of soil origin. Consequently, it is a very useful tool in the Sanitary Survey of surface waters where numerous differing sources can contribute to the total coliform content. The California Department of Health has used both the total coliform test and the fecal coliform test in its surface water studies since 1960. It is their opinion that if a single indicator test for bacteriological quality were to be used, the fecal coliform test would be the best for *fresh water* areas because of its selective ability.[7] While the use of a fecal coliform number as a river water quality objective is appropriate, the use of a fecal coliform standard for a measurement of disinfection would not be appropriate. All available data indicate that the fecal coliform strains are more fragile than the total coliform group and can be more easily destroyed or inactivated by disinfection or natural purification processes. It has been amply demonstrated that on the basis of chlorine demand tests, fecal coliforms can be completely destroyed while significant numbers of total coliforms remain.[5]

There is only sketchy information on the relative resistance of pathogenic agents to chlorination. The data suggests that bacterial agents may be as hardy as coliforms while most viruses are more resistant. The total coliform group is a more conservative indicator of effective disinfection and is more numerous in wastewater than is the fecal coliform group. It is conservatively estimated that the ratio of fecal coliforms to total coliforms in saline waters is 1:70. In San Francisco Bay it has been observed that when the ratio of fecal coliform to total coliforms, is 1:70, that area is generally out of the known pollution areas.[8] However, there is a considerable difference in the fecal to total coliform ratio from saline to fresh water areas. Instead of the 1:70 ratio observed above it is estimated that a 200–400/100 ml fecal coliform requirement in fresh water is comparable to a 2000–4000/100 ml concentration of total coliforms.[9] This comparison is most significant when disinfection is related to a final coliform count rather than a log reduction in the total coliforms present before disinfection.

Disinfection Efficiency: Bacteria Survival Versus Percent Kill

The effectiveness of a disinfectant dose for a given contact time is usually expressed as a ratio of logs reduction of initital to final bacteria count, or as a percent destruction of the initial bacteria count. In order to keep the proper perspective when evaluating reports of disinfecting procedures, the minimum acceptable surviving number of total coliforms should not be in excess of 1000/100 ml MPN. For comparing disinfectants of secondary effluents the final MPN should be no more than 230/100 ml, total coliform concentration. Let us see what this means when looking at studies reporting log reduction versus percent destruction.

Assuming a well oxidized secondary effluent, the total coliform concentration before disinfection is probably on the order of 1×10^6 (1,000,000), so a 4-log reduction would produce a final MPN of 100/100 ml, and a 99 percent reduction would yield a final count of 10,000/100 ml and a 99.99 percent reduction would yield 100/100 ml. So for this magnitude of initial count a 99.99 percent kill is comparable to a 4-log reduction.

A well oxidized and filtered tertiary effluent would probably have an effluent coliform count of 50,000/100 ml before disinfection. The coliform requirement for a disinfected tertiary effluent is usually 2.2/100 ml. Therefore, a log reduction greater than 4 is required, and the percent kill must be greater than 99.99. The point to be made is that disinfection efficiencies reported as 99 or 99.9 percent are meaningless. Actually, disinfection efficiency studies based on coliform destruction should specify the range of initial coliform as MPN/100 ml in addition to the log reduction. Then the mathematical model developed by

Collins et al. in 1970[10] can be used for verification. This model is as follows

$$y/y_0 = [1 + 0.23\, ct]^{-3} \qquad (1\text{-}1)$$

where

y_0 = initial coliform MPN/100 ml
y = final coliform MPN/100 ml
c = chlorine residual, mg/liter, at the *end* of contact time t
t = contact time in min.

This model has been verified by several practitioners and researchers since it was first published.

Bacterial Indicator Concepts

The parameter of major importance that would provide proof of disinfection is the resistance of the indicator organism to the disinfectant. Researchers over the years have expressed dissatisfaction with the coliform group as an indicator organism because it is not resistant enough to chlorine to allow any safety factor. For a group of organisms to be an ideal indicator, the following conditions should be met to demonstrate disinfection efficiency:

1. The indicator organism must be more resistant to disinfection than the pathogenic organisms.
2. The indicator must be present in the sample whenever pathogenic organisms are present.
3. The indicator must occur in greater numbers than the pathogens.
4. A simple, rapid, and unambiguous procedure must be capable of enumerating the indicator organisms.
5. The indicator should not regrow or otherwise increase in numbers in the aquatic environment after disinfection.
6. The indicator organism must be randomly distributed in the influent stream.
7. The presence of other organisms must not inhibit the growth of the indicator organism.
8. The indicator organism should be nonpathogenic to man.

The lack of disinfection resistance of the coliform group was clearly demonstrated when the organism *Klebsiella* was found in the City of Chicago's water distribution system.[11] The members of the *Klebsiella* genus can cause severe enteritis in children, and pneumonia and upper respiratory tract infection, septicemia, meningitis, peritonitis, and urinary tract infection in adults. Discovering such a hazardous organism in a distribution system, carrying water which had been coagulated, filtered, and chlorinated sufficiently to carry a small residual in the system, was indeed unsettling. In this instance it was concluded that *Klebsiella* was the organism responsible for most of the positive samples occurring in the routine coliform sampling procedures. The *Klebsiella* group are encapsulated organisms, and once in the distribution system, may be harbored in protective slime and sediment. For these reasons they are more resistant to disinfection than the coliforms. In 1974, Engelbrecht et al.[12] evaluated two promising groups of organisms believed to be resistant to chlorine in the range necessary to inactivate both bacillary pathogens and waterborne viruses. These groups are the acid-fast cultures *Mycobacterium fortuitum* and *M. phlei*, and a yeast *Candida parapsilosis*.

However, it must be recognized that the concept of proof of disinfection for wastewater discharges and water reuse situations is entirely different from that of water to be used for potable purposes. With the situation as it exists in 1976 with respect to the organics in wastewater, it is unlikely that it will be possible to pursue the reclamation of wastewater for potable use. Therefore, proof of disinfection for wastewater and water reuse should always be on the basis of destruction of the indicator organisms. This could conceivably be on the basis of a chlorine residual contact time envelope provided this combination of criteria could be developed for a "consensus" organism.[13,14]

WASTEWATER REUSE

Historical Background

Wastewater reclamation, in the United States, has been practiced since about 1920. One of the first systems to use the total discharge for beneficial use was the activated sludge sewage treatment plant in the Golden Gate Park of San Francisco. This operation began in about 1930. In 1935, there were 62 communities using treated wastewater for crop irrigation and regulations for this use had been in existence for about 15 yr. in California. As of 1970, there are about 600 reclamation projects operating on a continuous basis in the United States. About half of these are in the State of California. The first use, of course, was for crop irrigation. These crops were fiber, fodder, and seed crops, and there was little opportunity for public contact. Use of reclaimed water for landscape irrigation (i.e., irrigation of parks, playgrounds, golf courses, freeways, right-of-ways, and so forth) also has a fairly long history, but it wasn't until the last 10 or 15 yr. that there has been a sudden increase in the number of such installations. Where crop irrigation installations have increased 40 percent, landscape irrigation installations have multiplied nine-fold. There is a trend, then, and a general swing toward the uses of reclaimed water where the public may have more exposure and more contact. Public attention has been drawn to many outstanding reclamation systems in California such as the Santee project which includes swimming in addition to boating and fishing; and to the Contra Costa project of supplying industry cooling water needs from reclaimed wastewater.

Moreover, it is a matter of fiscal responsibility for any agency producing water and treating wastewater to have plans for wastewater reclamation. Otherwise such an agency would not qualify for Federal funds.

California Requirements

In 1968, the California State Board of Public Health adopted standards for the quality of reclaimed water used for crop irrigation, landscape, and for recreational impoundments. It is important to realize that scientists from all fields acknowledged that the establishment of specific quality limits for reclaimed water in terms of BOD, suspended solids, ether soluble materials, or other commonly used criteria, was not practical. The monitoring cost would be too great for smaller reclamation operations. The group decided to use as the keystone of the standards, the coliform bacteria concentration in the reclaimed water. The coliform bacteria concentrations which are allowable for different reclamation uses are supplemented in the standards by terms such as "oxidized wastewater," "filtered wastewater," and other descriptive terms which broadly identify the type of reclaimed water that is required without specifically identifying numerous quality limits.

Briefly these adopted standards are as follows:[15]

Primary Effluent. It can be used for surface irrigation of processed food crops, orchards, and vineyards; irrigation of fodder, fiber and seed crops. Virtually no public contact or ingestion is possible.

Oxidided Effluent (secondary). With a median coliform MPN of 23/100 ml it can be used for landscape irrigation, spray irrigation of processed food crops, landscape impoundments, and milk-cow pastures. Public contact is possible but ingestion is very unlikely.

Oxidized Effluent (secondary). With a median coliform MPN of 2.2/100 ml it can be used for surface irrigation of produce (makes this operation impractical) and restricted recreational impoundments. Public contact and minor ingestion is possible.

Filtered Effluent (tertiary). With a median coliform MPN 2.2/100 ml it can be used for spray irrigation of produce and unrestricted recreational impoundments. Public contact and minor ingestion are likely.

Other Applications. At present there are many cooling water applications using secondary effluent meeting the 2.2/100 ml MPN coliform requirements. Water reuse is here to stay, at least in California. It is used extensively to recharge groundwater supplies which accomplish two other major objectives: 1) provide a salt water barrier; and 2) prevent land subsidence due to withdrawal of natural gas and oil reserves.

Importance of Disinfection. Disinfection is the most important link in this chain of treatment. Chlorine must be relied upon to do the bulk of the disinfecting and to provide a persisting residual. Ozone may be called on to provide the assurance of viral inactivation.

The Environmental Health Laboratory at the Hebrew University, Jerusalem, Israel has been studying the health risks resulting from spray irrigation of non-disinfected wastewater. Their studies were undertaken to obtain data about the number and types of enteric bacteria dispersed into the air as a result of spray irrigation. Katzenelson and Teltch[47] discovered that coliform bacteria were found in the air at a distance of 400 yd. downwind from the irrigation line, and in one case a *Salmonella* bacterium was isolated 65 yd. from the source of irrigation. These findings inspired an epidemiological survey of the incidence of enteric communicable diseases in 77 agricultural communal settlements practicing spray irrigation with nondisinfected partially treated oxidation pond effluent as compared with those in 130 similar settlements not practicing any form of wastewater irrigation. Katzenelson, Buium, and Shuval[48] reported in 1976 that the incidence of shigellosis, salmonellosis, typhoid fever, and infectious hepatitis was two to four times higher in communities practicing wastewater irrigation. Moreover, it was found that there were no differences in the incidence of enteric diseases between these two communities during the winter nonirrigation season. *The result of these studies has brought about recommendations for strong wastewater-treatment measures including effective bacterial and virucidal inactivation by disinfection to prevent the spread of enteric diseases due to airborne contamination of communities adjacent to spray irrigation projects.*

VIRUSES

The Virus Hazard

It is of some concern that the frequency of infectious hepatitis has remained at a static level of 50,000–60,000 cases per yr. in the United States whereas the incidence of typhoid fever has dropped from approximately 2000 cases in 1955 to only 300 cases in 1968.[13] This indicates that although waterborne bacterial infections have been all but eliminated, water utility people may have a more severe task when dealing with waterborne viral infections. The problem of viruses in water supplies has received a great deal of attention in recent years and is well documented.[16-21] In evaluating the virus hazard, two factors should concern the protectors of our water supplies, whether for wastewater disinfection or potable water treatment. These factors are the origin of infectious hepatitis and the origin of the significant increase in gastroenteritis. Gastroenteritis is not a reportable disease as distinguished from relatively well-defined illnesses, e.g., infectious hepatitis, shigellosis, salmonellosis, and typhoid. Yet it can be estimated that the number of gastroenteritis cases occuring per year is hundreds of thousands and possibly millions. During the period 1961–1970 a total of 26,546 cases were definitely attributed to contaminated water.[19] Of 52 waterborne-disease outbreaks in the United States in 1971–1972, there were 22 outbreaks of gastroenteritis, amounting to 5615 cases from a total of 6817 cases of waterborne illnesses.[20] The concern here is that these cases may be the result of some yet unidentified viruses. There are more than 100 viruses excreted in human feces that have been reported to be in contaminated water. Any of these could cause a waterborne disease.

Other factors of viral infections that are great cause for concern are included in the evidence researched by McDermott who points out that poliomyelitis virus has hurdled the technical barriers of water treatment of the Paris, France, water supply which consists of coagulation, filtration, and disinfection by both chlorine and ozone.[5]

In the same discussion McDermott refers to the work by Plotkin and Katz that the minimum infective dose by a virus is 1 pfu.[18] Although this statistic may be open to question, the contemplation of this situation and the possibility of 100 or more viruses capable of contributing to a waterborne disease should be of extreme concern to water producers.

All of these concerns are confirmed and substantiated by the 1970 report of the Committee on Environmental Quality Management, ASCE Sanitary Engineering Division.[16] Some of the important conclusions of this report are as follows:

1. There is no doubt that the virus of infectious hepatitis can be transmitted by drinking water and epidemiological opinion uniformly supports this conclusion.

2. Although evidence is scanty, it should also be assumed that the enteric viruses and other possible causative agents of viral gastroenteritis can be transmitted by drinking water.

3. There is no doubt that a positive coliform index means that virus may be present; however, absence of coliform may not mean that virus is absent. The coliform index, therefore, while a good laboratory tool, is not a reliable index for viruses. Greater assurance of the absence of virus would be a turbidity of less than 0.1 Jackson Unit and an HOCl residual of 1 mg/liter after a contact period of 30 min.

4. The evidence available indicates that a risk of hepatitis infection results from the consumption of raw or steamed undepurated shellfish taken from sewage polluted waters and that the Public Health Service Coliform Standard (70 coliforms/100 ml) has been shown by experience to be a reliable indication

of risk-free shellfish waters. The Committee of Environmental Quality Management believes that a high level of protection would be provided by activated sludge treatment and chlorination of the effluent to a level producing an amperometric chlorine residual of 5+ mg/liter after 30 min. contact.*

5. Virus multiplication in polluted water appears not to be a significant possibility, and from the control point of view it can be disregarded.

6. Viruses are present in certain river waters and failure to isolate them results presumably from their low concentrations and the relatively ineffective sampling and concentration procedures employed.

7. Enteroviruses and the virus of infectious hepatitis can survive for prolonged periods under conditions prevailing in drinking-water reservoirs. Long detention times therefore cannot be considered as a safety factor.

8. Enteric viruses differ in resistance to free chlorine. Adenovirus 3 is less resistant than *E. coli* while poliovirus 1 and Coxsackie virus A2 and A9 appear to be more resistant than any of the other enteroviruses studied.

Viruses in Sewage Contaminated Supplies

In the United States the enteric virus concentration in raw sewage probably ranges from two or three to more than 1000 infectious virus units/100 ml with peak levels occurring in late summer and early fall.[17] If the coliform bacteria concentration of raw sewage is estimated at 10^7-10^8/100 ml organisms then the enteric virus concentration is perhaps 5–7 orders of magnitude lower.[22,23]

Although available evidence indicates that enteric virus concentrations in drinking water are likely to be very low it is important to be aware of the fact that as little as one virus infectious unit is probably capable of producing an infection in humans.[18] Most enteric virus isolations have been made from heavily polluted surface waters, but Berg and coworkers detected enteric viruses in Missouri River water having fecal coliform concentrations as low as 60/100 ml.[22] While virtually nothing is known about enteric virus levels in United States potable water supplies, monitoring of the potable water supplies of Paris, France, in the 1960s revealed that about 18 percent of the 200 samples examined contained enteric viruses; and the average virus concentration was estimated at one infectious unit/300 liter.[25] The heavily polluted Seine River is one of the major sources of the Paris water supply. Nupen (1974) and coworkers have reported finding enteric viruses in 10 liter samples of drinking water in South Africa.[26]

While there is little quantitative information available on enteric virus levels in sewage-contaminated surface and ground waters, there is plenty of evidence

*Amperometric residual is stipulated to be the total chlorine residual with no reference made to any free chlorine residual fraction.

that wastewaters are a primary source of enteric virus contamination of man's environment. With the possible exception of a few poliomyelitis outbreaks, there is no evidence of waterborne outbreaks in the United States caused by other specific viruses. The most prevalent waterborne disease in the United States continues to be gastroenteritis of unknown etiology. Therefore, for reasons described above it is imperative that wastewater-treatment processes and wastewater-reuse systems address themselves to this virus hazard.

Virus Inactivation

To date, the most comprehensive study of virus inactivation on a pilot-plant scale has been the work by the County Sanitation Districts of Los Angeles County. The results of this two-yr. study are contained in the "Pomona Virus Study—Final Report" prepared for the California State Water Resources Control Board and the U.S. Environmental Protection Agency February 1977. This work has been summarized in a paper by Selna, Miele, and Baird.[27]

The Sanitation Districts of Los Angeles County have been active participants in various water reuse programs since the mid 1950's. Various discharges of disinfected secondary effluent have been spread in percolation basins for the purpose of replenishing groundwater which is used for domestic supplies. Conveyance of this water to the percolation sites is through open flood control channels and in transit is unintentionally used for recreational activities including body contact. About 80 mgd of chlorinated secondary effluent is subject to this unplanned recreational use in flood control channels. These channels are classified as "unrestricted recreational impoundments" by the California State Department of Health, therefore wastewater discharged to such channels must comply with Title 22 of the California Administrative Code. This document contains the effluent quality and treatment system requirements for recreational reuse as decreed by the California State Department of Health. In order to qualify for such use, secondary effluent must be coagulated, settled, filtered, and disinfected to achieve a median total coliform MPN of 2.2/100 ml or less, and to provide an effluent which will protect swimmers against viral illnesses. Since this required treatment is expensive, both from a capital and operational standpoint, the prime objectives of the Pomona study was to investigate alternate methods of tertiary treatment that might be more cost effective than the required Title 22 System and still produce an effluent with the required degree of public health protection.

Treatment Systems. This work investigated the following four systems:

1. This is the Title 22 treatment called for by the California State Department of Health. This was a 40 GPM secondary effluent (NH_3N, 20 mg/liter)

treated with alum and ionic polymer followed by flash mixing, flocculation, sedimentation, dual media filtration and disinfection.

2. This was a 25 GPM secondary effluent similar in quality to 1, treated with alum and ionic polymer followed by flash mixing, dual media filtration and disinfection.

3. This was 100 GPM secondary effluent also similar in quality to 1 and 2. Treatment consisted of two-stage carbon absorption followed by disinfection.

4. This was 25 GPM of a nitrified secondary effluent (NH_3—N, 0.1 mg/liter) similar in quality to the others, treated with alum and anionic polymer followed by flash mixing dual filtration and disinfection.

Disinfection Process:

Chlorine. Systems 1, 2, and 3 were capable of disinfection by either ozone or chlorine. System 4 was disinfected with chlorine only since nitrification is not considered to enhance the performance of ozone.

Chlorine was applied in systems 1, 2, and 3 to produce two levels of chlorine residual at the end of the contact chamber (i.e., 5 and 10 mg/liter). These residuals required chlorine dosages of about 10 to 15 mg/liter respectively.

In system 4 the free chlorine residual at the outlet of the contact chamber was maintained at 4 mg/liter. This required a chlorine dosage of 10 mg/liter.

All chlorine residuals were measured by the DPD, ferrous ammonium sulfate titrimetric method.

Careful attention was given to the degree of mixing at the point of application. A $1/3$ hp flash mixer was installed in a 32 gal. confined mixing chamber. This calculates to a G factor of 450. (See Chapter 3—Mixing). A conventional perforated-type chlorine diffuser was used discharging about three in. from the mixer impeller.

The chlorine contact chambers were designed to have plug flow characteristics. After an optimization study to achieve a coliform MPN of 2.2/100 ml a detention time of 120 min. was selected. A tracer study of the prototype contact chamber revealed a modal time of 98 min. and a minimum time of 58 min. So far the sake of comparison with laboratory studies it might be stated that the bacterial kills and virus inactivation were accomplished with a one-hr contact time and not a two-hr time.

Ozone. The ozone system was arranged to provide a *dosage* level of 10 mg/liter in system 1, and 10–50 mg/liter in system 2, and 6 mg/liter in system 3. In each case the ozone contact time was 18 min. The ozone contactors consisted of six 14-inch diameter PVC columns 18 ft high.

Predisinfection Effluent Quality. The quality of the wastewater at the point of disinfection is of considerable interest. The nonnitrified effluent contained

20 mg/liter NH_3—N, with suspended solids on the order of 1.5 mg/liter and turbidities of 1–1.5 FTU. The pH was 7.5 and TDS was 580 mg/liter. At this pH the undissociated HOCl in system 4 was about 50 percent.

Results. The virus inactivation results from the Pomona study are illustrated in Fig. 1-2, 1-3, and 1-4. The effectiveness of combined chlorine residual is a real surprise, which advances a totally new concept: *that chloramines do in fact have virucidal efficiency potentially equal to that of free chlorine.* Until the revelation of the Pomona study it was believed that the only virucidal chlorine compound was free chlorine (HOCl). This is not seen to be the case for tertiary effluents.

Moreover, it was found that system 4 using free chlorine increased the pre-

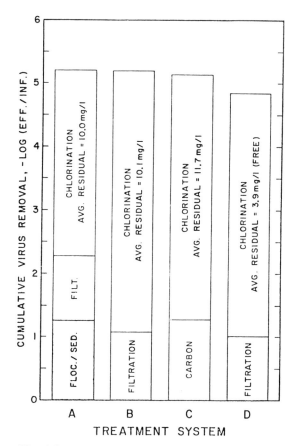

Fig. 1-2 **Virus removal at high chlorine residuals.**

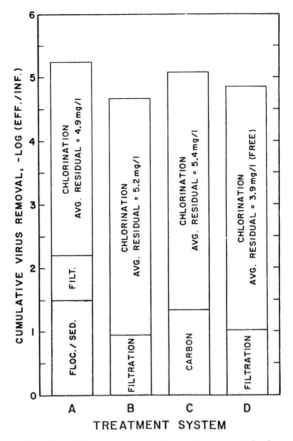

Fig. 1-3 Virus removal at low chlorine residuals.

disinfection effluent concentration of chloroform from 0.6 to 198 micrograms per liter, while the other systems using all combined chlorine increased the chloroform content on the order of 1.0 to 6.0 micrograms per liter (approximately). This observation plus the demonstrated effectiveness of chloramines as a virucide leads to the conclusion *that there is no benefit from free residual chlorination of tertiary effluents which are required to* meet the Title 22 requirements of the California State Department of Health *where indirect potable reuse is called for.*

A dominant feature of Fig. 1-2 is that the majority of virus removal occurs in the disinfection step. The data presented in this figure indicate that reliance must be placed on the disinfection step rather than the filtration or carbon absorption steps prior to disinfection.

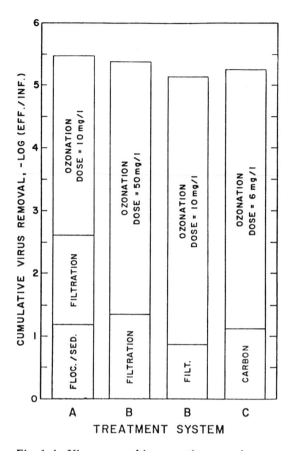

Fig. 1-4   Virus removal in ozonation experiments.

Other Conclusions. Other important conclusions drawn from the Pomona study are:

1. Virus inactivation in tertiary treatment systems employing combined chlorine residuals of 5-10 mg/liter ranged from 4.7 to 5.2 logs. These results were obtained in poliovirus seeding experiments. The additional virus removal at the 10 mg/liter residual over the 5 mg/liter residual was minimal but the coliform kill was consistently better at the higher residual.

2. The system employing free residual chlorination produced about 4.9 logs virus removal when the average free chlorine residual was 4 mg/liter.

3. All of the chlorination studies indicated consistent capability for attainment of the 2.2/100 ml MPN coliform requirement by all of the systems.

4. In the experiments using ozonation, virus removal ranged from 5.1 to 5.5 logs (see Fig. 1-4); however, attainment of the 2.2/100 ml MPN coliform standard was hampered by water quality variations.

5. Based on the results of the virus experiments it was concluded that system 2 (direct filtration) or system 3 (carbon adsorption) tertiary treatment systems are more cost effective than system 1 (the one required by the California State Dept. of Health).

6. The direct filtration system (2) is the lowest cost with chlorination and is estimated at 13.7¢/1000 gal. for total capital and operating cost. This compares to system 1 at 21.5¢/1000 gal. with chlorine, 17.2¢/100 gal. for system 3 with chlorine and 19.9¢/1000 gal. for system 4 with free chlorine. The systems using ozonation were more costly.

Comparison with Other Investigations. It is difficult at best to compare virus inactivation studies owing to evaluation of detection limits, and seeding and analytical procedures; however, it is worthwhile to compare the Pomona study which is an outstanding investigation with other noteworthy virus inactivation studies as follows:

Ludovici et al.[28] performed a large number of pilot-plant experiments using the tertiary effluent from the Tucson, Arizona wastewater treatment plant. This effluent ranged in pH from 7.3 to 7.6; NH_3-N from 2.8-5.5 mg/liter; organic N. 1.4-5.6 mg/liter; BOD 1.5-8.5 mg/liter; and COD 15-33 mg/liter. No turbidity information was given, but obviously this is a high quality effluent except for the 5.6 mg/liter organic N which could have some inhibiting effect on the chlorination process (see Chapter 2). The effluent was seeded variously with polio 1, Coxsackie B1, and Coxsackie B2. The chlorine dosages were 2 and 4 mg/liter which means that all the residuals reported at the end of 30 min. contact time had to be combined chlorine. These were 0.31-1.2 mg/liter for the 2 mg/liter dose and 0.99-2.76 mg/liter for the 4 mg/liter dose in the polio 1 experiments. The mean reduction for the 2 mg/liter dose was 96.19 percent and a mean of 99.3 percent for 4 mg/liter—all at 30 min. contact time. The Coxsackie B1 and B2 viruses were much more susceptible to the combined chlorine residuals where a mean of 99.8 percent reduction was achieved with a 4 mg/liter dose for Coxsackie B1 and a 100 percent reduction in the Coxsackie B2 experiments with the same chlorine dose. Concurrently with these experiments Ludovici et al investigated destruction of total coliforms. With a maximum Y_0 coliform MPN of 19,100/100 ml,* the maximum final coliform MPN during the Coxsackie B2 seeding experiment was 1.1/100 ml with a 4 mg/liter chlorine dose, and a 30 min. contact time. The chlorine residual range was 1.5-2.94 mg/liter. The same results occurred in the polio 1 seeding experiments

---

*Low-initial coliform count (Y_0 in the Collins model) is anything less than 50,000/100 ml. By comparison some raw potable water supplies exceed 3000/100 ml at the intake.

demonstrating the great enhancement of the chlorination process by a high-quality effluent having a low-initial coliform count.

These virus studies by Ludovici et al.[28] are noteworthy and deserve some comment. For virus inactivation, reporting destruction as a percent of the original number of organisms is a good comparative method, but may not be compatible with what a regulatory agency is likely to concede as appropriate disinfection. We need to have more discussion on a standard of disinfection where there is a public health threat from viruses in certain wastewaters.

Other recent virus inactivation studies include the Louisville, Kentucky, experiments by Pavoni and Tittlebaum.[29] They applied ozone to a 40,000 GPD activated sludge plant effluent seeded with F_2 virus (bacteriophage) at a concentration of 10^{11} plaque units per ml and a rate of 1 ml to 1 liter of sewage. They reported "virtually 100 percent efficiency (inactivation) after a contact time of 5 min. with a total ozone dosage of approximately 15 mg/liter and a residual of 0.15 mg/liter."

Nupen et al. treated a high-quality effluent for water reuse and found that a free chlorine residual beyond the breakpoint, which produced a pH of about 6.0 and a residual of not less than 0.6 mg/liter, inactivated polio virus in 35 min.[26]

One of the major difficulties in the inactivation of viruses is their variable sensitivity to disinfectants. Liu and co-workers tested 16 types of human enteric viruses for resistance to free chlorine in treated Potomac River water. The criteria used was time in minutes required for a 99.99 percent inactivation by a 0.5 mg per liter free chlorine residual at pH 7.8 and 2°C.[21] These are listed below in descending order:

polio type II (36.5 min.)
Coxsackie B5 (34.5 min.)
E. coli type 29 (18.2 min.)
E. coli type 12 (16.7 min.)
polio type III (16.6 min.)
Coxsackie B3 (15.7 min.)
adenovirus 7a (12.5 min.)
polio type I (12.0 min.)
Coxsackie B1 (8.5 min.)
adenovirus 12 (8.1 min.)
Coxsackie A9 (7.0 min.)
E. coli 7 (6.8 min.)
adenovirus 3 (4.3 min.)
reovirus 2 (4.2 min.)
reovirus 3 (4.0 min.)
reovirus 1 (2.7 min.)

So the picture developed to date by virus inactivation studies is a bit murky.

Future Considerations of Virus Destruction

The results of the experiments described above clearly illustrate the importance of predisinfection processes when treating wastewater effluents. Raw water for potable water supplies is hopefully of much higher quality than the well-oxidized and filtered effluent of a tertiary wastewater plant. Therefore, we have to recognize the existence of two different standards of disinfection related to virus inactivation: one for potable water and the other for wastewater. For the latter, it seems hopeless to expect significant virus destruction unless the effluent is of tertiary quality. As for raw potable water, it is the consensus that a 1.0 mg per liter free chlorine residual at the end of 30 min. contact time at a pH not to exceed 8.0 will destroy nearly all pathogenic viruses.[1]

For tertiary effluents, based on the Pomona study it appears that a *combined* chlorine residual between 5 and 10 mg/liter with a two-hr contact is as good or better than a 4 mg/liter *free* chlorine residual for the same contact time. This is a new concept. Other disinfectants, e.g., chlorine dioxide, might prove to be better virucides than either chlorine or ozone. There is no question of the efficacy of ozone as a virucide, but on a cost-effective basis it seems to be very little better than chlorine with the disadvantage that it may not produce an effluent to meet the coliform limit. All of this gives impetus to the future investigation of chlorine dioxide as a combination virucide and bactericide.

One other item of significance when considering the treatment of waste waters for virus inactivation is the report by Boardman and Sproul[30] that none of the particulate systems investigated—clay, hydrated aluminum oxide, and calcium carbonate—protected the T7 phage from inactivation by chlorine. Therefore, viral absorption on an exposed surface provides negligible protection from disinfection, which raises the question of how much turbidity might be tolerated in a high quality effluent. Total encapsulation of the virus would appear to be the major mechanism by which a particle may be afforded protection due to adsorption.

THE REGROWTH PHENOMENON

Significance of Regrowth

Numerous studies have demonstrated the regrowth phenomenon of coliform and fecal coliform organisms after disinfection. This may be due to the destruction of bacterial predators and may depend upon the presence of certain nutrients in the wastewater or the receiving waters.[7] It has been observed both in wastewaters and in receiving waters downstream from a disinfected sewage discharge. It is the judgment of authorities such as Geldreich that pathogenic bacteria like *Salmonella* and *Shigella*, of the same family as the coliform group, also regrow. If the disinfection process has a comparable effectiveness against such pathogens as with coliforms, a disinfection requirement resulting in a reduction of the numerous coliform bacteria down to a level of 23 or 230/100 ml would virtually assure the absence of the pathogenic bacteria present in sewage in smaller numbers. This would eliminate the pathogen regrowth potential, whereas a more liberal criteria of 200 or 400 fecal coliforms would not. Therefore, the phenomenon of regrowth after disinfection, whether it be a chlorine compound or ozone as the disinfectant, is not of any adverse public health significance.

The Clumping Phenomenon

It has been suggested over the years that "clumping" of organisms in the suspended solids of a wastewater effluent contributes to irrational data on the efficiency of disinfection after the break up of these clumps. Recent investigations by White[31] confirm an extremely small margin of disinfection reliability of a primary effluent. The extension of this observation to storm water overflows leads White to believe that it is not possible to adequately disinfect primary effluents or storm water overflows. (This is exemplified by the fact that a primary effluent could be chlorinated with reasonable dosages of about 200 lb/mg to achieve a 230 MPN/100 ml in the effluent at a one-hr plus contact time, but if the sample for this accomplishment were taken in the same place but on the discharge side of the sample pump, the MPN would rise to 30,000/100 ml.) This clearly demonstrates the futility of attempting disinfection of primary effluents. The cause of this dramatic increase in MPN is the clumping phenomenon. The sample pump breaks up the clumps thereby releasing great quantities of organisms protected from the disinfectant by the clumps. All of these data lend credence to state regulatory agencies asking for better and more effective predisinfection unit processes such as secondary treatment.

TOXICITY OF CHLORINE RESIDUALS

Chlorine residuals of low concentration whether free or combined are toxic to most fish and other aquatic life depending upon the species and time of exposure. Most oxidants like chlorine and ozone are known to be irritants to both freshwater and saltwater fish.

The resistance of fish to toxic substances varies according to the size of the fish. Fingerlings are very susceptible whereas carp and other large fish are highly resistant to toxic agents. The resistance is also proportional to the size of the fish scales: the larger the scales the greater the resistance. This is why goldfish are hardier than trout.

A review of the literature reveals a general agreement by the investigators as to what constitutes a lethal chlorine residual.

In 1968, Tsai[32] reported that the effects of chlorinated domestic sewage effluents on a fish population may stem not only from indirect degradation of water quality and alteration of stream bottom, but also from direct action on the fish similar to that of industrial discharges. When wastewater is chlorinated other toxic compounds may also be formed. For example, if thiocyanate is present this can be converted to the highly toxic cyanogen chloride by the chlorination process. It is also well known that under certain conditions an array of chloro-organic compounds are formed during the disinfection of wastewater by chlorination, and some of these compounds may also be toxic to aquatic life.[33] Very little is known about how these complex reactions affect aquatic life.

In 1971, Esvelt, Kaufman and Selleck[34] reported in a comprehensive study that the average daily emission of toxicity to the San Francisco Bay system was about 56 percent from municipal sources and 44 percent from direct releases by industry. The studies were performed long before it was thought necessary to be concerned about the concentration of the chlorine residual in the effluent owing to the enormous dilution factor of the receiving waters. The total chlorine residuals encountered (as measured by the back titration procedure with an amperometric endpoint) ranged from 1 to 8 mg/liter. The test fish used throughout the project were golden shiners. Using continuous-flow on-line bioassays they reported a 96 hr TL_{50}* in municipal wastewaters to be 0.2 mg/liter total chlorine residual. Even though the fish bioassay procedure is not necessarily a good measure of toxicity,[35] this investigation revealed the following important conclusions:

1. Toxicity removal by biological treatment was 75 percent while that removed by lime precipitation was 40 percent. Both ion exchange removal of ammonia and sorption of organics when coupled to lime precipitation provided an overall 65 percent removal.
2. MBAS and NH_3-N represent significant toxicants in municipal wastewaters.
3. Chlorination increased the toxicity of treated municipal wastewaters in all instances.
4. A chlorinated and dechlorinated effluent (with a slight excess of sulfite ion) was less toxic than either the unchlorinated or chlorinated effluent.
5. Dechlorination completely removed the chlorine-induced toxicity.

Also in 1971, the Michigan Department of Natural Resources conducted four separate studies at different wastewater-treatment plants and reported that for rainbow trout the 96 hr TL_{50} concentration below two plants was 0.023

*This means that 50 percent of the fish subjected for 96 hr to the specified residual will die. TL = tolerance limit.

mg/liter, and that for fathead minnows concentrations less than 0.1 mg/liter were toxic in the plant effluents.[36]

Another recent investigation demonstrated how the use of the orthololidine method of measuring and control of wastewater chlorine residuals resulted in gross overchlorination beyond disinfection requirements.[37] This practice resulted in a major fish kill in the lower James River (Virginia).

Another case involving fish kills occurred in the Sacramento River (California) below a wastewater-treatment plant discharge. Chlorine-induced toxicity in the wastewater was immediately suspected. Consequently the regulatory agencies conducted bioassays to determine the source of the problem.[38] River water collected above, at, and below the discharge site was used in the static bioassays. King salmon fry were used as the test fish. The test fish were held in the receiving waters 150 ft upstream from the waste plume and in the waste plume 100, 200, and 300 ft downstream from the discharge where very little dilution of the waste discharge with the receiving water occurred. All of the test fish below the discharge died within a captive 14-hr period of exposure while all the fish upstream survived. The total chlorine residuals measured by the back titration amperometric endpoint procedure ranged from 0.2 mg/liter to 0.3 mg/liter during the test period. Therefore, it could only be concluded that the fish kills were caused by the toxicity in the wastewater—presumably the chlorine residual.

Probably the most comprehensive study of the toxicity of chlorine residuals to aquatic life in the receiving waters is the 1975 report by Arthur et al.[39] The disinfection system under investigation by both chlorine and ozone was able to produce an effluent with coliform levels less than 1000/100 ml. All of the studies were in freshwater systems. Both fish and invertebrates were included in the study. The chlorinated effluent was more lethal than the ozonated effluent or the chlorinated-dechlorinated effluent, which confirms the observations and conclusions of previous investigators. The fish were more sensitive than the invertebrates to the chlorinated effluent in the 94-hr tests. The respective 94-hr TL_{50} values of total residual chlorine to fish and invertebrates ranged from 0.08–0.26 and 0.21 to greater than 0.81 mg/liter respectively.

A discussion of the effect of chlorine residuals would not be complete without the consideration of the literature review and analysis of the aquatic life criteria as it relates to treatment of wastewaters provided by Brungs in his 1976 report.[40]

Need for Dechlorination

It was shortly after the Esvelt, Kaufman and Selleck report published in 1971[41] that the Water Quality Control Board, State of California, in cooperation with the Department of Fish and Game began issuing orders for certain wastewater-treatment plants to supplement the chlorination system with dechlorination facilities. The first wastewater-treatment plant to add dechlorination (by sulfur

dioxide) in California was the city of Burlingame circa 1972. The orders specified a chlorine residual not to exceed 0.1 mg/liter. For technical reasons explained in Chapter 3 this meant total dechlorination with a slight excess of
$SO_3^=$ ion. In the meantime the Sanitary Engineering Research Laboratory,
University of California at Richmond, under the direction of Dr. Warren
Kaufman, investigated the toxicity of the sulfite ion (a product of overde-
chlorination with sulfur dioxide) and found none up to more than 10
mg/liter.[42]

As of 1976, there were about thirty chlorination-dechlorination systems
operating in California wastewater treatment plants and many more in the
design-purchase stage. The California Water Resources Control Board and the
Fish and Game Department are in agreement by following the conclusion
from the bay studies on toxicity, that a chlorinated and dechlorinated sewage
effluent is less toxic to aquatic life than either a chlorinated or unchlorinated
effluent (has not been treated by a chlorination process).

Problems leading up to the chlorine toxicity dilemma have been gross overchlorination and poor control systems. It has been argued in some quarters
that optimizing the chlorination system and controlling the dechlorination
of the effluent to a comfortable 0.5 mg/liter chlorine residual would result
in a sewage plume that would lose this small residual quickly in the receiving
waters by the effects of chlorine demand and dilution.

It is interesting to note that a study by Stone[43] revealed some interesting
facts about the die-away of chlorine residuals in San Francisco Bay. When a
chlorinated secondary effluent was diluted with seawater there was no uptake
of chlorine residual. The residual depletion was strictly by dilution: the seawater exerted no apparent chlorine demand on the combined chlorine residual
which varied from 0.9 to 7.6 mg/liter at the end of 24 min. before dilution
with seawater.

Freshwater receiving streams are known to exert a chlorine demand which
in most cases could quickly absorb a 0.5 mg/liter chlorine residual. This should
be investigated on a case by case basis because dechlorinating to 0.5 mg/liter
has many more advantages than the use of excess sulfite ion.

AVAILABLE METHODS OF DISINFECTION

Introduction

The scope of possible methods of wastewater effluent disinfection is great and
includes natural processes (predatation and normal die-away), environmental factors (salinity, solar radiation) and methods having certain industrial applications
(ultrasonics, heat). Only those methods which appear to have possible general
application for wastewater and water reuse disinfection will be explored in detail
in this text. In this introduction the general features or characteristics which

have influenced overall use of the various disinfection methods will be identified and included in the design coverage except for gamma radiation, and iodine for reasons described below. The 1975 comparative cost analysis where available will also be shown.

Chlorine

Chlorine has and will probably continue to be the dominant disinfectant of wastewaters. It is available in different forms, and the characteristics of these forms greatly influence the system design. Because of the importance of chlorine in wastewater treatment the specific features of gaseous chlorine and chlorine compounds are discussed in detail in subsequent chapters of this text. (See Table of Contents).

Liquid-Gas Chlorine. This is the basic chlorine compound. It is used in large volumes by the chemical industry where derivatives of chlorine find use as pesticides for agriculture, plastics, food preservatives, and pharmaceuticals. It is used in large quantities as a bleaching agent for paper and textiles. Only about 4–5 percent of the total annual production in North America is used for sanitary purposes: household bleaches, restaurant sanitizers, potable water treatment, wastewater treatment, swimming pools, cooling waters, and other industrial process water treatment.

Liquid-gas chlorine is manufactured commercially by the electrolysis of a saturated salt solution. The gas collected in the process is moist and must be dried by passing it through a concentrated sulfuric acid solution to remove the moisture. It is then liquified by a combination of compression and cooling and is stored in steel containers from 150 lb cylinders to 90-ton tank cars. Liquid-gas chlorine is the principal form of chlorine used in waste water disinfection. It is also used in wastewater treatment for odor control, destruction of hydrogen dulfide, prevention of septicity, control of activated sludge, bulking, etc.

Hypochlorite:

Imported Hypochlorite. "Available chlorine" can be provided either in the form of sodium or calcium hypochlorite. The most popular form is the sodium hypochlorite. Calcium hypochlorite is much too difficult to manage owing to excessive maintenance problems resulting from the deposition of the calcium ion throughout the system. Therefore, calcium hypochlorite could only be considered an emergency alternative.

Sodium hypochlorite is a clear liquid available in concentrations of 5, 10, and, 17 percent by weight (trade strength, available chlorine—see Chapter 5).

Calcium hypochlorite is available either as a dry granular white powder or in tablet form in strengths of either 35 or 70 percent chlorine by weight.

Imported sodium hypochlorite is being used in some large wastewater treatment plants as a measure to avoid the potential hazard of liquid-gas chlorine delivered and stored in containers under vapor pressures of 80-110 psi.

On-site generation. Complete systems are available for the on-site manufacture of hypochlorite solutions by electrolysis which also avoid the potential hazard of handling the liquid-gas chlorine in pressurized containers. Systems are available for electrolytic cells using either seawater, brackish water, or concentrated salt brines. The hypochlorite is produced in much the same way the liquid-gas chlorine is manufactured, except that there is no need to separate the chlorine gas and the sodium hydroxide which are the products of the electrolysis. This also eliminates the necessity of the sulfuric acid drying step required in the manufacture of liquid-gas chlorine. In addition to the formation of sodium hypochlorite, hydrogen gas is also a product of the electrolytic action. This is diluted with air and vented to atmosphere in concentrations well below the combustible capability of hydrogen. Equipment for this on-site production includes the electrolytic cells, rectifiers, electrical switchgear, brinemaker, brine-treatment unit (where required), water-treatment system for cellwater, cooling equipment, storage tanks for brine and hypochlorite. For economy the process is operated at a constant rate, and excess hypochlorite solution is stored to meet the high demand periods. Experience with these systems at wastewater treatment plants is relatively limited.

Another method of on-site manufacture of hypochlorite of considerable merit is the use of tank car quantities of chlorine gas supplemented by either calcium hydroxide or sodium hydroxide solution to produce an 8000-9000 mg/liter hypochlorite solution. This system is economically appealing from an equipment cost consideration when the average daily chlorine feed rate exceeds 5-6 tons/day (see Chapter 5).

Chlorine Dioxide

Chlorine dioxide is an unstable gas similar to ozone so it must be generated at the point of use. It cannot be stored in steel containers as can chlorine. It is generated on-site as an aqueous ClO_2 solution by reacting a solution of sodium chlorite with the aqueous solution of a conventional chlorinator injector discharge. In wastewater treatment it has a distinct advantage in that it does not combine with the ammonia nitrogen normally present. Therefore, in a nitrogen-laden wastewater it is reputed to have a disinfection efficiency for both bacterial and viral destruction comparable to *free chlorine.* Experience with chlorine dioxide on wastewaters is limited and it is more expensive than chlorine (see Chapter 6).

Bromine, Bromine Chloride and Iodine

Bromine, bromine chloride, and iodine have been used in various ways as an alternative to chlorine. Bromine and bromine chloride are relatively soluble in water, more so than chlorine; however, bromine is much too hazardous a chemical to handle in the treatment of wastewater. Bromine chloride is much easier to handle because it has a vapor pressure of about 30 psi at room temperature. The materials required to meter bromine chloride are considerably different from those of chlorine so a separate species of equipment is required for feeding and metering this gas (see Chapter 7).

Bromine compounds have one distinct advantage over the various chlorine compounds concerning the toxicity of residuals to aquatic life in the receiving waters: bromine residuals die away very rapidly as compared to chlorine residuals.* This means that bromine compound systems might not need debromination facilities. However, this characteristic makes bromine or bromine chloride residual control virtually impossible. Moreover, the bromine compounds are more expensive than those of chlorine.

Iodine is a gray-brown crystalline solid which is only slightly soluble in water. It is derived from kelp or oil field brines, and is mined from deposits in South America. It has been used as an effective method of water treatment on an emergency basis. There are so many unknown factors about iodine as a wastewater disinfectant that this, coupled with its high cost and uncertain availability, conspires to eliminate it from consideration as a practical wastewater disinfectant. The higher molecular weight of both bromine and iodine puts them at a distinct competitive disadvantage to chlorine (see Chapter 7).

Ozone

Ozone is another unstable gas which must be produced at the point of use. It is produced commercially by the reaction of an oxygen containing gas (air or pure oxygen) in an electric discharge. It is a powerful oxidant and has been used since the early 1900s for odor and color removal as well as disinfection of potable-water supplies in Western Europe and Canada. It has been investigated recently as a process for polishing tertiary effluents for both color removal and disinfection. From these recent investigations it appears that ozone in combination with either chlorine or chlorine dioxide could solve the disinfection problem of both bacterial and viral contamination in tertiary wastewater effluents. This is particularly significant where there is consideration of wastewater reuse (see Chapter 8).

---

*The rapid die-away refers to bromamine residuals. Free bromine (HOBr) residuals persist for a signficant period of time.

Ultraviolet Radiation

Ultraviolet radiation (UV) from sunlight has a sterilizing effect on microorganisms but most of the radiation from this natural source is screened out by the atmosphere before reaching the earth's surface. Ultraviolet radiation can be produced by special lamps (mercury vapor) and is presently used in special situations for high quality potable water disinfection. The disinfection reaction occurs on the thin film surfaces of water where the microorganisms can be readily exposed to the radiation reaction. With wastewater, lethal action cannot be exerted through more than a few millimeters. Beyond this limiting distance the high absorption of the rays by the water and the suspended solids dissipates the ultraviolet energy. Consequently, the problems of providing effective exposure of sewage effluents containing varying amounts of interfering suspended solids and ordinary turbidity to the UV rays is such that the practical application must be to a very thin sheet of wastewater flow of nearly uniform thickness. This is an engineering problem of considerable magnitude. Furthermore, the monitoring requirements for proof of disinfection of a UV system are nonexistent.

The UV disinfection concept will be applicable to water treatment situations where the raw water is of highest quality and where the UV radiated effluent is carefully monitored as is the case in a large number of industrial water processes.

The use of UV in wastewater must be limited to special cases of water renovation and not for general use as a wastewater disinfectant unless it is of potable water quality (see Chapter 9).

Gamma Radiation

In addition to UV radiation, gamma radiation has been investigated recently as a method of wastewater treatment and disinfection. In sufficient dosages, gamma radiation is an effective sterilant and is used as a method of sterilizing surgical instruments. Unlike UV radiation, gamma rays are capable of great penetration. Gamma radiation has the ability to alter organic and inorganic molecules and this effect may benefit tertiary treatment processes. Ths most convenient source of energy for irradiation is cobalt 60 which is available in virtually unlimited quantity. The cost of radiation energy is high and gamma radiation as a disinfection process for wastewater is not economically or otherwise competitive with other methods.

The proponents of gamma radiation point out that cesium 137 is the major component of nuclear waste material, which is a by-product of nuclear power plants and is therefore available as a source of gamma rays for water purification, and thereby diminishes the amount of chlorine required for disinfection.

Woodbridge[44] claims six and seven orders of magnitude reduction in the concentration of microorganisms have been obtained by irradiation in less than five min. exposure. The reported advantages of this method include reliability, beneficial side effects and no residual effects. Disadvantages are principally associated with safety needs, excessive cost and virtually no operating experience. The application requires considerably more engineering design information than is now presently available. It should not be ruled out, however, as an adjunct to present methods. Because of lack of information as a wastewater disinfectant, nothing more will be said in this text on the subject.

COMPARATIVE COSTS OF DIFFERENT METHODS OF DISINFECTION

Cost comparisons are unfair at best because all things are not equal between the various methods. One has to weigh the advantages and disadvantages of each method; cost is only one of many factors in evaluating the various methods. The method which will predominate: is one which gets the job done easily; is known to have a minimum health and safety risk; is easy to apply, measure, and control; and for which the handling equipment is reliable and easy to operate. The chlorination-dechlorination method fits this description and will certainly remain popular for a very long time.

TABLE 1-1. Cost Summary

Plant Size (mgd)	1	10	100
Capital Cost	$K	$K	$K
Process			
Chlorine	60	190	840
Chlorine/SO_2	70	220	930
Chlorine/SO_2/aeration[b]	120	360	1,580
Chlorine/carbon	640	2,800	8,400
Ozone/air[a]	190	1,070	6,880
Ozone/oxygen[a]	160	700	4,210
Ultraviolet[a]	70	360	1,780
Bromine chloride	50	130	410
Activated Sludge	1,450	5,790	39,800
Disinfection Cost	¢/Kgal.	¢/Kgal.	¢/Kgal.
Process			
Chlorine	3.49	1.42	0.70
Chlorine/SO_2	4.37	1.75	0.89
Chlorine/SO_2/aeration[b]	7.66	2.39	1.19
Chlorine/carbon	19.00	8.60	3.28
Ozone/air	7.31	4.02	2.84
Ozone/oxygen[a]	7.15	3.49	2.36
Ultraviolet[a]	4.19	2.70	2.27
Bromine chloride	4.52	3.04	2.65
Activated Sludge	55.90	20.20	14.00

[a]Tertiary treatment stage is not included in these costs.
[b]Aeration is not required following dechlorination by SO_2 because a properly designed system will not remove any DO in the effluent. (*author's note*)

In 1976 the EPA published a report on the various methods of wastewater disinfection.[45] Tabulated below is the comparative capital and process cost of the various methods studied (see Table 1-1).

Missing from the above tabulation is chlorine dioxide. The capital cost increase over chlorine for a chlorine dioxide installation is small: probably 15 percent. The following is the process chemical cost for chlorine dioxide assuming chlorine at 15 ¢/lb and sodium chlorite at 70 ¢/lb; 2 mg/liter dose and 30 min. contact time.[46]

TABLE 1-2.

Design capacity (mgd)	1	10	100	150
Chlorine dioxide (¢/Kgal.)	4	2	1	1

Chapter 1—Summary

Many modern wastewater disinfection practices in the United States had their origin in California. In the early 1920's, health officials were alarmed at the possible long-term deleterious effects of raw or poorly treated sewage discharging into the surf waters along the Pacific coast and freshwater streams throughout the state. A comprehensive investigation of the consequences of sewage discharges as related to public health indices resulted in a law suit against the City of Los Angeles. The presiding judge upheld the State Health Department contention that wherever samples taken from the surf exceeded a 1000 MPN/100 ml statistical coliform concentration, it constituted sewage contamination injurious to public health and that the offending dischargers must be made to provide proper sewage treatment.

During the last thirty-five years, the California State Department of Health has formulated a set of requirements for all receiving waters. These include confined saline waters, estuaries, surface waters, ephemeral streams, and shellfish growing areas. An additional constraint is considered whenever the receiving water is used for bathing or water contact sports. The numbers applied for these situations vary from 230 MPN/100 ml for confined saline waters down to 2.2/100 ml for sewage discharging into ephemeral streams or negative estuaries.

This concept of receiving water quality based upon coliform concentration carries with it the hypothesis that certain degrees of treatment are imperative to achieve the various numbers specified. In other words, the designer should not attempt to depend wholly upon a disinfection system to achieve the desired coliform count in the plant effluent. For example, disinfection of a sewage discharging into a confined saline water or estuary cannot consistently accomplish the required 230/100 ml MPN coliform without secondary treatment. Attempts to "disinfect" raw sewage or stormwater overflows are a waste of chemicals, time, and energy. Similarly tertiary treatment is almost always a necessity to meet a 2.2/100 ml MPN standard. When such a severe requirement is placed upon an effluent, there is a further implication of virus destruction. Virus destruction can only be accomplished on high quality effluents, regardless of the disinfectant used.

Nitrification of an effluent is no longer considered a necessity when a low coliform count is required. It was once thought that if an effluent was nitrified then the practice of free residual chlorination could be assured which would result in a more reliable and efficient disinfection system. Through the years 1970-1977, it has been found that nitrification to produce a free chlorine residual is not necessarily worth the effort.

Disinfection studies should represent the degree of disinfection based upon the total coliform concentration in the plant effluent after disinfection.

A mathematical model has been developed which relates the coliform concentration before and after disinfection with chlorine residual at the end of a specified contact time. This model clearly demonstrates that the higher the quality of effluent results in lower numbers of coliforms to be destroyed. This translates to higher efficiency of disinfection.

To date there does not seem to be a better indicator organism for proof of disinfection than the total coliform group.

Wastewater reuse requires a different approach. In these cases the sewage discharge is being used directly for land irrigation, spraying of crops, industrial cooling water, other makeup water requirements, groundwater recharge, prevention of salt water intrusion, and prevention of land subsidence due to underground withdrawals. All of these applications depend heavily upon the unit process of disinfection. It becomes the most important link in this chain of treatment for these applications.

Whenever wastewater is used in situations where there is the possibility of human contact or ingestion, the problem of widespread virus infection becomes the most serious concern of public health officials everywhere.

A recent study by the Los Angeles County Sanitation Districts revealed that contrary to previous beliefs, combined chlorine residuals could be nearly as effective as comparative free chlorine residuals in the destruction of viruses, provided that the disinfection facility is properly designed.

All of the data currently available demonstrate conclusively what the Collins mathematical model tells us: that disinfection efficiency is related directly to the quality of the effluent, and that for any quality of effluent the degree of disinfection is directly related to the total chlorine residual and contact time; provided that mixing is rapid and that the contact chamber demonstrates plug flow conditions.

The phenomenon of regrowth which occurs temporarily in some cases downstream from the point of disinfection is not considered as significant to public health. The public health practitioners recognize this phenomenon. They firmly believe that all pathogenic organisms are destroyed in the disinfection process.

A source of constant worry to any practitioner of disinfection is the clumping phenomenon. It is theorized that clumps of suspended or colloidal-like particles such as are present in raw sewage, primary effluents and poorly treated secondary effluents, may be able to pass through the disinfection system only to break up downstream and spew into the effluent gross amounts of coliforms and possibly pathogens which were sheltered from the disinfectant.

This concept is of grave concern to public health people and is a major reason for assuming that raw sewage and/or primary effluents should not be considered as candidates for the disinfection process.

The toxicity of chlorine residuals to aquatic life is well documented and has given rise to the addition of the dechlorination step to complete the disinfection process, whenever chlorine (or chlorine dioxide) is the disinfectant.

A chlorinated-dechlorinated effluent has been proved to be less toxic than either a chlorinated or a nonchlorinated effluent.

Available methods of disinfection of wastewaters include all of the halogens (Cl_2, I_2, Br_2, BrCl, ClO_2), ozone, ultra violet radiation, gamma radiation, and possibly some combinations with sonics.

Cost effective analyses of these various methods always place the chlorination-dechlorination method in the most favorable position.

From evaluations of the art of disinfection it is clear the process will not provide the desired results unless the other unit processes of the wastewater-treatment system are performing properly. Therefore, a disinfection system is a protective device for public health as well as a sensitive monitor of the entire waster-treatment process.

REFERENCES

1. White, G. C. *Handbook of Chlorination.* Van Nostrand Reinhold Co., N.Y., 1972.

2. *Standard Methods for the Examination of Water and Wastewater.* 14th Ed. American Public Health Assoc. 1975.

3. "Report on a Pollution Survey of Santa Monica Bay Beaches in 1942." California State Board of Health (June 26, 1943).

4. Stevenson, A. H. "Studies of Bathing Water Quality and Health." *Am. J. Public Health* 43: 529 (1953).

5. California State Department of Health Symposium: "Fecal Coliform Bacteria in Water and Wastewater." Berkeley, Calif. (May 21, 1968).

6. Unz, R. F. "Fecal Coliforms and Fecal Streptococci in the Bacteriology of Water Quality," *Water & Sewage Works* 115: 238 (1968).

7. Jopling, W. "Statement in Support of California Recommended Disinfection Requirements." Paper presented by the California State Department of Public Health at a joint meeting of Arizona, California, Nevada, and EPA regarding waste discharge requirements and water quality objectives for the Colorado River, Las Vegas, Nev. (Oct. 28, 1975).

8. "National Symposium on Estuarine Pollution." Stanford Univ. (1967).

9. Jopling, W. and Young, C. Private correspondence. California State Dept. of Health, Berkeley, Calif. (1976).

10. Collins, H. F., Selleck, R. E., and White, G. C. "Problems in Obtaining Adequate Sewage Disinfection." *ASCE J. San. Eng. Div.* **97**: 549 (Oct. 1971).

11. Ptak, D. V., Ginsburg, W., and Willey, B. F. "Identification and Incidence of *Klebsiella* in Chlorinated Water Supplies." *J. AWWA* **65**: 604 (Sept. 1973).

12. Engelbrecht, R. S., Foster, D. H., Masarik, M. T., and Sai, S. H. "Detection of New Microbial Indicators of Chlorination Efficiency." Paper presented at the AWWA Water Technology Conference, Dallas, Tex. (December 1–3, 1974).

13. White, G. C. "Disinfection: The Last Line of Defense for Potable Water." *J. AWWA* **67**: 410 (Aug. 1975).

14. White, G. C. "Disinfection Committee Report." A paper presented at the AWWA Annual Conference, Minneapolis, Minn. (June 1975).

15. Jopling, W. F. "Water Re-use Standards for the State of California." A paper presented at the annual WPCF Conference, Anaheim, Calif. (May 9, 1969).

16. Committee on Environmental Quality Management. "Engineering Evaluation of Virus Hazard in Water." *ASCE J. San. Engr. Div.* **96**: 111 (Feb. 1970).

17. Sobsey, M. D. "Enteric Viruses and Drinking Water Supplies." *J. AWWA* **67**: 414 (Aug. 1975).

18. Katz, M. and Plotkin, S. A. "Minimal Infective Dose of Attenuated Poliovirus for Man." *Am. J.* Public Health **57**: 1837 (1967).

19. McDermott, J. H. "Virus Problems and Their Relation to Water Supply." Presented at Virginia Sect. Meeting, AWWA, Roanoke, Va. (Oct. 25, 1973).

20. McCabe, L. J. "Significance of Virus Problems." Presented at AWWA Water Qual. Conf., Cincinnati, Ohio (Dec. 3 and 4, 1973).

21. Effect of Chlorination on Human Enteric Viruses in Partially Treated Water from Potomac Estuary. Proc. Congr. Hearings, Proc. Serv. No. 92–94. Washington, D.C. (1973).

22. Clarke, N. A. and Kabler, P. W. "Human Enteric Viruses in Sewage." *Health Lab Sci.* **1**: 44 (1964).

23. Geldereich, E. E. and Clarke, N. A. "The Coliform Test: A Criterion for the Viral Safety of Water." (V. Snoeyink and V. Griffin, editors), Proc. 13th Water Qual. Conf., Univ. of Ill., Urbana, Ill. (1971).

24. Berg, G. "Reassessment of the Virus Problem in Sewage and in Surface and Renovated Waters. *Prog. Water Technol.* **3**: 87–94 Pergamon Press, N.Y. (1973).

25. Coin, L. et al. "Modern Microbiological Virological Aspects of Water Pollution." Ad. Water Pollution Research, Proc. 2nd International Conf., Pergamon Press, N.Y., pp. 1–10 (1966).

26. Nupen, E. M., Bateman, B. W., and McKenny, N. C. "The Reduction of Virus by the Various Unit Processes Used in the Reclamation of Sewage in Potable Waters." A paper presented at the Virus Symposium, Austin, Tex. (April 1974).

27. Selna, M. W., Miele, R. P., and Baird, R. B., "Disinfection for Water Reuse." A Paper Presented for the Disinfection Seminar at the Ann. Conf. AWWA, Anaheim, Calif. (May 8, 1977).

28. Ludovici, P. P., Philips, R. A., and Veter, W. S. "Comparative Inactivation of Bacteria and Viruses in Tertiary Treated Wastewater by Chlorination." *Disinfection: Water and Wastewater*, J. D. Johnson (editor), Ann Arbor Science, Ann Arbor, Mich., p. 359 (1975).

29. Pavoni, V. L. and Tittlebaum, M. D. "Virus Inactivation in Secondary Wastewater Treatment Plant Effluent Using Ozone." Water Resources Symp. No. 7, Viruses in Water and Wastewater Systems, J. F. Malina Jr. and B. P. Sagik (editors), Univ. of Texas, Austin, Tex. (April 1974).

30. Boardman, G. D. and Sproul, O. V. "Protection of Viruses During Disinfection by Absorption to Particulate Matter." A paper presented at the 48th Annual Conf. WPCF, Miami, Fla. (Oct. 1975).

31. White, G. C. "Disinfection Facility Evaluation City of San Francisco." Unpublished report (1976).

32. Tsai, C. F. "Effects of Clorinated Sewage Effluents on Fishes in Upper Patuxent River, Maryland." *Chesapeake Sci.* **9**: 83 (June 1968).

33. Jolley, R. J. "Chlorine-Containing Organic Constituents in Chlorinated Effluents." *J. WPCF* **47**: 601 (Mar. 1975).

34. Esvelt, L. A., Kaufman, W. J., and Selleck, R. E. "Toxicity Assessment of Treated Municipal Wastewaters." Paper presented at the 44th Annual Conf. of the WPCF, San Francisco, Calif. (Oct. 4–8, 1971).

35. Esvelt, L. A. Private Correspondence. (Oct. 1971).

36. "Chlorinated Municipal Waste Toxicities to Rainbow Trout and Fathead Minnows." Water Pollution Control Research Series No. 18050 GZZ 10/71,

Bur. of Water Mngmt., Mich. Dept. of Nat. Resources for the EPA (Oct. 1971).

37. Bellanca, M. A. and Bailey, D. S. "A Case History of Some Effects of Chlorinated Effluents on the Aquatic Ecosystem in the Lower James River in Virginia." Paper presented at the 48th Annual Conf. WPCF, Miami Beach, Fla. (Oct. 5–10, 1975).

38. Collins, H. F. and Deaner, D. G. "Sewage Chlorination Versus Toxicity—A Dilemma?" *ASCE J. Environ. Eng. Div.*, 761 (Dec. 1973).

39. Arthur, J. W., Andrew, R. W., Mattson, V. R., Olson, D. T., Glass, G. E., Halligan, B. J., and Walbridge, C. T. "Comparative Toxicity of Sewage Effluent Disinfection to Freshwater Aquatic Life." EPA Report 600/3-75-012. Research Lab., Duluth, Minn. (Nov. 1975).

40. Brungs, W. A. "Effects of Wastewater and Cooling Water Chlorination on Aquatic Life." EPA Report 600/3-76-098. Research Lab., Duluth Minn. (Aug. 1976).

41. Esvelt, L. A., Kaufman, W. J., and Selleck, R. E. "Toxicity Removal from Municipal Wastewaters." SERL No. 71-8. Sanitary Engineering Research Lab., Univ. Of Calif., Berkeley (Oct. 1971).

42. Kaufman, W. J. Private Correspondence. (1972).

43. Stone, R. W., Kaufman, W. J., and Horne, A. J. "Long-Term Effects of Toxicity and Biostimulants on the Waters of Central San Francisco Bay." SERL Report 73-1. Univ. of Calif., Richmond, Calif. (1973).

44. Woodbridge, D. D. and Cooper, P. C. "Reduction of Chlorination by Irradiation." Paper presented at the AWWA Disinfection Seminar, Anaheim, Calif. (May 8, 1977).

45. A Task Force Report, "Disinfection of Wastewater." U.S. Environmental Prot. Agency No. 430/9-75-012, Washington, D.C. (March 1976).

46. Love, O. T. Jr., Carswell, N. K., and Symons, J. M. "Comparison of Practical Alternative Treatment Schemes for Reduction of Trihalomethanes in Drinking Water." A paper presented at the IOI Workshop on Ozone and Chlorine Dioxide, Cincinnati, Ohio (Nov. 17–19, 1976).

47. Katzenelson, E. and Teltch, B. "Dispersion of Enteric Bacteria by Spray Irrigation." *J. WPCF*, 48: 710 (April 1976).

48. Katzenelson, E., Buium, I., and Shuval, H. I. "Risk of Communicable Disease Infection Associated With Wastewater Irrigation in Agricultural Settlements." *Science* 194: 944 (Nov. 26, 1976).

2

Chemistry and kinetics of the chlorination - dechlorination method

---

FUNDAMENTALS OF CHLORINE CHEMISTRY

Introduction

Many reactions occur when chlorine is added to wastewater. Some occur immediately; others develop over a period of hours or days. All of these reactions are affected in some way by temperature, pH, buffering capacity of the wastewater, and the molar strength of the chlorinating agent. To get an insight into these diverse reactions, it is appropriate to review the fundamentals of chlorine-water chemistry.

Hydrolysis of Chlorine Gas

When chlorine gas is dissolved in water it hydrolyzes rapidly according to the equation:

$$Cl_2 + H_2O \rightleftharpoons HOCl + H^+ + Cl^- \qquad (2\text{-}1)$$

The rapidity of this reaction has been studied by many investigators.[1] The complete hydrolysis occurs in a few tenths of a second at 18°C; at 0°C only a few seconds is needed. This unusually rapid rate of reaction is best explained if the mechanism is a reaction of the chlorine molecule with the hydroxyl ion rather than with the water molecule. This can be represented as follows:

$$Cl_2 + OH^- \rightleftharpoons HOCl + Cl^- \tag{2-2}$$

The rate constant for this reaction is about 5×10^{14} indicating that the reaction occurs at almost every collision of ions.[2] This reaction is of great practical importance because it relates to the chemistry of aqueous chlorine solutions discharging from conventional chlorination equipment. The resulting solution in a chlorinator discharge is limited by design to 3500 mg/liter. At this concentration the most highly buffered injector water would result in a pH of no higher than 3. At this pH the amount of molecular chlorine in equilibrium with HOCl is substantial. Concentrations higher than 3500 mg/liter cause excessive chlorine gas release at the point of application which is extremely undesirable. Likewise if negative pressures exist in the chlorine solution piping, this contributes to the release of molecular chlorine at the point of application. In addition to the degassing effect, the release of gas in the solution piping has been known to adversely affect the hydraulic gradient between the injector and the point of application. Injector systems are usually designed to maintain at least 2 psig at the injector discharge. At this pressure and a temperature of 20°C, the solubility of chlorine in water is only about 7.5 g/liter.[3]

To demonstrate the relationship of the molecular chlorine-hypochlorous acid equilibrium for both buffered and unbuffered water, the following tables have been compiled from a computer printout provided by the Bioengineering Research and Development Lab, U.S. Army, Fort Detrick, Maryland.[4] The results are based upon the Cl_2-HOCl equilibrium; the Cl_3^- ion formation from Cl_2 and the chloride ion; a mass balance for all chlorine species; and an ion balance on Cl^-. Thus the mole percent for HOCl in the following tables is based upon a lengthy and complex cubic equation* which is best described as follows:

$$\text{Percent HOCl} = \frac{100 \times (HOCl)}{[(HOCl) + (Cl_2) + (OCL^-) + (Cl_3^-)]} \tag{2-3}$$

Table 2-1 illustrates what happens in the chlorine solution discharge from a chlorinator ranging in feed rates to produce concentrations varying from 500-3500 mg/liter. It also demonstrates the necessity for maintaining a constant high concentration of chlorine (e.g., 1500-2000 mg/liter at a low pH in the generation of chlorine dioxide). The molecular chlorine present in the solution coming in contact with the sodium chlorite provides the impetus for a fast and complete reaction.

Table 2-2 demonstrates the stability of a chlorine water solution buffered with either sodium hydroxide or calcium hydroxide. These figures are of interest for on-site manufacture as well as on-site generation of hypochlorite (see Chapter 5).

*This equation is shown in the Appendix.

TABLE 2-1   Percent Molecular Chlorine and Hypochlorous Acid in a Water Solution Buffered from pH 1-6 at 15°C at Atmospheric Pressure

	Solution Concentration (mg/liter)									
	500		1000		1500		2000		3500	
pH	Cl_2	HOCl	Cl_2	HOCl	Cl_2	HOCl	Cl_2	HOCl	Cl_2	HOCl
1	54.30	45.65	64.67	35.25	69.94	29.95	73.29	26.57	78.91	20.89
2	17.66	82.31	27.41	72.52	33.95	65.93	38.78	61.05	49.70	49.97
3	2.48	97.51	4.73	95.25	6.79	93.17	8.68	91.26	13.57	86.28
4	0.26	99.72	0.52	99.46	0.77	99.20	1.02	98.45	1.76	98.19
5	0.026	99.74	0.05	99.71	0.078	99.68	0.104	99.66	0.181	99.58
6	0.000	97.68	0.005	97.67	0.008	97.67	0.010	99.67	0.018	97.66

In any hypochlorite solution the active ingredient is always hypochlorous acid.

$$NaOCl + H_2O \longrightarrow HOCl + Na^+ + OH^- \qquad (2\text{-}4)$$

$$Ca(OCl)_2 + 2H_2O \longrightarrow 2HOCl + Ca^{++} + (OH)^= \qquad (2\text{-}5)$$

When a chlorine solution, such as the solution discharge of a conventional chlorinator (unbuffered), is subjected to negative pressure conditions, the solubility is reduced which usually results in the release of molecular chlorine at the point of application provided the diffuser is in an open body of water such as an open channel.

For example, at atmospheric pressure and 20°C, the maximum solubility of chlorine is about 7395 mg/liter. However, if the solution is subjected to a negative head of 9 in. Hg the solubility is reduced to about 5560 mg/liter.[4] There-

TABLE 2-2   Percent Molecular Chlorine, Hypochlorous Acid and OCl⁻ Ion in a Water Solution Buffered from pH 6-9 at 20°C

	Solution Concentration (mg/liter)								
	5000			7000			10000		
pH	Cl_2	HOCl	OCl^-	Cl_2	HOCl	OCl^-	Cl_2	HOCl	OCl^-
6.5	.0063	92.28	7.71	.0088	92.28	7.71	.0126	92.28	7.71
7.0	.0017	79.10	20.89	.0024	79.10	20.89	.0034	79.10	20.89
7.5	.0004	54.84	45.51	.0005	54.49	49.51	.0007	54.49	45.51
8.0	.0001	27.46	72.54	.0001	27.46	72.54	.0001	27.46	72.54
8.5	.0000	10.69	89.31	.0000	10.69	89.30	.0000	10.69	89.30
9.0	.0000	3.65	96.35	.0000	3.65	96.35	.0000	3.65	96.35

fore, all systems that are not closed should be designed to avoid negative pressure conditions in the chlorinator solution lines.

Chemistry of Hypochlorous Acid

The next most important reaction in the chlorination of an aqueous solution is the formation of hypochlorous acid. This species of chlorine is the most germicidal of all chlorine compounds with the possible exception of chlorine dioxide.

Hypochlorous acid is a "weak" acid which means that it tends to undergo partial dissociation as follows:

$$HOCl \rightleftharpoons H^+ + OCl^- \tag{2-6}$$

to produce a hydrogen ion and a hypochlorite ion. In waters of pH between 6.5 and 8.5 the reaction is incomplete and both species are present to some degree. The extent of this reaction can be calculated from the equation

$$Ki = \frac{(H^+)\,(OCl^-)}{(HOCl)} \tag{2-7}$$

Ki, the ionization constant, varies in magnitude with temperature. The values of this constant shown in Table 2-3 have been computed from the acid dissociation constant, pKa, based on J. C. Morris,[5] best fit formula developed in 1966 as follows:

$$pK_a = \frac{3000.00}{T} - 10.0686 + 0.0253T \tag{2-8}$$

where $T = 273 +$ degrees centigrade.

Table 2-4 shows the percent undissociated HOCl species for the various temperatures and pH values from 5–11.7. The percent OCl^- ion is the difference between these numbers and 100.

Reactions with Wastewater

There are numerous constituents present in wastewater which react immediately with the HOCl from the chlorinator injector discharge or the hypochlorite solution. Consequently, free chlorine $(HOCl + OCl^-)$ is probably consumed or converted to some form of chloramine in a matter of seconds after mixing with the

TABLE 2-3

Temperature (°C)	0	5	10	15	20	25	30
Ki $\times 10^{-8}$ (moles/liter)	1.488	1.753	2.032	2.320	2.611	2.898	3.175

TABLE 2-4[a]

	Percent HOCl						
pH	0°C	5°C	10°C	15°C	20°C	25°C	30°C
5.0	99.85	99.82	99.80	99.79	99.74	99.71	99.68
5.5	99.53	99.45	99.36	99.27	99.18	99.09	99.00
6.0	98.53	98.28	98.00	97.73	97.45	97.18	96.92
6.1	98.16	97.84	97.50	97.16	96.82	96.48	96.15
6.2	97.69	97.29	96.88	96.45	96.02	95.60	95.20
6.3	97.11	96.62	96.10	95.57	95.05	94.53	94.04
6.4	96.39	95.78	95.14	94.49	93.84	93.21	92.61
6.5	95.50	94.75	93.96	93.16	92.37	91.60	90.87
6.6	94.40	93.47	92.51	91.54	90.58	89.65	88.78
6.7	93.05	91.92	90.75	89.58	88.43	87.32	86.27
6.8	91.41	90.03	88.63	87.23	85.85	84.54	83.31
6.9	89.42	87.77	86.10	84.43	82.82	81.29	79.86
7.0	87.04	85.08	83.10	81.16	79.29	77.53	75.90
7.1	84.22	81.92	79.63	77.39	75.26	73.27	71.44
7.2	80.91	78.25	75.64	73.11	70.73	68.52	66.52
7.3	77.10	74.08	71.15	68.35	65.75	63.36	61.22
7.4	72.78	69.42	66.20	63.18	60.39	57.87	55.63
7.5	67.99	64.33	60.88	57.68	54.77	52.18	49.90
7.6	62.79	58.89	55.27	51.98	49.03	46.43	44.17
7.7	57.27	53.23	49.54	46.23	43.32	40.77	38.59
7.8	51.57	47.48	43.81	40.58	37.77	35.35	33.30
7.9	45.82	41.79	38.25	35.17	32.53	30.28	28.39
8.0	40.18	36.32	32.98	30.12	27.69	25.65	23.95
8.1	34.79	31.18	28.10	25.50	23.32	21.51	20.01
8.2	29.77	26.46	23.69	21.38	19.46	17.88	16.58
8.3	25.19	22.23	19.78	17.76	16.10	14.74	13.63
8.4	21.10	18.50	16.38	14.64	13.23	12.07	11.14
8.5	17.52	15.28	13.46	11.99	10.80	9.84	9.06
8.6	14.44	12.53	11.00	9.77	8.77	7.97	7.33
8.7	11.82	10.22	8.94	7.92	7.10	6.44	5.91
8.8	9.62	8.29	7.23	6.39	5.72	5.18	4.75
8.9	7.80	6.70	5.83	5.15	4.60	4.16	3.81
9.0	6.29	5.39	4.69	4.13	3.69	3.33	3.05
9.5	2.08	1.77	1.53	1.34	1.19	1.08	0.98
10.0	0.67	0.57	0.49	0.43	0.38	0.34	0.31
10.5	0.21	0.18	0.15	0.14	0.12	0.11	0.10
11.0	0.07	0.06	0.05	0.04	0.04	0.03	0.03
11.5	0.02	0.02	0.015	0.013	0.012	0.01	0.01
11.7	0.01	0.01	0.01	0.01	0.007	0.007	0.006

[a]Computer printout *courtesy* D. S. Cherry, N.C. State Univ., Raleigh, N.C.

wastewater stream, owing to the presence of ammonia nitrogen. Very little is known about this specific reaction. Simultaneously with the Cl_2-NH_3 reaction, are the other inorganic reactions of chlorine with reduced substances such as $S^=$, HS^-, $SO_3^=$, NO_2^-, Fe^{++}, Mn^{++}, etc. These substances react with both the available HOCl and combined chlorine (NH_2Cl, $NHCl_2$) to reduce these compounds so that the active chlorine compound is eventually reduced to the stable chloride ion which is nonbactericidal. At this point, there is no measurable residual. The chlorine consumption in the first minute of reaction is probably due to the reactions with inorganic substances. The reactions occurring in the following 2 or 3 min. are probably of organic chlorine demand and are much slower reactions. The 10 min. chlorine demand of a fresh domestic sewage may be as low as 5 mg/liter to produce a measurable residual; however, this figure may escalate to 40 mg/liter if the same sewage becomes septic. Ammonia nitrogen does not exert a chlorine demand on the chlorination process until the breakpoint reaction is reached. Ammonia nitrogen converts the applied chlorine to chloramines.

The most significant chemical reactions between chlorine and the various chemical constituents in wastewater effluents are those with the various nitrogenous compounds, either inorganic (ammonia nitrogen) or organic (proteins and their degradation products).

Ammonia Nitrogen. With the exception of highly nitrified effluents there is usually an appreciable amount of ammonia nitrogen in all wastewater effluents. The range is on the order of 10 to 40 mg/liter. The ammonium ion exists in equilibrium with ammonia nitrogen and hydrogen and the distribution is dependent upon pH and temperature. The relative distribution can be defined as follows:

$$NH_4^+ \rightleftharpoons NH_3 + H^+; \text{where } K = 5 \times 10^{-10} \text{ at } 20°C \qquad (2\text{-}9)$$

where

NH_3 = nondissociated ammonia
NH_4^+ = ammonium ion
H^+ = hydrogen ion
K = dissociation constant

According to the dissociation constant, the pH value at which nondissociated ammonia and ammonium ion are present in equal proportions (pK) is about pH 9.3 at 20°C. Above pH 9.3 nondissociated ammonia (NH_3) predominates; below pH 9.3 ammonium ion (NH_4^+) predominates.

At usual wastewater pH levels the predominant chlorine reaction proceeds as follows:

$$HOCl + NH_4^+ \rightleftharpoons NH_2Cl + H_2O + H^+ \qquad (2\text{-}10)$$

when the chlorine to ammonia nitrogen weight ratio is less than 5:1. If the pH drops below 7, dichloramine ($NHCl_2$) will begin to form, and at a much lower pH nitrogen trichloride will form. However, as soon as the 5:1 weight ratio of chlorine to ammonia nitrogen is exceeded a new set of reactions takes place and will be described below.

Germicidal Effectiveness of Chlorine Compounds. The predominant chlorine compounds formed in the chlorination of wastewater are those of combined chlorine. Free chlorine does not exist in sewage effluents, not even for a fraction of a second because at neutral pH in an environment of excess ammonia nitrogen the conversion of free chlorine to monochloramine is practically instantaneous. Free chlorine residuals may however exist in a highly nitrified effluent.

The inorganic chloramines initially formed in wastewaters tend to convert with time to less reactive chloro-compounds which are most probably organic chloramines. These latter compounds have much less germicidal efficiency than the inorganic chloramines; i.e., mono-, di-, or tri-chloramine.

It has been well documented that chloramines are inferior to free chlorine as a disinfectant.[1] However, recent work tends to disprove this. (See Chapter 1.) Under proper conditions a chloramine residual might be nearly as bactericidal or virucidal as free chlorine.[26]

In 1967 Morris[29] presented a tabulation of germicide concentrations giving 99% inactivation within ten minutes contact time. From this Morris derived a lethality coefficient:

$$\Lambda = 0.46/C_{99:10}$$

where C is the concentration of the chlorine compound in mg/liter. The values of Λ computed from this 1967 tabulation are shown in Table 2-5.

Monochloramine compared to free chlorine is a much less reactive compound; therefore, it persists over a much longer period of time in an environment of reducing substances. In wastewater it hydrolyzes to form organic chloramines owing to the presence of organic nitrogen compounds. This has an adverse effect upon the germicidal efficiency of the chloramine residual. This effect is discussed below.

TABLE 2-5   Values of Λ at 5 C.[29]

Λ in $(mg/liter)^{-1}$ $(min.)^{-1}$

Species	Enteric Bacteria	Amoebic Cysts	Viruses	Spores
$HOCl$ as Cl_2	20	0.05	1.0 up	0.05
OCl^- as Cl_2	0.2	0.0005	<0.02	<0.0005
NH_2Cl as Cl_2	0.1	0.02	0.005	0.001

Organic Nitrogen. In addition to reacting with free ammonia nitrogen in wastewater, chlorine may combine with amino acids, proteinaceous material, and many other sewage constituents. Very little is known about the extent of such reactions, their reversibility, or the disinfecting power of the products. Some organic chloramines have little or no germicidal power at pH values near 7 but they titrate as combined chlorine residual using the iodometric and DPD methods. Therefore, it is generally concluded that ammonia chloramines are far superior to organic chloramines as disinfecting compounds, and that the process of chlorination is hindered by the presence of organic nitrogen. However, it is generally conceded that the ammonia chloramines formed by the presence of the ammonia nitrogen in wastewater will slowly hydrolyze with time to react with organic-nitrogen compounds present to form organic-chloramine compounds. This may explain the phenomenon of decreasing germicidal efficiency of a combined chlorine residual with time.

It has been shown in water treatment that the nuisance residuals of combined chlorine which appear at and beyond the breakpoint may be organic chloramines with very little germicidal power. These chlorine residual fractions respond to quantitative determination, mostly as dichloramine. As early as 1966 Feng[6] discovered a great disparity in the germicidal efficiency between ammonia chloramines and those found in an environment of pure organic nitrogen compounds. He has reported that methionine, which is one of the indispensable amino acids for biological growth that is expected to be present in wastewater, forms a measurable chlorine residual with no germicidal power. Feng also investigated the lethal activities of the glycine, taurine, and gelatin chloramines. His work shows that taurine chloramines are as lethally active as ammonium chloramines at pH 9.5, but that their germicidal efficiency falls off as the pH decreases. The glycine chloramines are as germicidally active as monochloramine at pH 4 but are totally inert at pH 7, and the gelatin chloramines are active at pH 9.5 but are inert at pH 7 and 4. There are certain to be other such organic-nitrogenous compounds which contribute to the total chlorine residual which have little or no germicidal effect.

In 1973–1975 White made several chlorine residual profiles in secondary effluents using the forward amperometric titration procedure. These studies revealed two significant facts: 1) the combined chlorine residual titrated as much as twenty percent dichloramine; 2) the dichloramine fraction increased with time. The ammonia nitrogen content of these effluents was in the range of 15 to 30 mg/liter and the chlorine dosages were less than 20 mg/liter. So there was no possibility of the existence of free chlorine residual. It is reasonable to assume that the fraction titrating as dichloramine probably represents organic chloramines since it is not likely that a true dichloramine would form without an excess of free chlorine.

Sung[7] made a controlled laboratory study of fifteen organic compounds representing seven groups to evaluate their individual and combined effect upon the chlorination process. Nine of the fifteen compounds were found to interfere with the germicidal efficiency of the chlorination process. Of these nine compounds five were organic nitrogen compounds. Cystine and uric acid were the severest inhibitors of the nitrogen group. When five of the interfering compounds were mixed together, their combined effect was found to be more pronounced than any of their individual effects, but did not equal the sum of their individual effects. Sung compared the germicidal efficiency of a simulated wastewater with and without the interfering organic compounds. He found that the germicidal efficiency of wastewater containing the interfering compounds by themselves and the resulting chlorine residuals had little or no germicidal effect. The greatest interference was observed to be caused by cystine, tannic acid, humic acids, uric acids and arginine.

Cystine is an amino acid connected by two sulfur groups that is known to react with chlorine. Tannic, humic and uric acids are capable of exerting a significant chlorine demand when present in water or wastewater. *Arginine* is a basic amino acid. The reaction between chlorine and arginine is almost instantaneous.

The organic compounds that had little or no interfering effects on the chlorination process were: acetic acid, cellubiose, dextrose, glutamic acid, uracil, and lauric acid.

The significance of the above findings by Sung[7] confirms the theory of interference in the chlorination process by the presence of organic nitrogen. It further points to the fact that present analytical techniques do not provide for separating the chlorine residual fractions into those of equal germicidal efficiency.

It is also interesting to note that Esvelt, Kaufman and Selleck[8] found that the toxicity of combined chlorine residuals diminished with time. This finding together with those of Sung[7] demonstrates quite convincingly that there are a significant number of organic compounds in wastewater which will react with chlorine to form organic chloramines of little or no germicidal potential, and moreover, these compounds appear to increase in concentration with time. This apparent increase of this chlorine residual fraction with time would also explain the loss of germicidal efficiency of combined chlorine residuals in wastewater with time. The supposition is that the predominantly monochloramine residual present hydrolyzes to form free chlorine (HOCl), and both the monochloramine and HOCl react with the organic nitrogen compounds to form the ineffective organic chloramines.

It also appears that these chloramines titrate as part of the total combined residual as outlined in the 14th Ed. of *Standard Methods* using either the amperometric end-point of the iodometric procedure or the DPD-FAS titrimetric procedure. It would be helpful if future research could provide an analytic procedure for chlorine residuals which could differentiate between the anomalous low-germicidal efficient residuals and the pure mono- and dichloramine residuals in wastewater effluents.

Disinfection Efficiency in the Absence of NH_3-N. The chemistry of wastewater chlorination takes on a different aspect when a secondary effluent becomes nitrified. This situation occurs inadvertently in some plants when there is an overload due to cannery wastes. In one such plant, when the ammonia nitrogen disappears, the organic nitrogen concentration increases from about 6 to 10 mg/liter and simultaneously the disinfection efficiency of the chlorination system falls off suddenly and dramatically.[9] In such a situation one would expect the appearance of the free chlorine species and a subsequent improvement in the disinfection efficiency. At this particular plant free chlorine has never been detected using forward titration procedures. To correct for the drop in disinfection efficiency the plant personnel increased the chlorine dosage to the maximum capability of the system. While this increase in dosage represented almost double the dosage when ammonia nitrogen was present in the effluent, (20 mg/liter) it was never possible to achieve the 230/100 ml MPN coliform concentration in the effluent when the ammonia nitrogen disappeared. Whenever the ammonia concentration in the effluent exceeds 1 mg/liter there is never any problem in meeting the disinfection requirement with chlorine dosages below the breakpoint.

To overcome this disappearance of ammonia nitrogen during the canning season, the plant influent is seeded with the secondary digester supernatant which contains approximately 550 mg/liter ammonia nitrogen. There is sufficient quantity of this liquor to dose the plant influent with 10 mg/liter NH_3-N. This remedial action restores the disinfection efficiency.

Nitrified Effluents. The anomaly described above gives rise to speculation about nitrified effluents. From this evidence it might be advisable never to allow the ammonia nitrogen content to drop below 0.5 mg/liter. However, some plants that are now nitrifying to near zero ammonia nitrogen in the effluent and have a low-organic nitrogen concentration (1-2 mg/liter) produce free chlorine residuals, which achieve a high-disinfection efficiency. For example, these plants can achieve consistently 2.2 MPN coliforms/100 ml with chlorine dosages as low as 7-8 mg/liter providing a 1.5 mg/liter free residual at the end of 30 min. contact time.[10]

Plants producing nitrified effluents containing small amounts of ammonia nitrogen (less than 1.0 mg/liter) should experiment with dosages below and beyond the breakpoint if the coliform requirement is 2.2/100 ml in the effluent. There is no need to waste chlorine on the breakpoint reaction unnecessarily. A great many plants producing high-quality activated sludge effluents are capable of achieving coliform MPN concentrations of 23/100 ml with 100 percent combined chlorine residuals.

Any high-quality tertiary effluent which has not been nitrified can be disinfected by combined chlorine to produce an effluent showing a MPN coliform concentration of 2.2/100 ml with chlorine dosages as low as 6-8 mg/liter.[9] Assuming that the chlorination system is optimized, the results depend entirely upon the "quality" of the effluent as described below.

Effluent Quality as Related to Chlorine Dosage. The single most important effluent quality parameter as it affects the efficiency of the chlorination process is the coliform concentration in the treated effluent prior to the application of chlorine. This conclusion is based on the extensive unpublished field studies by White over a period of several years (1970-1976).

The basic concept of adequate wastewater disinfection is expressed in the mathematical model developed by Collins et al.[11] in 1970 based on a comprehensive pilot-plant study. This study resulted in the formulation of an equation which has been substantiated many times since its publication: *that if there is good mixing at the point of chlorine application and if there are plug flow conditions in the contact chamber (no short circuiting) one can expect a definitive coliform reduction with a given chlorine residual at the end of a specified contact time.* This is the *ct* relationship commonly referred to elsewhere as *the chlorine concentration–contact time envelope.*

The original model[11] has been subsequently fine-tuned by Collins based on plant-scale studies.[12] This recent work reinforces the practical aspects of his original model which is represented by the following equation:

$$Y = Y_o [1 + 0.23\, ct]^{-3} \qquad (2\text{-}11)$$

where

Y = MPN in chlorinated wastewater at end of time t
Y_o = MPN in effluent prior to chlorination
c = total chlorine residual, mg/liter at the end of contact time t
t = contact time, in min.

The validity of this equation has since been substantiated by others based on actual full-scale plant operation.

It should be pointed out that the mathematical model shown here was developed from a pilot-plant system that had excellent mixing in a highly turbulent regime and an ideal plug flow reactor for a contact chamber.

Now let us try some numbers to see its affect upon the three variables in order to appreciate the significance of the MPN concentration before and after chlorination. But first it should be pointed out that the minimum contact time for disinfection should be at least 30 min. unless the effluent is a highly oxidized 60-100 day pond effluent following secondary treatment, then the contact time can be reduced to a minimum of 15 min.

1. Primary effluent: Y_o is usually about $35 \times 10^6/100$ ml. A surf water requirement in California is 1000/100 ml MPN, so $Y = 1000$. Substituting these

values in the equation, c calculates to 4.6 when t is 30 min.:

$$1 + 0.23\ ct^3 = \frac{35 \times 10^6}{1000}$$

$$ct = 138 \qquad\qquad (2\text{-}12)$$

The residual die-away in a good domestic effluent would be about 6-8 mg/liter. This means the dosage required is on the order of 11-13 mg/liter for an optimized system and a normal chlorine demand.

2. Secondary effluent: Y_o is quite often on the order of 1×10^6 in a good secondary effluent. If the effluent were discharging into a confined body of water with plenty of dilution the coliform requirement might be 230/100 ml MPN. In this case c calculates to 2.2 mg/liter for $t = 30$ min. Such an effluent would require a dosage of about 7-9 mg/liter. If the disinfection requirement were 23/100 ml MPN total coliforms, then c calculates to 4.95.

3. Tertiary Effluent. The ease with which these effluents can be treated depends a great deal upon whether or not the filtered effluent has been preceded by coagulation and sedimentation. A conventional water reuse situation would consist of filtration of secondary effluent preceded by coagulation and sedimentation. The Y_o of such an effluent would probably range between 3000 and 10,000 coliform per 100 ml. In California, whenever tertiary effluent is required the coliform requirement is always the same as that for potable water, namely, 2.2/100 ml. Assuming $Y_o = 10,000$ and $Y = 2.2$ then c calculates to 2.25 mg/liter for $t = 30$ min. A tertiary effluent with the predisinfection processes described above would require chlorine dosages on the order of 5-7 mg/liter.

A filtered effluent with chemical coagulation but without sedimentation produces an effluent with coliform concentrations considerably higher, on the order of 50,000/100 ml. In this case using $Y = 2.2/100$ ml, c calculates to 3.96 mg/liter for $t = 30$ min. This is almost twice the residual required in the preceding case.

These examples clearly demonstrate the effect of effluent quality as it relates to the Y_o coliform concentration. The easiest effluents to disinfect that White[9] has investigated to date is a secondary effluent followed by 100 day ponds. In these particular cases, the Y_o rarely exceeds 4000/100 ml and the final coliform is usually less than 3/100 ml using chlorine dosages on the order of 3.5 to 4.0 mg/liter and 15 min. contact time. The chlorine residuals at the end of 15 min. are usually on the order of 2 mg/liter. This is what an optimized system can do when Y_o is a low figure.

Anomalies of Wastewater Chlorination

There are always situations which do not conform to predictable patterns. The anomaly described above which can occur when the ammonia nitrogen disappears from the effluent is a case in point. Another is the recent findings by Selna et al.[26] (see Chapter 1) which demonstrated that combined chlorine if given sufficient time was almost equivalent to free chlorine as a virucide. This is a new concept.

White has found that some tertiary effluents have great difficulty in responding to conventional chlorination practices, even though free chlorine residuals of abnormal concentrations are unable to achieve consistent MPN coliform concentrations of 23/100 ml. Fortunately these cases are a rarity, but they do deserve extensive investigation because they cannot be dismissed lightly.

In 1976 Collins et al. reported the following anomalies.[27] A comparison was made of free chlorine residuals in local potable water (seeded with settled sewage organisms) with a tertiary effluent consisting of a well oxidized activated sludge subjected to high pH lime precipitation (pH 11), sedimentation, recarbonation, and sand filtration. The tertiary effluent was dosed past the breakpoint and it was found that the rate of kill was about 150 times slower in the tertiary effluent. Collins hypothesized that this gross difference in disinfection efficiency might be the direct result of bacterial encapusalation in the calcium compounds resulting from the lime precipitation process. His assumption was validated when Ingols[28] discovered this phenomenon while studying the effect of calcium ion on the disinfection efficiency of swimming pool-treatment methods.

The other anomaly is again the comparison of free versus combined chlorine disinfection efficiency. Collins et al.[27] show in their Figure 11 that if given enough time, combined chlorine will eventually "catch-up" to free chlorine in bactericidal efficiency. Up until now this has never been considered a possibility. A quick answer to this phenomenon is that combined chlorine residuals (mono and dichloramine) might simply take longer to diffuse through the cell walls of the bacteria. Again, this is worthy of further investigation as it corroborates the findings of Selna et al.[26] on their virus studies (see Chapter 1).

Furthermore, Collins et al.[27] made a tentative conclusion, which by their own admission needs additional experimental verification, that there may exist a limit on the bactericidal action of chlorine beyond which it makes no difference whether free residual chlorination is practiced or not.

THE BREAKPOINT REACTION

In order to fully understand the chemistry of wastewater disinfection, it is desirable to review the breakpoint reaction; although this is not a primary goal to be achieved in the disinfection of wastewater. This reaction occurs when sufficient chlorine has been added to the wastewater to cause the chemical oxidation of the ammonium ion in solution to nitrogen gas and other end products. Some disagreement still exists in the literature concerning the actual chemical pathway to the end products.[1] However, a recent (1976) comprehensive study by

Saunier[13,14], with the latest sophisticated methods, under the guidance of Prof. Robert E. Selleck, Univ. of California, Berkeley, Calif., has developed a computer model based on more than 80 experimental runs, within a pH range of 6–9; ammonia nitrogen from 1 to 20 mg/liter; and chlorine to nitrogen molar dose ratios of 1.6:3.5; and temperatures between 12°C and 21°C. The model as developed by this research can be used to compute the production of the various species of chlorine compounds formed during the breakpoint reaction; the pH of the water immediately after chlorine addition; the amount of chemical required for buffering the reaction to the optimum pH for the greatest speed of the reaction, as well as the decrease in pH as the reaction proceeds. Based on Saunier's[13] research the following set of reactions appears to be the most reasonable:[15]

$$NH_4^+ + HOCl \Longrightarrow NH_2Cl + H_2O + H^+ \qquad (2\text{-}12)$$

$$NH_2Cl + HOCl \Longrightarrow NHCl_2 + H_2O \qquad (2\text{-}13)$$

$$0.5\ NHCl_2 + 0.5\ H_2O \Longrightarrow 0.5\ NOH + H^+ + Cl^- \qquad (2\text{-}14)$$

$$0.5\ NHCl_2 + 0.5NOH \Longrightarrow 0.5N_2 + 0.5\ HOCl + 0.5H^+ + 0.5\ Cl^- \qquad (2\text{-}15)$$

and finally:

$$NH_4^+ + 1.5\ HOCl \longrightarrow 0.5N_2 + 1.5\ H_2O + 2.5\ H^+ + 1.5\ Cl^- \qquad (2\text{-}16)$$

The above reactions are based on the formation of NOH, apparently a catalytic intermediary compound which Saunier and Selleck believe is a result of the formation of hydroxylamine (NH_2OH) as an intermediate reaction.[13,14] These reactions are as follows:

$$NHCl_2 + 2H_2O \longrightarrow NH_2OH + HCl + HOCl \qquad (2\text{-}17)$$

$$NH_2OH + HOCl \longrightarrow NOH + HCl + H_2O \qquad (2\text{-}18)$$

$$NOH + NHCl_2 \longrightarrow N_2 + HOCl + HCl \qquad (2\text{-}19)$$

The formation of the hydroxylamine and the resulting formation of the catalytic compound NOH greatly increases the speed of the breakpoint reaction as the ammonia nitrogen concentration increases with the greater NOH production.

The breakpoint occurs through the sequential formation of monochloramine and dichloramine with the subsequent catalytic decomposition of dichloramine to produce an end product of nitrogen gas, with a partial return of free chlorine residual (HOCl) to the solution. These reactions confirm that 1.5 moles (gram molecular weight) of chlorine are required to oxidize 1.0 mole of ammonia to nitrogen gas.

Stoichiometrically, the breakpoint reaction requires a weight ratio of chlorine to ammonia nitrogen $(Cl_2 : NH_4{}^+)$-N) at the breakpoint of 7.6:1 as shown below:

$$\text{Molecular wt HOCl} = 70.9 \text{ (as } Cl_2)$$

$$\text{Moles HOCl required} = 1.5$$

$$\text{Molecular wt HH}_4{}^+ = 14 \text{ (as N)}$$

$$\text{Moles NH}_4{}^+ \text{ required:} = 1.0$$

Therefore, $Cl_2 : NH_4{}^+$-N = (1.5) (70.9):(1.0) (14.0) = 7.6:1; so for each mg per liter of $NH_4{}^+$-N, 7.6 mg per liter of chlorine is required to reach the breakpoint. In actual wastewater-treatment practice, as was demonstrated by the Rancho Cordova project, it required 10 mg per liter of chlorine for each 1.0 mg per liter ammonia nitrogen present in the process imfluent.[13] Approximately 70 percent of this breakpoint dosage was consumed to produce nitrogen gas (N_2) from the ammonium ion $(NH_4{}^+)$ at pH set points between pH 7 and 8. The oxidation of $NH_4{}^+$ to $NO_3{}^-$ consumed 8 to 19 percent of the total chlorine dosed to the system. Overall, about 96 percent of the total chlorine dosage was accounted for in reactions between chlorine and nitrogeneous species in specific chemical pathways and free chlorine residual remaining in solution following breakpoint.

The Breakpoint Curve

The breakpoint chlorination curve is a graphic representation of chemical relationships which exist as varying amounts of chlorine are added to dilute solutions of ammonia nitrogen. The theoretical breakpoint curve as shown in Fig. 2-1 has several characteristic features. The characteristics of the B-P curve shown in Zone 1 include principally the reaction between chlorine and the ammonium ion indicated in Eq. 2-12. The hump of the breakpoint curve occurs, theoretically at a chlorine to ammonia nitrogen weight ratio of 5:1 (molar ratio of 1:1). That ratio corresponds to the point at which the reacting chlorine and ammonia nitrogen molecules are present in solution in equal numbers.

The chemical equilibria of Zone 2 favor the formation of dichloramine and the oxidation of the ammonium ion according to Eqs. 2-13, 2-14, and 2-15. These reactions proceed in competition with each other to produce a breakpoint at a theoretical Cl_2 to $NH_4{}^+$ wt ratio of 7.6:1. At the breakpoint the ammonium ion is at a minimum.

To the right of the breakpoint, Zone 3 chemical equilibria require the build-up of free chlorine residual.

In practical applications of breakpoint chlorination, reactions occur which result in the formation of nitrogen gas, nitrate, nitrogen trichloride, and other end products. These reactions consume chlorine and cause the $Cl_2 : NH_4{}^+$-N ratio to exceed the stoichiometric value of 7.6:1 and affects the shape of the breakpoint curve.

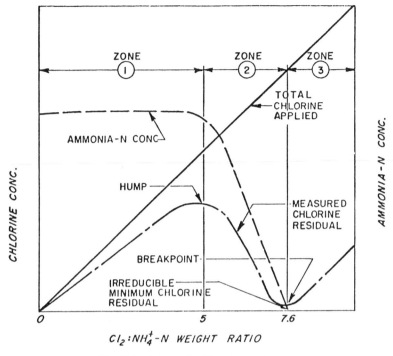

Fig. 2-1 Theoretical breakpoint curve.

Side Reactions of Breakpoint Chlorination

The following are some of the chemical reactions other than the direct oxidation of ammonia to nitrogen gas. The reaction products and chlorine consumption for such reactions are governed by factors such as the type and degree of pre-treatment, initial $Cl_2 : NH_4^+$-N ratio, pH, and alkalinity. These reactions are as follows:

Description	Reaction Stoichromerity
Breakpoint reaction	$NH_4^+ + 1.5\ HOCl \longrightarrow 0.5N_2 + 1.5H_2O + 2.5H^+ + 1.5Cl^-$
	(2-20)
NCl$_3$ formation	$NH_4^+ + 3HOCl \longrightarrow NCl_3 + 3H_2O + H^+$ (2-21)
Nitrate formation	
1-From ammonia	$NH_4^+ + 4HOCl \longrightarrow NO_3^- + H_2O + 6H^+ + 4Cl^-$ (2-22)
2-From nitrite	$NO_2 + HOCl \longrightarrow NO_3^- + H^+ + Cl^-$ (2-23)

pH and Alkalinity Considerations

The nature and concentration of the breakpoint chlorination end products, the chlorine dosage required to reach breakpoint, and the rate of the breakpoint reaction are all affected by the initial pH (following chemical addition) and the pH change which occurs as the breakpoint reaction proceeds. The initial pH in the reaction zone and pH change through breakpoint depends upon the pH and alkalinity of the process influent stream, the ammonia concentration, the chlorine dosage, and the amount of alkalinity supplementation.

Acidity is generated in breakpoint chlorination applications (nitrogen removal) from both the hydrolysis and dissociation of chlorine gas (when Cl_2 gas solutions are used), and the oxidation of ammonia nitrogen.

When the acidity generated is from the hydrolysis and dissociation of chlorine gas the following reaction occurs:

$$1.5\ Cl_2 + 1.5\ H_2O \longrightarrow 1.50\ OCl^- + \underline{3H^+} + 1.5\ Cl^- \qquad (2\text{-}24)$$

If acidity is from the oxidation of ammonia the following two reactions prevail:

$$NH_4^+ + 1.5\ OCl^- \longrightarrow 0.5\ N_2 + 1.5\ H_2O + \underline{H^+} + 1.5\ Cl^- \qquad (2\text{-}25)$$

and

$$NH_4^+ + 1.5Cl_2 \longrightarrow 0.5\ N_2 + \underline{4\ H^+} + 3Cl^- \qquad (2\text{-}26)$$

To counteract the acidity generated in the above reactions either lime or caustic can be used as follows:

$$Lime:\ \ 2\ CaO + H_2O \longrightarrow 2Ca^{++} + 4OH^- \qquad (2\text{-}27)$$

$$Caustic:\ \ 4\ NaOH \longrightarrow 4Na^+ + 4OH^- \qquad (2\text{-}28)$$

Stoichiometrically, three moles of hydrogen ions are liberated in the hydrolysis and dissociation of chlorine gas to provide sufficient chlorine for the oxidation of one mole of ammonia nitrogen; assuming the initial pH in the reaction zone is alkaline. One mole of hydrogen ion is liberated in the oxidation of ammonia to nitrogen gas.

NITROGEN REMOVAL

Introduction

The concept of nitrogen removal by the use of chlorine is an extension of the *Breakpoint Reaction*. Owing to factors which may favor this procedure to remove small amounts of ammonia nitrogen, it is appropriate to summarize the conclusions of the Rancho Cordova project[15,16] and the comprehensive study of the chlorine–ammonia nitrogen chemistry by Saunier.[13]

Rancho Cordova Project

A full-scale demonstration of nitrogen removal by breakpoint chlorination was carried out at the Rancho Cordova secondary treatment plant, Sacramento Co., California, between December 1975 and March 1976.[15,16] During this time the automatic chlorination facility was operated 24 hr. a day 5 days a week. The process flow rates varied from about 0.1 to 1.2 mgd. Influent ammonia nitrogen concentrations were ordinarily in the range of 15 to 25 mg per liter. The breakpoint process succeeded in a consistent removal of about 97 percent of ammonia nitrogen.

A number of specific observations and conclusions made as a result of the Rancho Cordova breakpoint chlorination demonstration program are enumerated below.

1. The dosage of chlorine at Rancho Cordova required to reach breakpoint and maintain a controllable free residual in the process stream averaged 10 mg per liter for each 1.0 mg per liter ammonia nitrogen present in the process influent which is the plant effluent.

2. Approximately 70 percent of the breakpoint chlorine dosage was consumed to produce nitrogen gas (N_2) from ammonia (NH_4^+) at pH set points between pH 7 and 8. The oxidation of NH_4^+ to NO_3^- consumed 8 percent to 19 percent of the total chlorine dosed to the system. Overall, about 96 percent of the total chlorine dosage was accounted for in reactions between chlorine and nitrogenous species in specific chemical pathways and free chlorine residual remaining in solution following breakpoint.

3. Nitrate (NO_3^-) production in breakpoint chlorination was not found to be pH sensitive, with about 1.0 mg per liter NO_3^- (as N) produced from NH_4^+ across a final system pH range of pH 6.5 to 8.5. The production of NO_3^- from NO_2^- was wholly dependent upon influent NO_2^- concentration.

4. Nitrogen trichloride (NCl_3) production was observed to be fairly insensitive to pH across a range of final system pH values from pH 7 to 8. The median value for NCl_3 production was about 0.4 mg per liter (as N) when breakpoint effluent was used as the source of chlorine injector water. While the amount of chlorine consumed in the formation of NCl_3 was relatively small (4-6 percent of total dosed), NCl_3 generation affects the minimum ammonia concentration that can be achieved in breakpoint, since its concentration decays slowly in dilute solution and it is converted to ammonia upon dechlorination with sulfite ion ($SO_3^=$).

5. If the breakpoint process influent is used as injector water, reactions between chlorine and ammonia do occur in the injector water and can consume chlorine in undesirable side reactions. Therefore, injector water should always come from the reacted process effluent. At Rancho Cordova when secondary effluent (process influent) was used as injector water source, the NCl_3 formed in the injector discharge increased the NCl_3 (NH_4^+ remaining) in the process effluent by about 0.2 mg liter.

6. The concentration of organic nitrogen compounds was not affected by breakpoint chlorination.

7. The rate of reaction for breakpoint chlorination was found to vary depending upon the pH control point (final system pH), with fastest rates observed at a pH set point of pH 7.0. The time to completion was found to be between 60 sec and 90 sec at pH 7.0. The reaction rate slowed considerably at pH set point 6.5, with gradual reductions in rate observed as pH increased from pH 7.3 to pH 8.5.

8. Variations in the amount of mechanical mixing intensity in the zone of breakpoint chemical application had no effect upon overall system chemical consumption and effluent quality. Mechanical mixing, to facilitate a rapid and thorough blending of process chemicals and influent stream, was important in damping free residual oscillations for control purposes.

9. Sodium hydroxide (NaOH) was used throughout the study as an alkalinity supplement. The amount of NaOH required to neutralize all breakpoint acidity (1.53 lb NaOH/lb Cl_2) was essentially identical to that predicted from chemical stoichiometry.

10. The small amount of NCl_3 formed did not present any odor problem partly due to the closed pipe reactor and partly because of the reliability of the sophisticated control system.

11. Although it was built into the control system, the continuous NH_3-N process influent analyzer signal to the chlorination system was not necessary for the successful operation of the process.

12. The key control parameters, in addition to the plant flow signal, were the chlorine dosage trim signal from the free residual analyzer and the ability of the sodium hydroxide feed system to maintain a set point of the process pH within ±0.2 pH units.

13. The DPD, FAS titrimetric method for the determination of the chlorine residual species proved most reliable and a time saving method of analysis.

Saunier's Research

Saunier studied the kinetics of the breakpoint reaction in both tap water and tertiary effluent.[13] Only the facets of this research will be discussed here as they may relate to tertiary effluents, because it is doubtful that the application of chlorine to achieve breakpoint proportions will be practiced in other than tertiary effluents containing ammonia nitrogen concentrations of less than 2.0 mg per liter.

One of the most important findings by Saunier was the confirmation that there is a considerable time difference required to complete the breakpoint reaction as related to the ammonia nitrogen concentration, other factors being equal.

Fig. 2-2 illustrates the distribution of the chlorine residual species (as predicted by Saunier's model) when the NH_3-N is 0.5 mg per liter and the contact time is 2.5 min.; Fig. 2-3 is the same situation except that the contact time is 20 min.

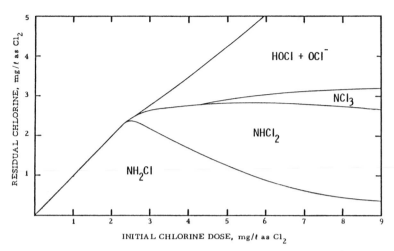

Fig. 2-2   Chlorine dose residual curves predicted by the model after 2.5 min. contact time (pH = 7.4, NH_3-N = 0.5 mg/liter, Temp. = 15°C).

Fig. 2-4 is the predicted breakpoint kinetics but with an ammonia nitrogen concentration of 2.5 mg per liter.

Compare these curves with Fig. 2-5 which is a tertiary effluent with an NH_3-N content of 18 mg per liter and molar ratio of Cl:N of 1.77. The breakpoint reaction in Fig. 2-5 occurs in less than 3 min. (at the disappearance of NH_2Cl) while it takes about 20 min. for water containing less than 1.0 mg per liter NH_3-N and about 8-10 min. when the NH_3-N concentration is 2.5 mg per liter, other factors such as pH and temperature being equal. The model prediction of Fig. 2-5 is in close agreement with the Rancho Cordova project findings[13] where molar ratios of Cl_2-NH_3-N of 2:1 produced the breakpoint reaction in 60-90 sec. at NH_3-N initial concentrations of 15 to 20 mg per liter. So for tertiary effluents nitrified to NH_3-N concentrations of less than 2 mg per liter, it will require 15-20 min. to complete the breakpoint reaction and produce a controllable, stable, free chlorine residual.

Further findings by Saunier showed that after the breakpoint is reached, the resulting *dichloramine fraction* of the *residual is nearly as germicidal as the free HOCl residual*, and furthermore the HOCl residual is 30 times more germicidal than the OCl⁻ ion. These findings are indeed different from previous findings.[1]

Nitrogen gas and nitrate are the final end products of the breakpoint reaction. Nitrogen trichloride* formed during the reaction decreases slowly with time. While NCl_3 does not appear to be an end product of the reaction, for all practical purposes it is an end product; because in the time frame of the process in practice,

*Nitrogen trichloride is a most effective germicide.[1]

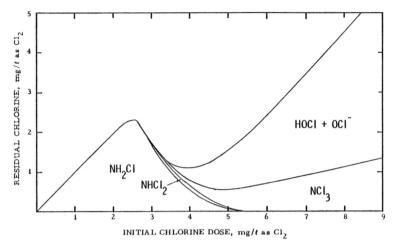

Fig. 2-3 Chlorine dose residual curves predicted by the model after 20 min. contact time (pH = 7.4, NH$_2$-N = 0.5 mg/liter, Temp. = 15°C).

the remaining NCl$_3$ (at the end of the contact time) will revert to ammonia nitrogen after dechlorination.

Saunier also confirmed the speed of the breakpoint reaction to occur most rapidly at pH 7.5. He also confirmed that the NCl$_3$ concentration increased with increasing Cl to N ratios at all levels of pH. Also that the organochloramines interfered mainly with the dichloramine and indirectly with the NCl$_3$ DPD-FAS

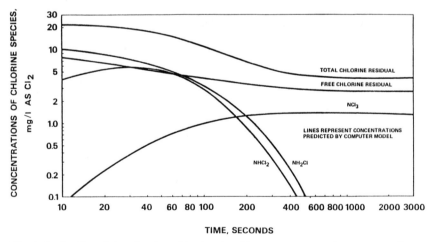

Fig. 2-4 Predicted breakpoint chlorination kinetics in a plug flow reactor (pH = 7.50, NH$_3$-N = 2.5 mg/liter, Temp. = 15°C and Cl$_2$/N = 9.0 by wt). The lines represent concentrations predicted by a computer model.

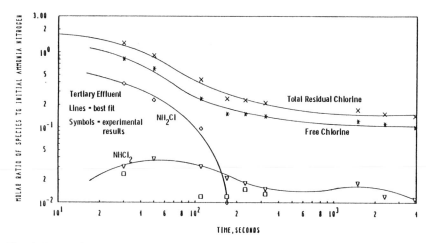

Fig. 2-5 Breakpoint chlorination kinetics of a tertiary effluent in a plug flow reactor (pilot-plant experiment). pH = 7.5 after 150 sec., Temp = 18.8°C, N = 18.03 mg/liter, and Cl_2/N = 1.77 molar ratio. Lines = best fit; symbols = experimental results.

analysis, primarily at pH values above 7.5. Traces of organic nitrogen (0.02 mg per liter) formed organochloramines in sufficient amount to give false DPD-FAS dichloramine readings at pH values above pH 7.5. The oxidation of the organic nitrogen proceeded faster at pH values of less than 7.5.

In summarizing Saunier's findings, particularly as they relate to tertiary effluents, it can be concluded that if it is desired to achieve a free chlorine residual for bacteria and virus destruction, when the ammonia nitrogen concentration is less than 2.0 mg per liter, the process should be carried out at a control pH of 7.0-7.2 with a minimum contact period of 30 min. Moreover, Saunier's work shows some instability of the free chlorine residual in the usual control sampling time of 1 to 3 min. after the point of application. Therefore, residual control in situations where the initial ammonia nitrogen concentration is less than 2.0 mg per liter should be based on the total chlorine residual measurement.

At this level of chlorination with most wastewater effluents, pH adjustment due to the addition of chlorine would probably not be required. Most effluents have an alkalinity of 150 to 200 mg per liter which is easily capable of maintaining the pH at status quo with chlorine dosages as high as 25 mg per liter.

THE DECHLORINATION PROCESS

Chemistry of Dechlorination

Dechlorination is the practice of removing all or part of the total combined chlorine residual remaining after chlorination. This is usually accomplished by sulfur-bearing compounds which hydrolyze in water to produce the sulfite ion ($SO_3^=$). Granular activated carbon can also be used but it is much more expensive than the sulfur compounds and does not lend itself to predictable process control.

Sulfur Compounds. The chemical reaction of dechlorination amounts to the conversion of all positive chlorine atoms to negative chloride atoms. The reaction of sulfur dioxide with both free and combined chlorine residuals is nearly instantaneous and as follows:

$$SO_2 + H_2O \longrightarrow H_2SO_3 + HOCl \longrightarrow H_2SO_4 + HCl \qquad (2\text{-}29)$$

and with chloramine:

$$NH_2Cl + H_2SO_3 + H_2O \longrightarrow NH_4HSO_4 + HCl \qquad (2\text{-}30)$$

Similar products are formed in the dechlorination of dichloramine and nitrogen trichloride. From Eq. 2-29 it can be seen that the stoichiometric relationship requires 0.9 ppm SO_2 to dechlorinate 1 ppm chlorine. Some installations might not be large enough to warrant the use of sulfur dioxide in cylinders. The best chemical choice in such cases is either sodium sulfite (Na_2SO_3) or sodium metabisulfite ($Na_2S_2O_5$ —pyrosulfite).

Sodium sulfite reacts with chlorine as follows:

$$Na_2SO_3 + Cl_2 + H_2O \longrightarrow Na_2SO_4 + 2HCl \qquad (2\text{-}31)$$

This reaction requires 1.775 lb of pure sodium sulfite per lb of chlorine removed, whether it is a free or combined residual. The speed of this reaction is similar to that of sulfur dioxide—a matter of seconds for completion. The use of sodium thiosulfate as a dechlorinating agent *is not recommended* for chlorination systems.

The above reactions of the sulfur compounds to reduce the chlorine residual to chlorides does produce small amounts of acid; however, these would be neutralized by the buffering capacity of the wastewater.

In the neutral pH range of application, 2.8 parts of alkalinity as $CaCO_3$ are consumed for each part of chlorine removed.

The sodium sulfite reaction of Eq. 2-31 consumes 1.38 parts of alkalinity as $CaCO_3$ for each part of chlorine removed.

Sulfur dioxide is used in some special industrial processes as a dissolved oxygen scavenger. Therefore, chemically an excess of sulfite ion could remove dissolved oxygen in a dechlorinated effluent. However, the reaction between the sulfite ion and dissolved oxygen at the usual temperatures of wastewater is so slow that dilution would take place in the receiving waters, or some other chemical reaction would probably dominate to such an extent that there would not be any detectable oxygen depletion unless there was a gross excess of sulfite ion (i.e., 5-10 mg per liter.)*

Activated Carbon. The process of chlorine removal by activated carbon is not a pure adsorption** process. It also involves a chemical reaction between chlorine and the water where the carbon acts as a catalyst. The use of activated carbon can serve several functions other than the removal of chlorine residuals. Some removal of soluble organics is accomplished through adsorption by the carbon.

Activated carbon reacts with both free and combined chlorine residuals in the following manner.

Reaction with free chlorine residual:

$$C + 2HOCl \longrightarrow CO_2 + 2H + 2Cl^- \qquad (2\text{-}32)$$

Reactions with combined chlorine residual:

$$C + 2NH_2Cl + 2H_2O \longrightarrow CO_2 + 2NH_4{}^+ + 2Cl^- \qquad (2\text{-}33)$$

$$C + 4NHCl_2 + 2H_2O \longrightarrow CO_2 + 2N_2 + 8H^+ + 8Cl^- \qquad (2\text{-}34)$$

Carbon dioxide is formed in each case following the reaction with the chlorine residual. Ammonia is returned to solution following the reaction of carbon and monochloramine, but dichloramine has been observed to decompose to nitrogen gas following contact with activated carbon. These chemical pathways have been confirmed by Stasiuk et al.[17]

While HOCl is absorbed by the carbon it has been demonstrated that nitrogen trichloride is adsorbed by carbon.[1] Studies by Stasiuk et al.[17] showed complete dechlorination of both free and combined residuals following carbon contact times of 10 min. The use of activated carbon as a continuous dechlorinating agent is not competitive with compounds producing sulfite ion in solution. The only time activated carbon can be considered as a dechlorinating agent of choice is when it is to be used for the removal of both nitrogen and soluble organics. Installations using activated carbon for these purposes must necessarily be equipped to reactivate the granular carbon.

FORMATION OF ORGANOCHLORINE COMPOUNDS

These compounds enter surface waters from many sources, both point and non-point. Little is known about the chemical stability, biological degradation, distribution, and ecosystem behavior of the majority of these compounds. Most of

---

*For these reasons it is not necessary to aerate a chlor-dechlor effluent to restore or enhance the dissolved oxygen concentration.
**There is a significant difference between the phenomena of adsorption and absorption. For example, NCl_3 is adsorbed by carbon while HOCl is absorbed by carbon.

the surface water pollutants identified thus far are present in microgram per liter concentrations or less. At this concentration level, the highly chlorinated pesticides and hydrocarbons probably represent a more serious problem with respect to potable water treatment than those of the chloro-organics formed during wastewater disinfection. These organochlorine species of compounds are formed by the chloramine reaction in chlorinated wastewater effluents which discharge into surface waters.[23]

The formation of a wide variety of organochlorine compounds as a result of wastewater chlorination practices has been well documented by Jolley[19,23] and others.

The importance of the formation of these halogenated-organic compounds is the possible public health risk when they appear in potable water. The carcinogenic nature of several of these compounds has been demonstrated in the laboratory, using high concentrations and laboratory test animals. However, no direct cause-and-effect link with cancer in man has yet been established, and probably never will be.

The EPA is making a concerted effort to accumulate as much data as possible on the public health risk of these compounds. Their primary concern is the formation of the trihalomethanes (chloroform and bromoform) as a result of chlorination practices.[20] These compounds are known carcinogens so it is prudent to pursue a course of practice which would reduce or eliminate the formation of these compounds.

It is interesting and important to observe that there is no evidence that chlorination practices using chloramines will form any trihalomethanes; therefore, sewage discharges not treated beyond the breakpoint will not form trihalomethanes.[20] *This makes the practice of free residual chlorination of wastewater effluents highly speculative.*

The Rancho Cordova Investigation

A recent comprehensive study by Stone[22] covered the halomethane formation in the following situations: activated sludge effluent containing 20 mg/liter ammonia nitrogen chlorinated to the breakpoint with dosages of 200 mg/liter or more and resulting in substantial free chlorine residuals (7-12mg/liter); same effluent but with chlorination for disinfection (3-4mg/liter residual at 30 min. contact); same effluent except nitrified and filtered and chlorinated; a primary effluent dosed with 11 mg/liter chlorine; and an oxidation pond (nitrifying) effluent downstream from the activated sludge effluent.

The results of this study reveal a rapid formation of chloroform resulting from breakpoint chlorination as was expected. At a pH of 7, 88 μg/liter formed in 2 min. and 123 mg/liter at 5 min.

The activated sludge effluent chlorinated for disinfection purposes formed only 4 μg/liter of chloroform in 4 hr.

The highly polished activated sludge effluent dosed with chlorine at 11 mg/liter led to the formation of three times the amount of halomethane compounds than formed by disinfection chlorination of the activated sludge effluent. This is undoubtedly a result of the formation of free chlorine residuals in the highly polished effluent. In this environment of free residuals there is a steady growth of halomethane formation.

Chlorination of the primary effluent to 14 mg/liter caused no change in chloroform concentration, a slight decrease in dichloroethane, and an appreciable increase in dibromochloromethane. Overall, a steady increase in total halomethane concentration was noted to a level of approximately 0.210 μm/liter at 4 hr contact time. This is almost the same as that measured for the polished activated sludge effluent at the same contact time, even though differences in the form of chlorine residuals exist. The residual in the primary effluent is all combined chlorine.

Chlorination of the oxidation pond effluent to a dosage of 6.7 mg/liter yielded no change in THM concentration from unchlorinated levels through a 4-hr chlorine contact period even though chlorine residuals of 2.5–3.5 mg/liter were maintained throughout that time. Carbon tetrachloride and dibromochloromethane concentrations in the pond effluent were unaffected by chlorination. The low-ammonia nitrogen concentration of the nitrifying pond effluent insured that the chlorine residual exisiting throughout the 4-hr contact period was a mixture of both free and combined residuals.

A sample of the chlorine injector water was analyzed for THM concentration to determine if the high-chlorine concentrations (1000–3000 mg/liter) present in the injector water could cause THM formation prior to the injection of the chlorine solution into the process liquid. The injector water source was activated sludge effluent which had not been subjected to any previous application of chlorine. At the time of this particular sampling, the injector water chlorine concentration was on the order of 1800 mg/liter at a pH of 2.5. The residence time in the injector water system from the time of chlorine addition to the point of application was about 20 sec. Approximately 0.23 μm/liter* of THM were observed to be formed in the injector water under the above conditions. However, in the overall system, dilution of the injector water into the process stream results normally in a 200 or 300 to 1 dilution so that the net impact of THM formation due to the chlorination system itself is negligible. It is likely that the low pH value and the short contact time in the injector water system may be responsible for keeping THM concentrations below that observed for breakpoint chlorination of the process stream.

The above observations of THM formation in sewage treatment processes were also compared with data on municipal tap water and raw surface water. Analysis

---

*μm = micromoles.

of potable water taken from a municipal water system (surface water) in northern California showed THM levels almost identical to those reported as a median value of the NORS report,[24] with measured values of 25 μg/liter chloroform and 4.3 μg/liter bromodichloromethane. The total halomethane concentration was observed to be 0.235 μm/liter for the northern California potable water sample as compared to 0.218 for the median value of the NORS report. Analysis of samples collected above and below the wastewater discharge of the Rancho Cordova plant into the American River demonstrated that with the dilution factor of 250:1, current analytical techniques would be unable to detect the contribution of THM concentrations from this discharge to background THM levels in the American River.

The Occoquan Watershed Survey

A yearlong study of watershed runoff in this water service area located in northern Virginia near Washington, D.C. was reported by Hoehn et al. in 1976.[21] This report involves a system owned and operated by the Fairfax County Water Authority which provides potable water to more than 600,000 inhabitants in this rapidly urbanizing area. The Occoquan watershed collection system receives both argicultural and urban runoff as well as treated sewage from eleven plants in the area. The purpose of the report was to determine the effects of various factors in a given runoff area which might contribute to the formation of trihalomethane in the potable water supply at the downstream end of this runoff system. The findings of this comprehensive study support other evidence that chloroform concentrations appearing in surface waters do not necessarily have their origin in chlorinated wastewater discharges.

For example, the upstream control at Catharpin on Bull Run had a mean concentration of 2.2 μg/liter chloroform in all samples; while at Bull Run 2.3 miles below the last of eleven chlorinated sewage treatment plant discharges the samples showed a mean concentration of 3.2 μg/liter of chloroform. However, while the intake of the water treatment plant which is downstream from the Bull Run sampling point showed only 3.0 μg/liter chloroform (mean) the treated potable water showed a mean of 232.6 μg/liter chloroform. *The latter must be attributed to the practice of free chlorination at the treatment plant.*

Pomona, California Virus Study[26]

This comprehensive investigation of various tertiary effluent treatment processes is described in Chapter 1. The chloro-organic compounds investigated confirm other findings concerning the formation of chloroform due to chlorination procedures. It was found that free residual chlorination of a tertiary effluent increased the average chloroform concentration 330 fold (0.6-198 μg/liter) compared to a six-fold increase during combined chlorine residual chlorination. The tertiary effluent before chlorination contained chloroform concentrations on the order of 0.6 to 1.5 µg/liter.

Montgomery Maryland Simulation Study

A planned construction of a major advanced wastewater treatment (AWT) facility in the vicinity of Dickerson, Maryland has raised some concern about the discharge into the Potomac River at a point approximately 22 miles upstream from the Washington, D.C. water supply intake and 8 miles upstream from the proposed Leesburg, Virginia, water supply intake.[25] This AWT facility is designed to remove ammonia nitrogen by breakpoint chlorination. This will be followed by activated carbon adsorption and dechlorination. The effluent disinfection requirement is 2.2 MPN total coliform per 100 ml or less. A major consideration in the selection of this process was its projected ability to obtain the desired disinfection with free chlorine residuals which also should provide virus inactivation.

This study clearly demonstrated that *breakpoint chlorination results in the formation of appreciable amounts of chloroform and bromodichloromethane.* And that excessive amounts of free chlorine resulting from pursuing the breakpoint procduces increased concentrations of organochlorine compounds none of which are desirable in potable water.

The results of this study confirm other investigations, particulary the comprehensive work by Stone.[22]

Summary

Wastewater chlorination practices should be limited to the formation of combined residuals in order to prevent, as much as possible, the formation of trihalomethanes in surface waters. This practice will be able to achieve the desired wastewater disinfection without contributing to the formation of not only undesirable compounds in potable-water supplies, but compounds which may have carcinogenic properties.

Chapter 2—Summary

The Chemistry of chlorine in wastewater practices is not yet fully understood. There are anomalies.

Chlorine, whether an aqueous solution or a buffered hypochlorite solution, hydrolyzes to form hypochlorous acid which is the most powerful and effective species of the various chlorine compounds.

The hydrolysis of chlorine forms two species: 1) hypochlorous acid and 2) hypochlorite ion. The relationship of the formation of these two species is pH dependent. Hypochlorous acid has a germicidal efficiency about 100 times that of the hypochlorite ion (OCl^-) as defined by the lethality coefficient.

Hypochlorous acid reacts almost instantaneously with the ammonia nitrogen in wastewater to form inorganic chloramines. These chloramines are predominantly monochloramine in the neutral pH range. It is believed that these chloramines will in time react with the organic nitrogen content of wastewater to form various organic chloramine compounds shown to be less germicidal than inorganic chloramines. This would explain why the germicidal efficiency of combined chlorine residuals diminishes with time.

It has been found that when ammonia nitrogen in the treated effluent disappears, leaving only organic nitrogen, the germicidal efficiency of the resulting combined residual is greatly diminished. Moreover, in these instances where organic nitrogen is present in the absence of ammonia nitrogen, free chlorine residuals have not been detected. This is contrary to all expectations. A reasonable explanation is that the hypochlorous acid reacts rapidly with the various organic nitrogen compounds in the wastewater to form organic chloramines of poor germicidal efficiency, to the complete exclusion of any reaction by free chlorine.

Therefore, it is possible that nitrified effluents containing significant amounts of organic nitrogen (>5.0 mg/liter) can be more difficult to disinfect than nonnitrified effluents.

The governing factors of wastewater disinfection by chlorination are clearly demonstrated by the Collins mathematical model. This model is dependent upon rapid mixing of the chlorine and plug flow conditions in the contact chamber. When these conditions have been satisfied the coliform kill in a wastewater effluent is predictable for a given total chlorine residual (as measured by the back titration procedure) at the end of a specified contact time. The coliform concentration before disinfection must also be known.

The use of chlorine for nitrogen removal in wastewater effluents has been thoroughly investigated. As a chemical process it is cost-effective and readily controllable.

However, this process is dependent upon the formation of *free chlorine residuals* and while these residuals may not seem excessive (<5.0 mg/liter) they contribute heavily to the formation of trihalomethanes in the receiving waters. This is undesirable because these ccompounds are known carcinogens and should be eliminated as far as is possible from our drinking water supplies.

Waters containing residual chlorine discharging into receiving waters containing aquatic life have been found to be toxic with total chlorine residuals as low as 0.02 mg/liter.

Dechlorination has been used successfully to detoxify these residuals. Dechlorinated effluents are less toxic than nonchlorinated effluents.

The most practical of all dechlorination methods is by the application of sulfur dioxide or those chemicals which produce the sulfite ion $(SO_3^=)$ in solution. It requires about one part of sulfur dioxide to neutralize (dechlorinate) one part of total chlorine residual.

Sulfur dioxide is attractive because it can be metered and controlled with conventional chlorination equipment.

Activated carbon is another effective means of dechlorination. It is not competitive with the sulfur compounds when used solely for dechlorination. Activated carbon has many roles to play in the polishing of wastewater effluents so it should always be considered for its overall possibilities for soluble organics removal and nitrogen removal.

Organochlorine compounds enter surface waters from many sources, however, since *combined chlorine residuals* in sewage discharges do not contribute to the formation of trihalomethanes, wastewater chlorination practice should be limited to the formation of combined residual. This practice will be able to achieve desired wastewater disinfection levels without contributing to the formation of undesirable compounds or compounds having carcinogenic properties.

REFERENCES

1. White, G. C. *Handbook of Chlorination*. Van Nostrand Reinhold Co., N.Y. (1972).

2. Morris, J. C. "The Mechanism of the Hydrolysis of Chlorine." *J. Am. Chem. Soc.* **68**: 1692 (1946).

3. Adams, F. W. and Edmonds, R. G. "Absorption of Chlorine by Water in a Packed Tower." *Ind. Eng. Chem.* **29**: 447 (1937).

4. Rosenblatt, D. H. and Small, M. J. A Private Communication. U.S. Army Medical Bioengineering Research and Development Lab., Fort Detrick, Md. (1976).

5. Morris, J. C. "The Acid Ionization Constant of HOCl from 5 to 35°C." *J. Phys. Chem.* **70**: 3798 (Dec. 1966).

6. Feng, T. H. "Behavior of Organic Chloramine in Disinfection." *J. WPCF* **38**: 614 (1966).

7. Sung, R. D. "Effects of Organic Constituents in Wastewater on the Chlorination Process." Ph.D. Thesis, Univ. of Calif., Davis, Calif. (1974).

8. Esvelt, L. A., Kaufman, W. J., and Selleck, R. E. "Toxicity Assessment of Treated Municipal Wastewaters." A paper presented at the 44th Ann. Conf. of WPCF, San Francisco, Calif. (Oct. 4–8, 1971).

9. White, G. C. Unpublished field studies. San Jose–Santa Clara Water Pollution Control Plant, San Jose, Calif. (1973–1976).

10. Stokes, H. W. Private communications regarding the Las Virgenes Wastewater Treatment Plant, Calabasas, Calif. (1974).

11. Collins, H. F., Selleck, R. E., and White, G. C. "Problems in Obtaining Adequate Sewage Disinfection." ASCE *J. San. Engr. Div.* **97**, SA 5, Proc. #8430 (Oct. 1971).

12. Collins, H. F., White, G. C., and Sepp, E. "Interim Manual For Wastewater Chlorination and Dechlorination Practices." California State Dept. of Health (Feb. 1974).

13. Saunier, B. M. "Kinetics of Breakpoint Chlorination and Disinfection." Ph.D. Thesis, Univ. of Calif., Berkeley, Calif. (1976).

14. Saunier, B. M. and Selleck, R. E. "The Kinetics of Breakpoint Chlorination in Continuous Flow Systems." A paper presented at the AWWA Ann. Conf., New Orleans, La. (June 22, 1976).

15. Stone, R. W. "Rancho Cordova Breakpoint Chlorination Demonstration." A report prepared by Sacramento Area Consultants (Sept. 1976).

16. Stone, R. W., Saunier, B. M., Selleck, R. E., and White, G. C. "Pilot Plant and Full-Scale Ammonia Removal Investigations Using Breakpoint Chlorination." A paper presented at the 49th Ann. Conf. WPCF, Minneapolis, Minn. (Oct. 5, 1976).

17. Stasiuk, W. N., Hetling, L. J., and Shuster, W. W. "Removal of Ammonia Nitrogen by Breakpoint Chlorination Using an Activated Carbon Catalyst." New York State Dept. of Env. Conservation, Tech. Paper No. 26 (April 1973).

18. Stone, R. W. Unpublished results of Rancho Cordova Project: Chloro-organic formation, Private communication (1976).

19. Jolley, R. L. "Chlorine Containing Organic Constituents in Chlorinated Effluents." *J. WPCF* **47**: 601 (1975).

20. Symons, J. M. "Interim Treatment Guide for the Control of Chloroform and Other Trihalomethanes." A report by the EPA Env. Res. Lab., Cincinnati, Ohio (June 1976).

21. Hoehn, R. C., Randall, C. W., Bell, F. A. Jr., and Shaffer, P. T. B. "Trihalomethanes and Viruses in a Water Supply." A paper presented at the ASCE National Conf. on Env. Eng. Res. Dev. and Design, Seattle, Wash. (July 12–14, 1976).

22. Stone, R. W. "The Formation of Halogenated Organic Compounds in Wastewater Chlorination." An Unpublished Report by Brown and Caldwell Cons. Engrs. Walnut Creek, Calif. (1977).

23. Jolley, R. L. and Pitt, W. W. Jr. "Chloro-Organics in Surface Water Sources for Potable Water." A Paper Presented for the Disinfection Symposium at the Ann. AWWA Conf. Anaheim, Calif. (May 8, 1977).

24. Symons, J. M., Bellar, T. A., Carswell, J. K., Demarco, J., Kropp, K. L., Robeck, G. G., Seeger, D. R., Slocum, C. J., Smith, B. L., and Stevens, A. A. "National Organics Reconnaissance Survey for Halogenated Organics in Drinking Water." *J. AWWA* **67**: 634 (Nov. 1975).

25. Breidenbach, A. W. "Interim Report on Montgomery Simulation: Study of Formation and Removal of Volatile Chlorinated Organics." EPA MERL Cincinnati, Ohio (July 8, 1975).

26. Selna, M. W., Miele, R. P., and Baird, R. B. "Disinfection for Water Reuse." A Paper Presented in the Disinfection Seminar at the AWWA Annual Conf. Anaheim, Calif. (May 8, 1977).

27. Collins, H. F., Selleck, R. E., and Saunier, B. M. "Optimization of the Wastewater Chlorination Process." A Paper Presented at the National Conference on Environment and Des. Sponsored by EE Div. ASCE, Seattle, Wash. (July 12, 1976).

28. Ingols, R. S. Private communication. Georgia Inst. of Tech. Atlanta, Ga. (July 1977).

29. Morris, J. C. "Aspects of the Quantitative Assessment of Germicidal Efficiency," Disinfection: Water and Wastewater, J. D. Johnson (Editor), Ann Arbor Science, Ann Arbor, Mich., p. 1 (1975).

3

Chlorination-dechlorination facility design

FACTORS AFFECTING PROCESS EFFICIENCY

Introduction

The strict requirements set forth by the California regulatory agencies for wastewater disinfection focused on the performance of every wastewater chlorination system. Therefore, having been involved in every aspect of many wastewater chlorination systems, White decided in 1972 it was time to evaluate the performance of these systems and to determine the elements of an optimum system.[9] Over a six-month period in 1972, White made a personal investigation requiring at least three and sometimes four visits to 34 plants. From 1972 to 1976 inclusive another 12 plants were investigated to see if the findings of the 1972 survey could be corroborated. The treatment plant effluents investigated in the 1972 survey included eleven primary, fifteen activated sludge, four secondary using high-rate-recirculating biofilters, and four secondary followed by oxidation ponds. The second group included only secondary or tertiary effluents, and some of these were nitrified either on purpose or inadvertently.

Elements of an Optimum System

These investigations revealed that regardless of the treatment process the most effective chlorination systems were those that included the following elements:

1. A proper and workable automatic chlorine residual control system.
2. Excellent initial mixing.
3. Adequate contact time (at least 30 min. at peak flow) in an outfall conduit or a well-designed basin.
4. Competent operating personnel.

Let us examine each of these elements since these factors are the ones under the control of the design engineer and the operators. The chlorine demand, pH, and temperature of the wastewater is beyond their control but also has to be reckoned with. This becomes the duty of the operator but he has to have an optimum facility to accomplish the job.

Chlorine Control System

The chlorine control system consists of a supply system (150-lb, cylinder ton containers, tank cars, storage tanks), metering equipment (chlorinators), solution discharge equipment (injectors, pumps, diffusers) and control equipment (flow meters, residual analyzers).

Chlorine Supply System:

100- and 150-lb Cylinders. Although most packagers make it available, the 100-lb cylinder is seldom used. The most popular size is the 150-lb cylinder. The gross weight of a full 150-lb cylinder varies from 250 to 285 lbs, and they are best handled by a special two-wheeled hand truck. Each cylinder has a single outlet valve equipped with a fusible plug that will melt at about 158°F.

The most important design considerations are as follows:

1. Direct sunlight must never reach the cylinder.
2. The maximum withdrawal rate should be limited to 40 lb/day/cylinder.
3. Minimum allowable room temperature is 50°F.
4. Heat must never be applied directly to the cylinder.
5. Sufficient space should be allowed in the supply area for at least one spare cylinder for each one in service.

Either weighing scales or an automatic cylinder switchover device is an essential accessory, so that the operator can be sure of never running out of chlorine. In the situation where only one cylinder is in use at one time there is available a two-cylinder readout scale. This plus an automatic switchover device provides the operator with maximum confidence of a continuous chlorine supply.

If the situation requires a bank of chlorine cylinders the best choice for the operator is to group each of two banks of 150-lb cylinders to separate headers and utilize an automatic cylinder switching device. This eliminates the necessity of scales. These switchover devices are described under the heading of ton containers.

In general, the chlorine supply area should be kept cooler than the chlorinator. This reduces the possibility of reliquefaction. Installations using 150-lb cylinders are more often than not subject to low-withdrawal rates which makes them susceptible to reliquefaction in the piping between the cylinder and chlorinator. This phenomenon is accompanied by the deposit of chlorine impurities in the small orifices of the chlorinator mechanism resulting in excessive chlorinator maintenance. If the cylinder can be kept cooler than the chlorinator, reliquefaction will not occur. This situation suggests the need for insulation to prevent the chlorine supply area from being subjected to wide diurnal fluctuations of ambient temperatures. If this is not possible or if the reliquefaction phenomenon is anticipated for other reasons, then it is desirable to install an external chlorine pressure reducing valve between the cylinders and chlorinator. This valve should be as close to the cylinders as possible.

Filters and traps ahead of all chlorinator control apparatus are highly desirable. Most small chlorinators have built-in filters which should be preceded by a trap. The latter is an accessory that should be specified.

Ton Containers. Unlike the 150-lb cylinders, either gas or liquid may be withdrawn from ton containers; consequently each container has two outlet valves. Also unlike the small cylinders, the ton containers have six fusible plugs— three in each of the two dished heads.

Ton containers are transported by truck or multiple-unit tank cars (TMU). A truck can carry a maximum of fourteen containers; a TMU, fifteen.

The gross weight of these containers (3500 lb) dictates that proper handling equipment be used. The container is designed for use in the horizontal position. It must be positioned before connection to the supply header is made so that the two outlet valves line up vertically, one over the other. In this position, the eductor tube connected to the top valve is in the gas withdrawal position, and the eductor tube connected to the bottom valve is in the liquid withdrawal position.

Proper handling equipment includes the following:

1. Two-ton capacity electric hoist.
2. Cylinder lifting bar.
3. Cylinder trunnions.
4. Monorail for hoist.

The trunnions are used primarily to easily position the cylinder outlet valves. In addition, the trunnions provide support and proper spacing for the cylinders.

The practice of using parallel rails to store ton containers on a concrete pad should not be allowed because of the inherent danger of mashing an operator's fingers between two cylinders.

One of the critical design dimensions is the height of the monorail above the floor. The primary criterion is that the monorail be high enough to pick up a cylinder off the truck and have sufficient clearance to lift one cylinder over another one which is connected to the header. Figure 3-1 illustrates how to calculate the minimum distance of the monorail above the floor. For installations where seismic forces are of significance, the designer can prescribe the use of plastic cinch straps to prevent the cylinder from becoming dislodged from the trunnions.[1]

At room temperature the allowable continuous gas withdrawal rate from a ton cylinder is 400 lb/day. Theoretically, gas withdrawal can be used up to any capacity if enough cylinders are connected to the supply header. However, when

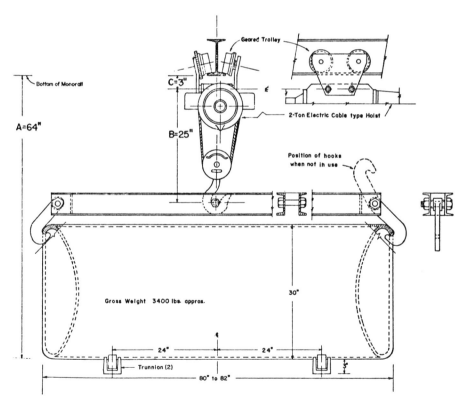

Fig. 3-1 Monorail height location. To determine monorail height above grade add truck bed height to distance "A" as illustrated. Check dimensions "B" and "C" with hoist manufacturer (*courtesy* Chlorine Specialities, Inc.).

the system withdrawal rate reaches 1500 lb/day, it is logical to change over to liquid withdrawal which requires the use of an evaporator. However, the following is one exception.

In hot climates where the "in shade" summer temperatures exceed 95 to 100°F on a consistent basis it is desirable to consider the use of an evaporator. Normally those climates experiencing summer temperatures of 100°F usually experience winter temperatures of 15°F. Therefore, the evaporator concept takes care of both extremes of climatic temperature. Liquid withdrawal of chlorine to an evaporator from a supply system is least affected by ambient temperature. The only precaution is to prevent direct sunlight on the cylinders. For winter operation no special precautions are required (such as artifical heating) because the gas temperature at the outlet of the evaporator will usually *exceed* 100°F regardless of room temperature and the temperature of the gas in the vacuum line to the injector will never fall below the critical temperature (35°F) where chlorine hydrate occurs.

For gas withdrawal rates up to 1500 lb/day, cylinder space should be provided for four cylinders in service, four standby cylinders, and four empty spaces for the next delivery. A further consideration for providing storage space is the cost reduction for quantity delivery of containers.

Gas withdrawal systems using ton containers require the same considerations as smaller cylinders: never let the sun shine directly on the cylinders and never apply heat directly to the cylinders. If it is not possible to keep the cylinders cooler than the chlorinators, then it is highly desirable to install an external pressure-reducing valve immediately downstream from the last cylinder. Just upstream from this valve a combination filter and sediment trap should also be installed. This is shown in Fig. 3-2.

The chlorine header system for liquid withdrawal is somewhat different from that for gas withdrawal. The piping and support system are the same, except that the flexible connections to the auxiliary header valve are connected to the bottom cylinder outlet valve (the top valve is for gas withdrawal). Because of the evaporator in the system, the filter is located immediately downstream from the evaporator outlet. The filter must always be installed in the gas phase. It is not possible to filter out chlorine impurities in the liquid phase, because the impurities are in solution. This places the filter just upstream from the chlorine pressure-reducing valve, which now becomes an automatic shut-off valve as well.*

A complete liquid-withdrawal system is illustrated in Fig. 3-3, which also shows proper spacing of the evaporator-chlorinator system.

It should be noted that some engineers prefer to use duplicate header sys-

---

*This pressure-reducing and shut-off valve is always a part of the evaporator system. It is electrically interlocked with the evaporator water bath temperature, and automatically shuts off the chlorine supply in the event the water bath temperature falls below 150°F.

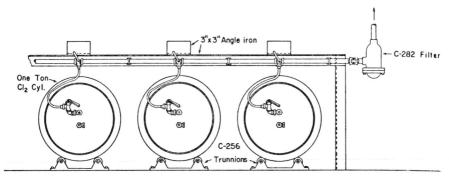

Fig. 3-2 Gas withdrawal system for ton containers.

tems from the containers to the evaporators. In this way one header can be taken out of service for the required periodic cleanings without interrupting the entire facility. Most systems rely on a single header; however, duplicate headers have advantages from an operator's viewpoint.

When contemplating the use of liquid withdrawal versus gas withdrawal, the following advantages of a liquid-withdrawal system should be considered:

1. The danger of reliquefaction of chlorine between the containers and the chlorinator is all but eliminated.
2. Fewer cylinders need to be connected at one time. Liquid-withdrawal rates of a ton cylinder can be as high as 10,000 lb/twenty-four hr.
3. The evaporators, although insulated, do give off some heat in the equipment room.

Liquid-withdrawal systems do not have the critical design problems with regard to temperature considerations as do the gas-withdrawal systems, except for inadvertent trapping of liquid in the header system which constitutes a temperature-pressure hazard. If liquid chlorine is trapped between two shut-off valves and the ambient temperature rises a few degrees, the liquid chlorine will try to expand. Since it cannot expand it will exert a pressure in accordance with the vapor pressure temperature curve shown in the Chlorine Institute Manual. For example, suppose a container connection full of liquid chlorine is trapped by yoke shut-off valves and at each end a ton container or a tank car (at an ambient temperature of $80°F$) is removed, then the vapor pressure would be 100 psi. Now if the connection full of liquid were allowed to reach a temperature of $90°F$ by placing it in the sun, the hydrostatic pressure of the contained liquid would probably exceed the tensile strength of the already fatigued flexible connection resulting in a rupture of the flexible connection.

A note of caution about manifolding ton containers withdrawing liquid: always be sure that the temperature of the cylinders is about the same; never

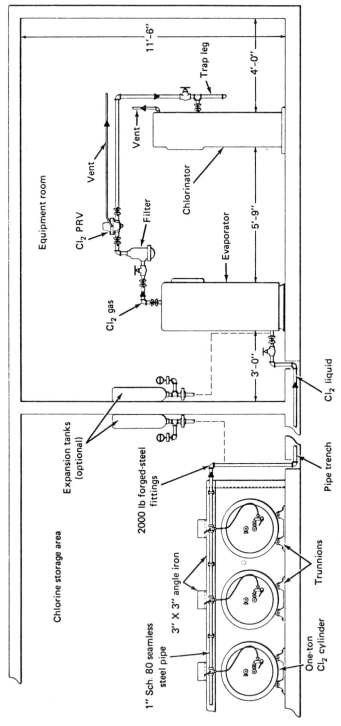

Fig. 3-3 Liquid withdrawal system showing chlorinator and evaporator spacing and ancillary equipment.

connect "hot" cylinders to the manifold simultaneously with cylinders already in use.

The cylinders can be placed in carport-type open structures with only a sun-shield when the system is operating entirely on liquid withdrawal (Fig. 3-4). If the cylinder storage area is remote from the chlorine control system, it is desirable to install an expansion tank in the header system using a frangible disc and pressure alarm in series with the expansion tank. (See Fig. 3-5). This device is necessary to protect the system against the condition where an operator might close both the outlet valve on the header and the inlet valve to the evaporator, thereby trapping liquid in the header. Any subsequent ambient temperature rise would result in a pressure rise in the liquid chlorine sufficient to rupture the header piping. This is a direct result of the hydrostatic pressure due to the expansion of the liquid chlorine (or SO_2) which is greater than that of the steel pipe or copper flexible connection as a function of a rise in ambient temperature.

Since chlorine headers must be cleaned occasionally, it is desirable in most cases to install duplicate headers between the cylinder area and the chlorination equipment.

Fig. 3-4 Typical chlorine storage with trunnions for rotation and spacing of ton containers (*courtesy* Chlorine Specialities, Inc.).

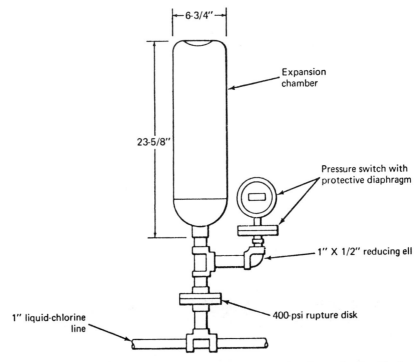

Fig. 3-5 Liquid chlorine expansion chamber. Connection is to the liquid phase only. All fittings are 2000 lb CWP forged steel. No bushings permitted. All piping is seamless carbon steel Sch 80. Teflon tape should be used as a lubricant for all threaded joints. Welded joints not advised as system should be designed for disassembly.

Traditionally, chlorine supply systems have been supplied with cylinder weighing scales. Since about 1965, more and more systems are using automatic cylinder switchover systems in lieu of scales. There are two types of these systems. The pressure differential-type system is illustrated in Fig. 3-6 and operates as follows. The two pressure reducing valves are each connected to a different supply cylinder or bank of cylinders. The outlet pressure of one reducing valve is set at about 20 psi lower than the other. When the cylinder operating through the reducing valve with a higher outlet pressure becomes empty, the other cylinder takes over. The operator replaces the empty cylinder, then by appropriate valving reroutes the chlorine flow through the reducing valve set with the higher outlet pressure. Figure 2-4 on page 45, *Handbook of Chlorination*[2] clearly illustrates an automatic pressure differential switchover system. The 1976 cost of the valves, gauges, and reducing valves illustrated in

Fig. 3-6 Automatic chlorine cylinder switchover system, pressure differential type.

Fig. 3-6 is approximately $1200. The withdrawal rate is limited to 8000 lb/day.

The vacuum differential switchover system is illustrated in Fig. 3-7.

This is a proprietary item which is listed as optional equipment. This arrangement achieves switchover by special vacuum regulator-check valve units. The unit on standby is held closed by a mechanical detent-type lockout. When the on-line supply is exhausted, system vacuum rises to a higher than normal level. This increased vacuum overcomes the latching force of the detent, and the standby supply comes on the line.[3] The 1976 cost of this system is approximately $400. It can be used with either 150-lb cylinders or ton containers but is limited to a 500 lb/day maximum withdrawal rate. A comparable pressure differential system is about $750.

Automatic switchover of liquid-chlorine systems can be done with the use of a specially designed liquid-chlorine chamber fitted with a level sensor which activates an electrically operated lever-type chlorine line valve, either ball or plug type. The liquid chamber should be about 4 in. in diameter and 18 in. high fitted with a sonic-type liquid level sensor.

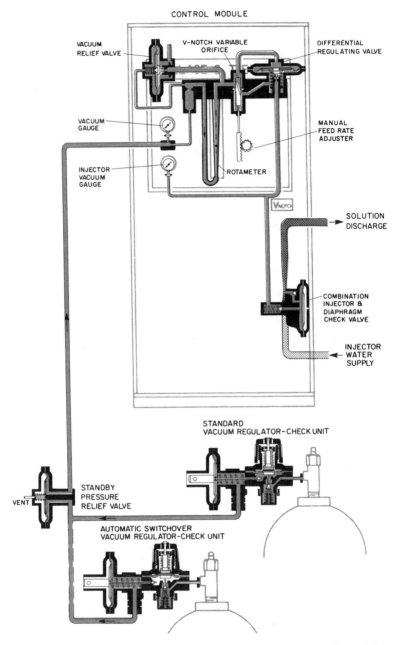

Fig. 3-7 Automatic cylinder switchover system—vacuum differential type (*courtesy* Wallace and Tiernan Division of Pennwalt Corp.).

An alternate method of automatic switchover on a liquid system is the "Reserve Tank" concept described below. This concept is usually limited to tank-car installations.

If scales are preferred, there are two types available: The load cell type with dial or electronic readout; and the lever type with beam and counterweight or dial readout. The supply system layout must be arranged so that the scales weigh only the cylinders in use. This may never be more than four per scale. In some ways a system using scales is not as flexible or reliable as the pressure differential switchover system, particularly for large installations.

Pressure gauges, switches, and alarms are desirable in the chlorine supply system. A gas-withdrawal system that is close coupled to the chlorination equipment could get by with the gas pressure gauge on the chlorinator; however, an additional gauge somewhere in the header system is always useful. A gas withdrawal system that uses an external chlorine pressure reducing valve should have one gauge upstream and one gauge downstream of this valve. An alarm for high pressure is useless on a gas withdrawal system. A low pressure alarm for loss of supply is optional. A vacuum alarm switch on the chlorinator will provide the same information.

A liquid-withdrawal system should have a pressure gauge somewhere in the header system upstream from the evaporator. It is also desirable to have a high-pressure switch to alarm overpressure conditions. This should be in addition to the alarm on the expansion tank if the latter is used. Low-pressure alarms on the liquid-supply system are optional.

Single-Unit Tank Cars. Chlorine is available in five sizes of single-unit cars: 16, 30, 55, 85, and 90 tons. Each car is equipped with an outlet dome that is made identical for all tank cars approved by the Chlorine Institute. This dome contains two liquid outlet valves that are in line with the longitudinal axis of the car and two gas outlet valves on an axis at right angles to the liquid valves. In the very center of the dome is a safety relief valve that will expel gas to atmosphere under overpressure conditions. In each liquid outlet line, there is installed a safety check valve described as an excess flow valve. In case the car is in an accident and the valves are sheared off, the check valves will jam in a shut tight position due to the momentary velocity of the liquid exiting through the sheared off valve. This is why any tank-car withdrawal system should be equipped with a rotameter to see that the unloading rate is less than that which will cause the check valve to jam closed. This rate is usually 7000 lb/hr.

For the layout of a single-unit car unloading site, the designer should consult the latest edition of the *Chlorine Institute Manual* which includes recommended unloading facilities, precautions, and federal regulations.

Two separate headers should be provided between the tank-car platform and the chlorination equipment. One line is for liquid and one is for gas. These headers are connected to the car with specially made flexible connections either of an annealed copper loop, or flexible reinforced metal hose. One flexible connection is to connect the liquid lines and the other connects to either the gas header or the air-padding system.

An air-padding system should be part of a tank-car unloading operation. While it is not always necessary, there are times when it is imperative. The air-padding system should include an air-drying system capable of providing air with a dew point of $-40°F$. This system should include a humidity alarm switch and a dew point sensor. Whenever both chlorine and sulfur dioxide are present at the disinfection facility, the air-padding system must never be used interchangeably. Each gas supply must have its own air-padding system.

Whenever air padding is deemed unnecessary, then nitrogen purging facilities should be considered as mandatory. This consists of a sufficient number of nitrogen cylinders which contain nitrogen gas at 2000 psi and a common pressure regulator with an outlet pressure regulated at 150 psi. This system should be arranged to allow the operator to purge the entire header system back into the containers. This system is most advantageous in the handling of liquid chlorine.

In addition to purging and air-padding systems, tank-car systems should be provided with a liquid chlorine absorption tank. This tank should be able to absorb all the liquid chlorine in the header and evaporator system. The absorption tank should be made of reinforced fiberglass or rubber lined steel. It should have a vent, a special connection for caustic and one for makeup water. There should also be a sampling tap to determine the effective absorption capacity of the solution. The piping of the liquid chlorine to the tank must have a barometric loop (see reserve tank system). The down drop of this loop and the sparger (diffuser) should be made of Kynar pipe. The size of the absorption tank is based on the stoichiometric combination of caustic (NaOH) and chlorine which is 1.13 lb of caustic for 1.0 lb chlorine. The absorption tank must have a minimum depth of 8 ft and the caustic solution is usually kept between 0.5 to 1.0 lb/gal.

EXAMPLE: 1000 lb. chlorine requires 1130 lb. caustic for neutralization. One gal. of 50 percent caustic contains 6.38 lb caustic. Therefore, 1130/6.38 = 177 gal. of 50 percent caustic is required for each charge of the absorption tank to neutralize 1000 lb of liquid chlorine. So to provide a tank having about 1.0 lb/gal. caustic, the absorption tank should have a capacity of approximately 1200 gal.

Accessory Equipment. In addition to the usual high- and low-pressure liquid chlorine pressure alarms and the liquid chlorine flow meter, one other device is most helpful to alert operating personnel to the imminent loss of chlorine supply in a tank car. This is a recording thermometer with an alarm switch installed on the liquid line adjacent to the tank car connection. As the car runs out of liquid chlorine, the temperature drops dramatically. This condition will sound an alarm. After the liquid has been exhausted, a 55-ton car will still be able to provide about 650 lb of gaseous chlorine at 70°F before the pressure drops below 30 psi. This is only applicable when no air padding is used.

Gauges should be provided in both headers just downstream from the flexible connections to the car and on the loading platform. An expansion tank with a pressure alarm switch, as illustrated in Fig. 3-5, should be provided on the liquid header piping.

Evaporators. When the rate of chlorine withdrawal exceeds 1500 lb/day an evaporator should be installed. This changes the supply system to liquid withdrawal which has different characteristics than gas-withdrawal systems as described above.

Evaporators are available in capacities of 4000 lb/day, 6000 lb/day, and 8000 lb/day. In a pinch, one cylinder can discharge liquid to an evaporator at a rate as high as 12,000 lb/day. This means that an evaporator can be used to conserve space for cylinder storage if necessary. The optimum storage requirements should be based on the quantity-discount price break that is offered by the local chlorine supplier. This usually occurs at a quantity of five, thus dictating space for five in service, five empties, and a vacant space for the incoming five, or a total space for fifteen ton cylinders.

The most widely used evaporator is the electric heater type as made by Fischer and Porter, and Wallace and Tiernan; Div. of Pennwalt Corp. These units are equipped with G.E. Calrod heating elements of various sizes depending upon the vaporization requirement. It takes approximately 65,000 Btus to vaporize 8000 lb of chlorine. However, the evaporator must have a wide margin of safety to allow for the partial filling of the chlorine vessel with impurities inherent in the manufacture of chlorine. Therefore, to provide a sufficient safety factor the chlorine gas, which is vaporized, must contain at least 20°F of superheat to prevent "misting" or liquid chlorine "fallout" in the gas discharge piping.

Misting occurs when the evaporator is pushed beyond its capacity. This is detrimental to the chlorinator because the little globules of mist contain the various impurities inherent in the production of chlorine. These impurities will "plate out" at the various stages of pressure reduction. This is another reason why a chlorine gas filter is always installed just downstream from the evaporator.

Evaporators are also available for use with recirculated hot water. One type utilizes the intermittent recirculating flow of hot water pumped in a circuit between a heat exchanger and the evaporator water bath. A temperature probe actuates the recirculating pump to maintain the water bath between 170 and 180°F. In the other hot water arrangement, a treatment plant utilizes a closed-loop hot water system, in which water at 200°F is intermittently pumped through a coil in the evaporator hot water bath at approximately 10 GPM. This arrangement requires an independent water bath makeup system.

Another method of heating the water bath is with live steam at about 10 psi. In place of the electric heaters, perforated copper steam diffusers are installed, through which the live steam is injected directly into the water bath.

Integral with the evaporator is the electrically interlocked chlorine pressure shut-off valve on the discharge line of the evaporator. The circuit that operates this valve, whether an air solenoid for a pneumatically operated valve or an electric motor operator, is connected to the low-temperature alarm circuit. The alarm circuit sounds and deenergizes the CPRV circuit when the water bath temperature drops to 150°F. This protects the chlorinator against possibly receiving severely damaging liquid chlorine from the evaporator.

All evaporators should be equipped with a cathodic protection system that protects both the water bath tank and the outside of the chlorine container from aggressive water corrosion. This system is provided with an indicating ammeter on the evaporator instrument panel to verify cathodic protection.

The outside of the water bath should be insulated with a $1/2$-inch covering of urethane foam.

Other accessories that are consistent with good practice and should be standard equipment but which are still considered by the manufacturers as optional are a gas temperature gauge and automatic water level control of the water bath. Standard accessories include a gas pressure gauge, water bath temperature gauge, water level indicator and low-temperature alarm switch. A high-temperature alarm switch is optional.

The electrical requirements of the electric heater type is as follows:

1. A three-wire 240 or 480 V circuit for the heater elements in the evaporator water bath. The load requirement is 12 kW for 6000 lb/day and 18 kW for 8000 lb/day.
2. A two-wire 120 V circuit is needed for the following functions:
   a. air solenoid or electric operator on chlorine pressure reducing and shut-off valve downstream from the evaporator, interlocked with low-temperature alarm
   b. low-temperature alarm
   c. high-temperature alarm (optional)
   d. solenoid valve on makeup line to water bath
   e. water level pressure switch.
3. Alarms for each evaporator should include low temperature of the water bath, and low-water bath level.

Evaporator Pressure Relief Devices

Historical Background. Until as recently as 1974 there was no mandatory requirement to provide any pressure relief device for the liquid chlorine vessel in an evaporator. However, the manufacturers of chlorine evaporators have always been conscious of the possibility of liquid vessel ruptures. The prevailing belief was and still is that any pressure relief system creates more of a hazard than it might prevent. Therefore, evaporators have always been designed with enough strength to hold any vapor pressure that could conceivably be encountered. Furthermore, the overall system design mitigates against any possibility that might allow the liquid vessel to get "skin" full of liquid chlorine. As a final precaution, the classic design of an evaporator is to have the connections to the liquid vessel made with lead gasketed "ammonia-type" unions. These unions act like relief valves under extremely high pressures. However, since the liquid chlorine vessel is fabricated according to the ASME Boiler and Pressure Vessel Code, it is subject to rigid inspection regardless of who the manufacturer may be and must therefore be certified accordingly.

As of December 1975,[29] all pressure vessels manufactured in accordance with Division 1 Section VIII of the ASME code must be protected from overpressuring by means of a safety device. This safety device need not be provided by the vessel manufacturer, but must be provided prior to placing the vessel in service. This latest code defines the general requirements of pressure relief devices. The relief valve must be able to relieve the pressure in the liquid chlorine vessel when this pressure exceeds 110 percent of the rated working pressure of the vessel.

The Chlorine Institute specifies in Pamphlet No. 9 that chlorine vaporizing equipment must have a pressure relief device.[30] This can be either a rupture disc or a spring loaded relief valve or both; preferably discharging to an adsorption system. When both are used the section between should be equipped with a vent or pressure alarm.

The State of California safety orders for "Unfired Pressure Vessels" indicates that in addition to compliance with the ASME Code the following control is also required.

467. Controls: (a) Any pressure vessel not specifically covered or exempted elsewhere in these orders shall be protected by one or more safety valves or rupture discs set to open at not more than the allowable working pressure of the vessel* and by such other controlling and indicating devices as are necessary to insure safe operation.

Current Practice. Both Fischer and Porter[31] and Wallace and Tiernan, Div. of Pennwalt Corp.[32] provide as optional equipment a relief valve system for all their various types of evaporators. Both illustrate the location of the relief valve on the gas and *not the liquid* phase of the evaporator connections. This current arrangement is shown in Fig. 3-7A.

*Both Fischer and Porter and Wallace and Tiernan use a working pressure design of approximately 500 psi at 212°F and a hydrostatic pressure test of 1450 psi at 125°F.[33,34]

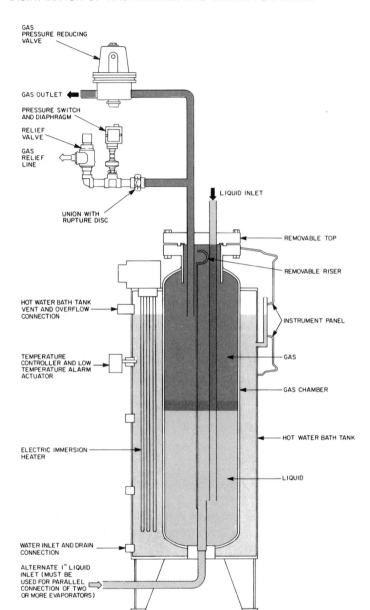

Fig. 3-7a   Evaporator showing relief system (*courtesy* **Wallace and Tiernan Division of Pennwalt Corp.**).

Relief Valves. The Fischer and Porter relief valve Model 71P1412 has been manufactured for over 20 yr. This valve opens at about 275 psig and seats tightly at about 200 psig. Fischer and Porter report a high-confidence level for this valve.

The Wallace and Tiernan valve carries Part No. U24733. It opens at 400 psig and closes at 380 psig. This valve is purchased from Crosby Valve Co. and other suppliers such as Dresser Industries and Ferris Valve Co.[33]

Safety Considerations. The discharge of the relief valve system (vent) brings up serious questions about the hazards of chlorine leaking to the atmosphere at high pressures. Whenever either of these valves open to relieve pressure they become subject to atmospheric corrosion. Maintenance routine should therefore require that the valve be overhauled after each opening or closing cycle. More-over, this valve must be protected at all times from chlorine vapor by installing a rupture disc as recommended by the manufacturer (upstream from the valve). This keeps the valve clean and dry during the periods of nonuse.

Owing to the potential hazard of a chlorine leak it is always advisable but not mandatory to discharge the vent from this system into a chlorine absorption tank. Therefore, all of these systems should be designed so that at a later date a barometric loop and absorption tank can be conveniently added to the relief valve vent. (See section describing "Reserve Tank.")

Absorption Tank for Relief Valve Vent System. In addition to a suitable size absorption tank (containing NaOH), the discharge piping between the relief valve discharge and the absorption tank must contain a barometric loop. (See Fig. 3-8). The barometric loop is mandatory with an absorption tank because it prevents the almost certain intrusion of moisture into the chlorine gas supply piping.

The absorption tank should be capable of neutralizing a maximum of 150 lb liquid chlorine from each evaporator during an overpressure crisis situation. This is about the amount of liquid chlorine which would be contained in each evaporator if the liquid vessel were full of liquid chlorine. While this is an im-probable situation it provides a generous safety factor to the size of the ab-sorption system. To this amount of chlorine should be added the amount in the liquid piping and expansion tanks* between the supply tanks and the evaporators.

Monitoring Relief System. As shown in Fig. 3-7A there will be a pressure switch to monitor a critical overpressure situation sufficient to rupture the

---

*It is reasonable to expect that the rupture discs on the expansion tanks might also be overpressured at the same time.

frangible disc. However, it is recommended that if this relief system is vented to the atmosphere and not to an absorption chamber the vent should be combined (but separated) in such a way that this vent and the usual chlorinator and chlorine pressure reducing valve vents be monitored continuously by a chlorine leak detector.

Chlorine Tank Car Supply Monitoring Systems:

Scales and Alarms. A continuous flow of chlorine, which must not be interrupted when a tank car has discharged all its liquid chlorine content, is a fundamental requirement of a tank car installation. Reliance on pressure drop is not the answer and where air padding is used on a continuous basis temperature drop monitoring is not reliable. The use of weighing scales for tank cars is an exorbitant expense, $30,000–$35,000 per car. Scales for chlorine storage tanks are much less expensive (i.e., $5000). Load cell installations are about $5000 and have certain advantages over scales.

The accuracy of tank-car scales and load cells is somewhat suspect because of the density change with ambient temperature variations, plus the fact that drag or tension exerted on the scale balance system by flexible tubing and/or rigid pipe connections introduce gross errors in the weighing mechanism. Unless a tank car or storage tank is heavily air padded (175–200 psi, see Chapter 5) it is not possible to measure and record the flow of liquid chlorine by conventional measuring devices with any reasonable accuracy. This is due to the characteristics of liquid chlorine. As the liquid flows past a point of slight pressure drop as in a rotameter, the liquid flashes to gas causing a great change in density. This makes calibration of a flow meter practically impossible. A magnetic meter will not respond to liquid chlorine; therefore, some alternative is necessary to provide a continuous monitor for the flow of liquid chlorine. A system which has been used for many years in the pulp and paper industry is the "reserve tank" concept as developed by the late Brian Shera of Pennwalt Corp., formerly Pennsylvania Salt Co.[4,5]

Reserve Tank. This is a liquid chlorine flow-through tank, as shown in Fig. 3-8, which provides an active reserve when the car has been emptied of liquid. The tank may also be installed vertically if preferred. These tanks have been made in sizes from 1000 to 16,000 lb of chlorine. The capacity should provide approximately 1–3 hr reserve chlorine at peak demand. Liquid chlorine enters and leaves the tank through pipes extending to the bottom of the tank. Venting of gas is necessary to provide a minimum of 20 percent gas volume as a safety factor against excessive filling of the vessel with liquid. This is accomplished by a dip pipe of an appropriate length to provide the gas volume. The vent piping is

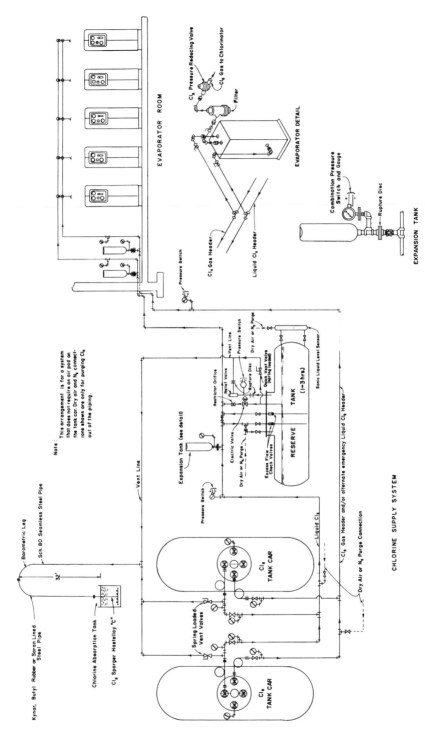

Fig. 3-8 Reserve tank concept for a chlorine tank car supply system.

fitted with an electric shutoff valve and a restrictor orifice. Upon leaving the restrictor orifice the vent pipe joins the liquid lines going to the vaporizers. This restrictor orifice allows a minor flow of gas that will flashover from the small amount of liquid being bled from the tank thus keeping the liquid level constant in the reserve tank so long as chlorine is flowing from the tank car. This vent also allows the liquid level to rise to its normal height when the reserve tank is being filled. When the liquid chlorine flow from the tank terminates, the chlorine level in the reserve tank will drop and this will be detected by the ultrasonic level sensor. The sensor sends a signal through a control box which will engage an alarm and also close the electric shutoff valve in the vent line. The operator now has two choices: shutoff the liquid flow from the reserve tank and bleed the remaining gas in the tank car to the process; or immediately switch to a full car. When the time comes to switch to a new car, the reset button on the electric valve is actuated so as to open the vent line and allow refilling of the reserve tank. The reserve-tank piping and valving accessories are equipped with both automatic and manual pressure relieving devices all discharging to the absorption tank. This prevents any overpressuring of the reserve tank. The pressure relief line has a rupture disc followed by a pressure switch alarm. This is followed by a spring loaded relief valve protected from moisture entry by a rupture disc which is installed backwards for ruptures at low pressure. In addition, there is a spring-loaded manual quick vent valve to the absorption tank. Fig. 3-8 illustrates all of the appropriate expansion tanks and pressure switches required to give the operator the necessary operating information.

The following are the recommendations of the Chlorine Institute for chlorine storage tanks as set forth in Pamphlet No. 5, "Facilities and Operating Procedures for Chlorine Storage," 2nd Ed., Jan. 18, 1962.

- Minimum design volume, 25.6 ft^3 for each 2000 lb of chlorine to be stored
- Minimum working pressure, 225 psig plus $1/8$-in. corrosion allowance
- Design and fabrication, compliance with the ASME-UPV Code, 300°F, 70 percent weld efficiency, spot X-ray or 100 percent X-ray longitudinal and circumferential seams
- Nozzle necks, Sch 160, seamless steel pipe
- Piping, Sch 80 seamless steel
- Flanges, 300 lb weld-neck, slip-on, or screwed
- Fittings, buttweld extra heavy
- Fittings, screwed, 3000 lb
- Valves, 300 lb, flanged, forged steel body
- Valves, 300 lb, screwed, forged steel body

The purge connections shown are essential for the inspection and repair of the system. They also prevent the emission of chlorine to the atmosphere and the entrance of moisture to the piping and tank system, thus accelerating corrosion and deteriorating the entire system. This is the only way to provide a completely closed system to the moisture in the atmosphere.

Materials of Construction:

Supply System. The chlorine supply system should consist of steel and cast iron products. The *supply system* is defined as that part of the system that begins at the chlorine containers and terminates at the inlet to the chlorinator. From the chlorinator and beyond, the materials of construction are entirely different and are discussed elsewhere.

The supply system piping must be Sch 80 black seamless steel and fittings must be 2000 lb forged steel. *Do not use bushings* (they cannot meet the 2000 lb criterion). Use reducing fittings instead.

All unions should be ammonia-type with a lead gasket joint. *Never use a ground joint union.* Filter bodies and reducing valve bodies are usually cast iron. Expansion tanks should be of welded steel construction, but can be a standard 100- or 150-lb chlorine cylinder. Valves for the chlorine supply system should be Chlorine Institute approved. Two types of valves are used, one is for main line shutoff purposes and the other for isolating cylinders (header valves). Header valves are identical to the outlet valves of ton containers; bronze bodies with monel seat and stem. Main line valves can be either the ball-type or rising stem-type. The ball-type is more popular because it utilizes a lever which not only indicates at a glance the position of the valve, but also makes it easier to operate the valve.

All gauges on the supply system must be equipped with a protector diaphragm. The diaphragm should be of silver and the diaphragm housing can be either Hastelloy "C" or silver cladded steel. Shutoff valves should not be used ahead of gauges. Gauges require a minimum maintenance so that when replacement is needed the entire supply line should be drained of pressure before replacing or removing the gauge. The value of a shutoff valve for this purpose is lost, because a valve in a chlorine supply system loses its reliability if it is not operated on a frequent basis.

In assembling the piping system, either welded or threaded construction can be used; welded is preferable. If threaded construction is used, the contractor must be cautioned to use sharp dies and all threaded pipe must be cleaned with solvent before assembly. Pipe dope should not be allowed; instead use teflon tape for thread lubricant.

Safety Equipment. This equipment consists of the following items; breathing apparatus, chlorine leak detectors, emergency chlorine container repair kits, and expansion tanks for liquid chlorine supply systems. These items are primarily used with the supply system because this system is under a positive vapor pressure.

There are two types of *breathing apparatus*, the canister-type gas mask and the oxygen or air-type breathing unit. The canister-type gas mask is limited in effectiveness such as changing chlorine cylinders or normal maintenance work. It is not satisfactory for use in repairing a leak. Therefore, either of the following types of equipment should be furnished: the air-type breathing unit (with thirty minute air supply as manufactured by Mine Safety Appliance Company or Scott Aviation Company) or the oxygen breathing apparatus (as manufactured by MSA). The latter is similar to a canister type. When the seal on the unit is broken, the unit manufactures its own oxygen which lasts for forty-five min. These canisters must be discarded after the seal is broken.

Every installation should have at least one *Chlorine Institute Container Emergency Kit*. The kit for ton containers is illustrated in Fig. 3-9. These kits are also available for 150-lb cylinders and tank cars; they are designed to seal off a

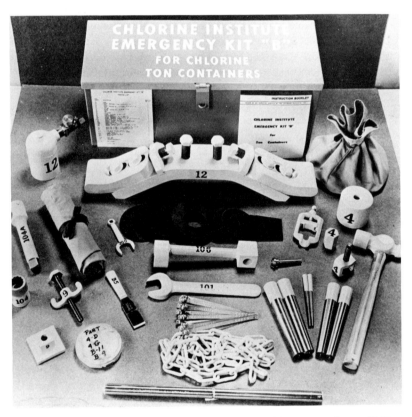

Fig. 3-9 Ton container emergency kit (*courtesy* Chlorine Institute, N.Y. and Chlorine Specialties, San Francisco, Calif.).

leaking fusible plug, a leaking outlet valve, or a moderate size rupture in the container shell.

A suitable *continuous chlorine leak detector** should be furnished for every chlorination station regardless of the size of the containers. This is not only desirable from a safety viewpoint, but is necessary to meet the OSHA personnel safety requirements. Leak detectors are equipped with circuitry to actuate external alarms. These detectors are of such nominal cost that more than one unit might be used for safety; one for the supply system and one for the control system. Leaks may develop in each of these separated areas.

Expansion tanks are mandatory for liquid chlorine supply systems. These chambers are necessary when there is danger of liquid chlorine becoming trapped in a length of pipe. If this situation is followed by a significant ambient temperature rise then hydrostatic pressure will develop. This may be sufficient to rupture the pipe because the coefficient of expansion of steel is much less than that of liquid chlorine. The recommended type of expansion tank with rupture disc and chlorine pressure switch as recommended by the Chlorine Institute is illustrated in Fig. 3-5. The frangible disc ruptures at pressures between 300 and 400 psi, thereby allowing the liquid trapped in the system to enter the expansion tank. This immediately produces some vapor pressure in the expansion tank, which actuates the pressure switch and sounds the alarm. Spring loaded relief valves discharging to the atmosphere should never be used on liquid chlorine lines. The chance of such a valve reseating properly is remote. Therefore, the discharge of any and all such valves should be to an absorption system.

Note: Additional information on chlorine supply systems can be found in Chapter 2, *Handbook of Chlorination.*

Chlorine Control System

Chlorinator Design Concepts:

Historical Background. In the years 1965–1975, there have been some significant changes in chlorinator operation concepts. These changes are much more subtle than those of ca. 1955 with the advent of plastics, particularly PVC. The famous bell jar line of chlorinators pioneered by Wallace and Tiernan was virtually replaced overnight by the introduction of PVC injection molding. So once again the original spring diaphragm concept was revived and its success was made possible by the research into plastics and corrosion resistant diaphragms, particularly by DuPont. The very first series of chlorinators marketed by Wallace and Tiernan were based on the spring-loaded diaphragm principle (ca. 1913). Owing to severe corrosion problems this concept was dropped as soon as the visible vacuum concept by C. F. Wallace was introduced, ca. 1920.

*This should include a SO_2 leak detector for the total chlor-dechlor system.

So when PVC injection molding and teflon diaphragms became available the chlorinator design approach reverted to the original spring loaded diaphragm principle but utilized the technology of the newly arrived plastics industry.

This design concept, pioneered by Fischer and Porter, was a much needed stimulus for the chlorinator industry. This was a small specialty manufacturing company at that time, but once the ingredient of quality competition was introduced the results for the consumer were dramatic. To counter the impact of the Fischer and Porter "all plastic" line, Wallace and Tiernan introduced their line of "plastic" chlorinators but with a revolutionary idea of chlorine feed rate control. Wallace and Tiernan incorporated the "V"-notch orifice concept, which they had been keeping under wraps for at least 10 yr. This concept of chlorine metering control proved to be unique. From this point forward Wallace and Tiernan developed unbelievably accurate and flexible control systems. Their success spurred their competitors to greater achievements some of which are described below.

Conventional Chlorinator Design.   Owing to the versatility of the combination V-notch orifice–spring-loaded diaphragm principle, Wallace and Tiernan were able to capitalize on even another idea that lay dormant for many years. This was their U.S. Patent 2,929,393 which is now known as the compound-loop control system, first introduced in 1960. Referring to Fig. 3-10 which is a flow

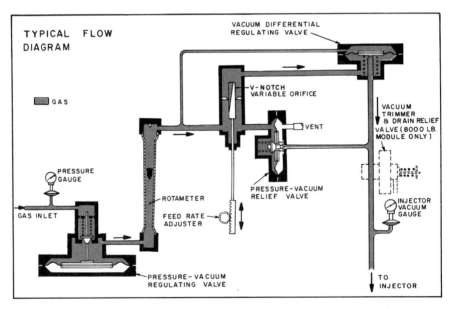

Fig. 3-10   Flow diagram V-800 series vacuum solution feed chlorinator (*courtesy* Wallace and Tiernan Division of Pennwalt Corp.).

diagram of a Wallace and Tiernan V-800 chlorinator: The chlorine gas enters the system through the pressure-vacuum regulating valve at which point the inlet chlorine pressure from the supply system is reduced to some constant level of negative (vacuum) pressure (the level of vacuum varies with each manufacturer). The gas then passes through the metering orifice (V-notch) to the differential vacuum regulator and then to the injector vacuum line.

The chlorine flow through the V-notch orifice is based on the classic flow formula $Q = AV$, where A is the area of the orifice opening (position of V-notch orifice positioner) and V, the velocity of the gas through the orifice. This gas velocity is best expressed in terms of the differential pressure across the orifice to produce a given velocity. This is equal to $C\sqrt{2gh}$ where h equals the differential vacuum across the differential vacuum regulator and C is the velocity coefficient of the V-notch orifice. The pressure-vacuum regulating valve illustrated in Fig. 3-10 is designed to maintain a constant pressure (usually a slight vacuum) upstream from the metering orifice and control device. The vacuum differential regulating valve is designed to maintain a constant pressure drop (h) across the metering orifice (V-notch orifice).

The metering orifice has a nominal range of 20:1 while that of the vacuum regulating valve (differential regulator) is about 10:1. If these two devices operate from independent signals—called "compound loop control"—then the overall automatic proportional chlorinator range is a product of the metering orifice range times the vacuum regulating valve range or 20:1 X 10:1 = 200:1. Figure 3-11 illustrates how this is accomplished. The flow proportional signal—a linear function—is sent to the chlorine metering orifice; the chlorine residual signal—a square root function—is sent to the differential regulating valve. The flow proportional signal is usually electric, either pulse duration, pulse frequency, milliamp, or potentiometer. This signal can also be pneumatic, 3–15 psi. The chlorine dosage signal to the differential-regulating valve is converted to variable vacuum, usually by means of a motorized vacuum valve. This is the basic concept of wide range chlorinator control, often referred to as the "compound loop system."

Sonic Flow Concept. The most recent development in chlorinator design which affects accuracy and control modes is the concept of sonic flow. Previously described concepts assumed that the gas flow through the metering orifice was a function of the differential pressure across that valve (h). This is true for a wide range of differential pressures; however, if the velocity through the valve is increased to the speed of sound in the gas flow at that point, a different set of conditions is encountered. Once the sonic velocity is reached, the flow through the valve is no longer a function of the pressure drop (h) across the valve. Under these conditions, gas flow is directly proportional to the area of opening in the control valve and is entirely independent of the downstream pressure which is a function of the injector vacuum. Therefore, when sonic flow conditions are

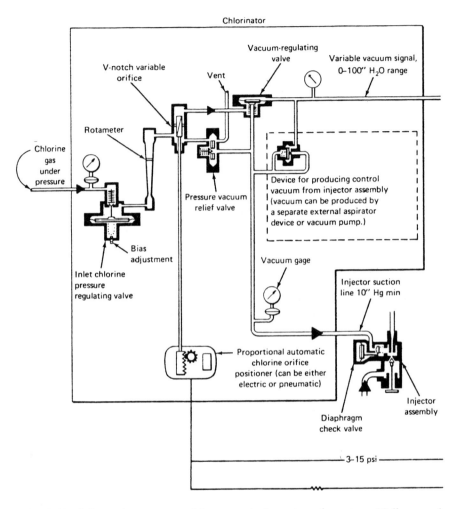

Fig. 3-11 Schematic compound-loop control system (*courtesy* Wallace and Tiernan Division of Pennwalt Corp.).

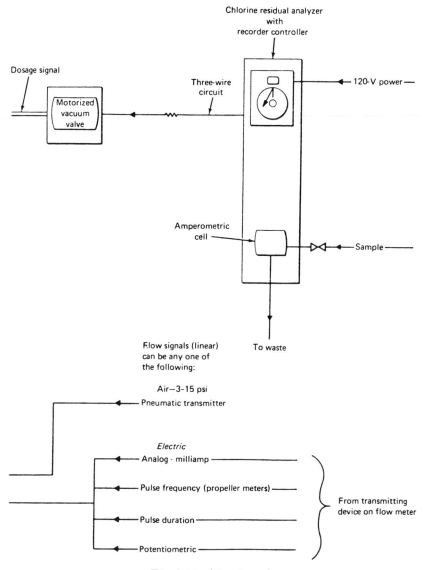

Fig. 3-11   (*Continued*)

attained, the differential vacuum regulator is no longer a necessary component of chlorinator design. Using this as a concept, Fischer and Porter have developed a high resolution characterized control valve which is available with a built-in computerized component which can accept either or both a flow signal and a residual signal (see Fig. 3-12). For single input applications the dosage signal originates from a dosage adjustment knob on the chloromatic valve. This valve and computer arrangement is designed so that at 100 percent flow signal and 50 percent residual signal, the chlorine gas flow will be at 100 percent of full

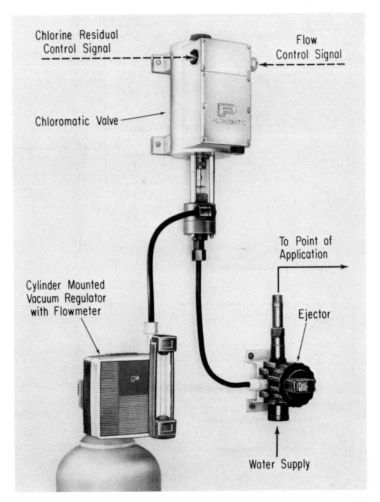

Fig. 3-12 Chloromatic valve for sonic flow chlorinators and other applications (*courtesy* **Fischer and Porter Co.**).

scale. Similarly at 20 percent flow signal and 50 percent residual signal, the chlorine gas flow will be at 20 percent of full scale.

This valve can only be used with chlorinators that are equipped with injectors that produce a high enough vacuum to achieve sonic velocities across the control valve. While the chloromatic valve is available in capacities up to 10,000 lb/day, it is only available in the sonic version up to 2000 lb/day chlorinators. Between 2000 and 10,000 lb/day capacities the chloromatic valve must be used with a vacuum differential regulating valve (see Fig. 3-10).

Chlorine Control System:

Capacity. The first step in the design of a chlorination facility is the determination of the maximum chlorine needed for the given situation, including prechlorination, intermediate, and postchlorination applications. While this text is specific for disinfection (postchlorination), total capacity must include the requirements for the other purposes stated above. Chlorinators should be divided into two groups, one group for prechlorination and intermediate points of application, the other group for postchlorination (disinfection). Equipment should be arranged so that the first group can provide standby service for the disinfection equipment. A third group may be desirable if both pre- and post-chlorination is to be continuous. The third group should be arranged for both intermediate chlorination and standby for either pre- or postchlorination.

Chlorinator capacity is a function of the flow signal and the chlorine demand of the wastewater. Attention is called here to the term "flow signal." Since it is imperative that the chlorine feed rate be controlled in proportion to the flow, the chlorinator capacity must be sized to the capacity of the primary flow meter. While it is true that chlorinators have over-riding dosage control, this feature must be reserved for variations in the chlorine demand of the wastewater. For example, at a proposed plant, the primary flow meter to be used has a range of 0-10 mgd which is the ultimate design capacity of the facility. Furthermore, the chlorine demand is to be a maximum of 150 lb/million gal. The chlorinator capacity for that plant must be 10 X 150 = 1500 lb/day *regardless of the expected initial low flows that may be expected for several years.* This capacity is required because the chlorinator needs the full range of the primary meter flow signal so that it can also provide a dosage range to meet the diurnal chlorine demand variation.

This concept applies to both pre- and postchlorination, but not to intermediate chlorination because the latter chlorine feed rate is usually on a manual control basis.

The required chlorine dosage for prechlorination is difficult to predict with any degree of accuracy in the absence of some historical evidence. If the sewage is moderately fresh, the dosage may be as low as 10-12 mg/liter. If sulfite wastes such as from a tannery are a factor, the chlorine demand might be as high as 40-50 mg/liter. If the sewage is septic upon arrival, the demand will most likely be from 30 to 40 mg/liter.

Intermediate points of application of chlorine require approximately the following equipment capacities:*

1. Secondary sedimentation tanks (10 mg/liter)
2. Return activated sludge—control of bulking (5 mg/liter)
3. Ahead of biofilters or trickling filters (10-15 mg/liter)
4. Sludge thickener line—odor control (50 mg/liter)

The use of chlorine at these points of application is rarely on a continuous basis, so the prechlorination equipment can be used for these purposes. If prechlorination is to be used continuously for odor control, additional equipment should be supplied for the intermediate points of application. This group of equipment should then be sized so as to provide standby service for either pre- or postchlorination.

Industrial wastes have a profound effect on the disinfection requirements of domestic wastes. This is further complicated by the seasonal nature of some industrial wastes and extent to which these wastes are pretreated before being discharged into the domestic collection system.

For domestic wastes with not more than 1-2 percent industrial waste, the following are minimum capacity guidelines for the designer:

Primary effluent	150-200	lb/mg
Secondary effluent	50-75	lb/mg
Secondary plus ponds	50	lb/mg
Tertiary (not nitrified)	30-50	lb/mg

Effluents that are treated to a free chlorine residual require about 10 parts of chlorine for each part of ammonia nitrogen.

If industrial wastes are present to the extent of 10 to 25 percent of the total wastewater flow, it may be necessary to increase the chlorine requirement for primary effluents by a factor of two and perhaps more depending on the nature of the wastes. This may apply to secondary effluents as well, depending on how effectively the treatment process can cope with the industrial waste. The more uncertain the factor of industrial waste, the more uncertain becomes the needed quantity of chlorine for disinfection. In this case, laboratory determination of chlorine demand will be necessary.

Methods of Control. Chlorinators for both pre- and postchlorination should be provided with flow proportional control. The prechlorinator should be con-

---

*All dosages are based upon treatment plant flows (PDWF) while sludge thickener dosage is for this specific flow rate.

trolled from a raw sewage flow meter and the postchlorinator from an effluent flow meter. Never attempt to control the ·postchlorinator from the raw sewage meter or vice-versa; it will not work because of the lag time between the two measuring points. As discussed previously, the flow proportional signal should be used to control the chlorine metering orifice in the chlorinator. This can be done either pneumatically or electrically. If it is to be pneumatic, the signal must be linear (3–15 psi). If it is to be done electrically, there are a variety of linear signals. The one most preferred is the analog milliamp signal, with outputs of 1–5, 4–20, and 10–50 mA. The most common is 4–20 mA.

In addition to the flow proportioning control, residual control for disinfection is necessary because of the diurnal variations in chlorine demand, and the necessity for close regulation of the disinfection process to meet stricter discharge standards.

Residual control is synonymous with automatic dosage control. This is accomplished automatically using an amperometric chlorine residual analyzer. The analyzer transmits a vacuum signal, proportional to the residual, to the differential regulator in the chlorinator. The chlorinator compounds this signal with the flow signal to achieve a wide range of operation in excess of 100 to 1. Such an arrangement of the two separate signals is known as "compound loop control." (See Fig. 3-13.)

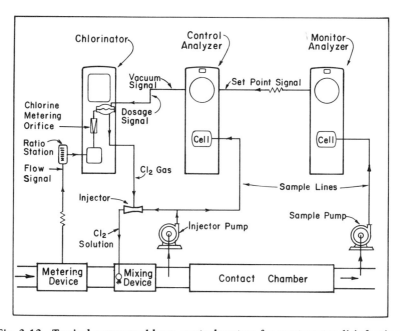

Fig. 3-13   Typical compound-loop control system for wastewater disinfection.

The dosage signal is transmitted from the chlorine residual analyzer via a vacuum transmitting device. The controls on the analyzer should be adjustable to accommodate loop times from 1 to 7 min. and length of correction times from 0 to 20 sec. Accessories should include a precision vacuum gauge to monitor the vacuum signal to the chlorinator. It is not good practice to interchange the signals (i.e, send the dosage signal to the chlorine metering orifice positioner and the flow proportional signal to the vacuum regulator).

An alternative to compound loop control is to summate the flow and dosage signals into one signal and send that signal to the chlorinator. The range on this type of system is limited to about 20 to 1. This system is illustrated in Fig. 3-14. Regardless of which method is used, flow proportional control signals should be equipped with ratio stations. These units serve as a precise and repeatable device for plant supervisory personnel to maintain an accurate method of logging shifts in basic chlorine dosage control.

In the present era of chlorination-dechlorination it is no longer practical to operate the chlorine control system solely by residual control because a flow signal is mandatory for the dechlorination system control.

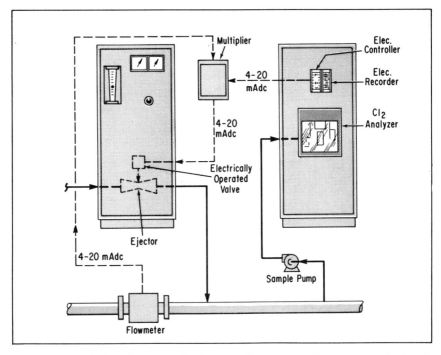

Fig. 3-14 Typical chlorine residual control system using summated signals (*courtesy* Fischer and Porter Co.).

The compound loop control system illustrated in Fig. 3-13 consists of two separate signals to the chlorinator, one for flow, which is modulated by a ratio station, and one for the residual as based on the immediate chlorine demand. In any chlorine control system the loop time is a critical factor. This should be kept as short as possible, preferably not more than 2 min.

Assuming proper mixing at the point of application, a representative sample for the analyzer should be taken downstream from the chlorine mixing device within 30 sec after the application of chlorine at average flow. At peak flow this time would be reduced to about 15 sec. However, it will take the sample another 45-60 sec to reach the cell since it has to pass through a filter and the internal piping within the analyzer; therefore, the earliest the sample can reach the measuring cell is about 1½-2 min. after the chlorine has been mixed with the wastewater. If the chlorine has been well mixed, the control residual will lie on the flat part of the residual die-away curve shown in Fig. 3-15. Another mandatory requirement of a well-designed system is to utilize the remote injector concept by installing the injector as close to the point of application as is physically possible. The average loop time should be on the order of 2 min. with a maximum not to exceed 5 min.

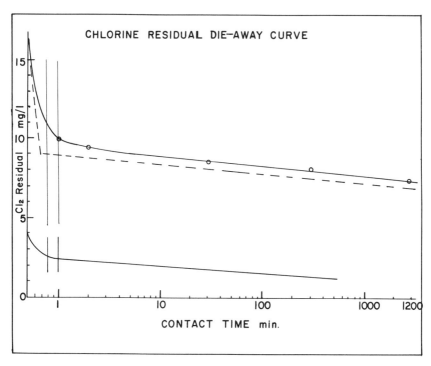

Fig. 3-15 Typical chlorine residual die-away curve.

Sample Lines. All sample lines to the various analyzers should be designed so that they can be easily purged of slime layers and other organic debris. Biofouling of sample lines introduces significant errors in the analyzer system by creating a false chlorine demand due to dechlorination in the line. The sample lines carrying a dechlorinated effluent are particularly vulnerable to biofouling.

Sample lines should be designed to provide velocities up to 10 ft/sec. This acts to decrease dead time in the control loop and produces some scouring action thereby inhibiting build-up of slime layers in the piping.

Alarms. Every chlorination facility should have an alarm system that adequately alerts the operators in the event of deficiencies, malfunctions, or hazardous situations related to chlorine supply, chlorine metering equipment, chlorine leaks, and chlorine residual.

The chlorine supply system should be monitored for chlorine leaks and low pressure which indicates loss of supply. High-pressure alarms are only appropriate for liquid withdrawal systems. Pressure alarms are triggered by high- and low-pressure switches. Chlorine leaks in the supply and control area should be monitored by a permanent chlorine leak detector and alarm that meets OSHA requirements. The detector should be capable of sensing atmospheric chlorine concentrations as low as one ppm (by volume).

The chlorinators should be equipped with high- and low-vacuum sensing devices. Low vacuum signifies failure of the injector system and high vacuum signifies loss of chlorine supply. Either of these is indicated by a vacuum switch.

Installations using evaporators require warning alarms of high- and low-water bath temperature and low-water bath level.

Chlorine residual control analyzers should be equipped with adjustable high- and low-residual alarms.

Systems using tank cars, storage tanks, or tank trucks that involve air-padding systems should have a high humidity alarm on the dried compressed air.

Monitoring Equipment. In addition to the chlorine leak detector previously described, two categories of monitoring equipment may be utilized. One is for monitoring chlorine residual at the end of the contact chamber, and the other is for recording of chlorine flow through the chlorinator.

An amperometric chlorine residual recorder should be installed to record the chlorine residual in the effluent from the contact chamber. This unit will be in addition to the analyzer used in the control system. The effluent monitor will insure against the loss of residual in the chlorine contact chamber and thus help to prevent improper disinfection and the proliferation of slime growths in the contact chamber.

The effluent monitor residual recorder and the control residual recorder should both be furnished with identical capabilities so that one can be a standby for the other.

Continuous recording of the amount of chlorine applied by each chlorinator provides valuable information, particularly for trouble-shooting the residual control process. Malfunctions in the system are readily identified when the chlorine feed rate can be compared with the chlorine residual chart at any given moment. Facilities utilizing chlorine residual control should have each chlorinator equipped with individual chlorine flow recorders. These recorders should be installed somewhere downstream from the chlorinator pressure-vacuum regulator as recommended by the chlorinator manufacturer. Chlorine flow recorders on the chlorine supply line upstream from the chlorinators are generally unsatisfactory.

Plant effluent flow meters should be provided with recorders so that the flow for any given moment can be compared with chlorine residual and chlorine dosage.

Injector Systems:

Description. The injector system is the heart of the entire chlorination facility. If this system is inoperable, no other part of the system can function. The various parts of this system include: 1) the operating water supply to the injector; 2) the injector; 3) the injector vacuum line from the chlorinator; 4) the injector discharge system described as the chlorine solution line; and 5) the diffuser at the point of application.

Water Supply. The injector must pull a specified vacuum in order to move the chlorine from the supply system throughout the chlorinator, dissolve it into the injector water supply, and carry it to the point of application. This requires a specified amount of water—about 40 gal. of water/day/lb of chlorine, more or less depending upon the conditions. The amount of water must be enough to maintain a chlorine solution strength not to exceed 3500 mg/liter. Therefore, the amount of water is a function of the amount of chlorine being fed to the system. A second factor is the pressure at the point of application of the chlorine. This is known as injector back pressure. A higher back pressure requires a higher injector inlet pressure and more operating water to make the injector function properly. The injector also has minimum operating water requirements. Chlorination equipment manufacturers use injector operating curves that specify how much water at what pressure is required for a given amount of chlorine to be applied against a given back pressure. It is the designer's responsibility to make an hydraulic analysis of the chlorine solution line between the injector and the point of application to establish the amount of back pressure to be expected. With this information and the maximum chlorine feed rate desired, the chlorinator manufacturer can then advise regarding the necessary water supply inlet pressure at the injector and the optimum injector operating water quantity.

The back pressure should never be allowed to fall below 2 psi at the injector discharge (see section on Diffusers).

Injector Pumps. In most systems, injector water supply pumps must be furnished. The choice of pump is between a turbine-type or a centrifugal-type. These two types of pumps have widely different characteristics insofar as chlorinator operation is concerned. Turbine pumps are usually the pumps of choice for chlorinators using the one-inch fixed throat type injector. These pumps should be equipped with an adjustable discharge pressure bypass assembly. It is standard practice to select a turbine pump capable of pumping about twice that of the injector requirement. This allows for impeller wear which is critical with a turbine pump. A centrifugal pump does not require a bypass assembly to allow for impeller wear. The pumping rate selection should be about 20–25 percent in excess of the injector requirement to provide the standard safety factor. Centrifugal pumps may be the preferred type at injector quantities in excess of 15 GPM provided that the total dynamic head requirements are not too high. Usually back pressures in excess of 30–40 psi will require turbine-type pumps.

High-back pressures are rare in wastewater practice and should be avoided. It requires a tremendous amount of hydraulic energy to apply 2000 or 3000 lb of chlorine against back pressures greater than 5–10 psi. In wastewater practice it is extremely desirable to locate the center line of the injector above the hydraulic gradient at the point of application. This creates a slight negative head which is undesirable but can be overcome by proper design of the diffuser (see the section on Diffusers).

Remote Injector Installation. Very often it is desirable to install the injector adjacent to the point of application, primarily for control purposes when using residual control. If so, certain features described below should be included. The vacuum line between the chlorinator and the injector should be sized so that the total pressure drop in this line is not more than one and one-half in. Hg. A shut-off valve should be provided at each end of this vacuum line. A vacuum gauge should be installed adjacent to, but downstream from, the shut-off valve at the injector end. The injector should always be installed in a horizontal position to achieve a low profile (better hydraulics) and to allow easy disassembly of the injector.

Long Vacuum Lines. The optimum pipe size for vacuum lines between the metering equipment (chlorinators or sulfonators) and the injector is subject to a great deal of scrutiny. White has investigated the hydraulic characteristics of three remote injector systems varying in distances from 750 to 8740 ft. The most comprehensive study was of a chlorination system with 3–2000 lb/day chlorinators remotely located from a single 3-in. injector connected by one 3-in. PVC vacuum line 8740 ft long.[6] Friction loss data were plotted showing

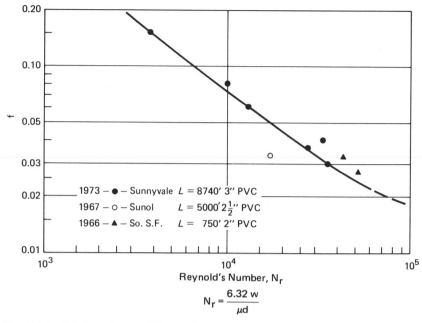

Fig. 3-16 Friction factor "f" as a function of Reynolds Number, N_r, for PVC injector vacuum lines.

the friction loss factor f as a function of Reynolds Number, N_r (see Fig. 3-16). The response time (lag time) between the chlorinators and the injector was determined by 30-sec. interval amperometric titrations of the chlorinated effluent immediately downstream from the chlorine diffuser, which was within 25 ft of the injector. As a check on these field observations Fischer and Porter Co. Engineering Department set up a laboratory experiment using a 400 ft $\frac{3}{4}$-in. vacuum line to a remote injector and a variable chlorine feed rate of 150–500 lb/day. The results of this experiment when extrapolated agreed closely with the field results by White and Stone described above.[6]

Several important conclusions were drawn from these observations: 1) for any given system the lag time appears to be almost constant regardless of magnitude of change of the gas feed rate; 2) the higher the vacuum level in the vacuum line the shorter the lag time because the lower density of the gas provides a higher velocity; 3) lag time is independent of total volume of the vacuum line; 4) friction factor f varies significantly with Reynolds Number (see Fig. 3-16); and 5) when the total pressure drop is greater than about 1.5 in. Hg there is a noticeable decay in the vacuum level of the entire line (see Fig. 3-17). Referring to this figure, the total pressure drop was about 1.5 in. Hg at flows less than 500 lb/day. Above this flow (i.e., 500-5500 lb/day), the pressure drop was practically constant. It varied from 2.88-3.06 in. of Hg, while the vacuum level varied from 5.25-22.25 in. Hg. Over this range of vacuum level and at a constant temperature of 68°F the density of gas varied from 1.55 lb/ft³ at the low vacuum to 0.048 lb/ft³ at the high vacuum—this is a range of 32-1. From these values it is obvious that the flow conditions of this vacuum line are highly unstable. The instability of this system is a result of the high-total pressure drop.

These observations bring up the question of the design procedure for an optimum size vacuum line for a given length and amount of gas flow. It would seem prudent to design for the severest conditions. Instead of designing for the minimum vacuum conditions allowable for the proper operation of chlorinators and sulfonators (about 8-10 in. Hg), the maximum injector vacuum level should be assumed (22-23 in. Hg). At this level the *total* pressure drop in the system, regardless of pipe length, should be limited to 1.50-1.75 in. Hg. One such system designed to include these factors has been in operation long enough to draw the following conclusion: it is a stable system with rapid response and therefore close to the optimum design. The maximum vacuum decay level from minimum to maximum feed rate is only about 3 in. Hg (i.e., from 24-21 in. Hg, vacuum).

The design procedure is to first choose a line size. This particular case was for a dechlorination system—3300 ft from sulfonator to injector, 8000 lb/day maximum feed rate. Let us try a 4-in. Sch 80 PVC pipe, $d = 3.826$ in. Using the following equation corrected to pressure drop in inches of Hg:

$$\Delta P = \frac{11.89 \times L \times f \times W^2}{10^9 \times \rho \times d^5} \tag{3-1}$$

where

ΔP = total pressure drop, in. Hg
L = length of line in ft
f = friction factor from Fig. 3-16
W = lb/day Cl_2 or SO_2
ρ = density of gas, lb/ft³ *
d = inside pipe diameter, in.

To find f from Fig. 3-16, it is necessary to calculate the Reynolds Number of the system:

$$N_r = \frac{6.32 \times \omega}{\mu \times d} \tag{3-2}$$

*For gas density values see Appendix.

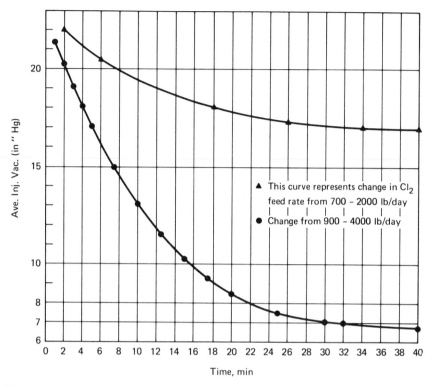

Fig. 3-17 Vacuum level decay in long vacuum lines as a result of excessive friction loss.

where

ω = lb/hr gas flow
μ = viscosity of gas in cp*
d = inside pipe diameter in in.

$$N = \frac{6.32 \times 8000}{0.0133 \times 3.826 \times 24}$$

$N = 41,399$

So from Fig. 3-16,

$$f = 0.027; \therefore \Delta P = \frac{11.89 \times 3300 \times 0.027 \times (8000)^2}{10^9 \times .05 \times (3.826)^5}$$

*For gas viscosity values see page 73, *Handbook of Chlorination*.

$\rho = 0.05$ lb/ft^3 at 22 in. Hg. vac. and 68°F.

Answer: $\Delta P = 1.65$ in. Hg.

At 6000 lb/day flow ΔP is about 0.16 in. Hg. Now to calculate the lag time in this system it would be appropriate to calculate it for 4000 lb/day gas flow, but at 25 in. Hg vacuum. So 4000 lb/day is 0.05 lb/sec. The gas density at 25 in. Hg vacuum and 68°F is 0.03 lb/ft^3 so;

$$Q = \frac{0.05 \text{ lb/sec}}{0.03 \text{ lb/ft}^3} = 1.54 \text{ cfs}$$

$$V = \frac{1.54}{.7986} = 19.32 \text{ ft/sec}$$

Therefore, the lag time for any change in gas feed rate is on the order of 3 min. To show that the lag time would be nearly the same at 8000 lb/day assuming a 22 in. Hg vacuum level, 8000 lb/day = 0.09 lb/sec and the density = 0.048

$$Q = \frac{.09 \text{ lb/sec}}{.048 \text{ lb/ft}^3} = 1.88 \text{ cfs}$$

$$V = \frac{1.88}{.07986} = 23.54 \text{ ft/sec}$$

So at double the gas flow the lag time decreased to about 2.35 min. Therefore, as the flow of gas changes, the density changes in a direction which provides an almost constant system lag time regardless of flow change.

It is to be noted that the physical properties of chlorine and sulfur dioxide gas under a vacuum are so similar that calculations for either are interchangeable; so for simplicity, long vacuum lines can be designed for either gas based on the physical characteristics of chlorine.

Flow Meters. Chlorinator installations that use either the 2-, 3-, or 4-in. adjustable injectors should be provided with some type of flow meter on the water inlet to each injector. Propeller meters with rate indication are quite satisfactory. This allows the operator to make the proper adjustment to the injector to insure maximum efficiency, proper chlorine solution strength, and adequate back pressure. The fixed throat injector systems do not need flow meters for this purpose.

Gauges and Alarms. Gauges should be provided to show the injector operating water pressure for each injector in the system. Each injector should be provided with a chlorine solution pressure gauge mounted immediately downstream from the injector discharge. These must be compound gauges reading to 30 in. Hg vacuum and to 15 psi pressure and equipped with a silver diaphragm protector. These gauges are necessary for the proper analysis of the injector operation. They provide the operator with the back pressure readings for various quantities of water passing through the injector. In some situations this is critical information that cannot be obtained in any other way. It also constitutes proof of proper or improper diffuser design and/or chlorine solution line design.

Every installation should be provided with a loss of water pressure alarm to indicate a failure in the water supply. Over pressure switches are not necessary for the injector system.

Chlorine Solution Lines. The piping downstream from the injector is the chlorine solution line. It is permissible to manifold the injector discharge from two or more chlorinators into one point of application, but a solution line to the point of disinfection should not be manifolded to any other point of application. The most desirable arrangement is for each injector (in a multiple injector system) to have its own solution line and diffuser. For pipe materials see "Materials of Construction." Chlorine solution flow proportioning systems are not worth the expense and most end in failure. Some waters, when carrying a chlorine solution, release a tremendous amount of carbon dioxide and other gases. The bubbles of gas in the solution passing through the rotameter can cause sufficient vibration so as to severely limit the accuracy of the reading. Glass tube rotameters should not be used because the vibration just described can cause the rotameter float to shatter the glass tube. The only rotameters that can be used on chlorine solution are the "straight through" metal (Hastelloy C) or PVC tube type with dial indication.

Diffusers. A properly designed diffuser is one of the most important elements of a successful chlorination facility. There are two basic types: one for discharging chlorine solution into a pipeline flowing full; and the other for discharging into an open channel. Illustrations of various types of diffusers are described in detail under the heading of *Mixing.*

Pipeline diffusers are of two general types. One discharges all the solution into the center of the pipe; this type is satisfactory for pipe sizes up to 30 in. in diameter. The other discharges the solution through perforations over the middle half of the diameter. These arrangements give the best possible dispersion of the chlorine solution. To insure rapid mix the diffuser should be either located in a highly turbulent flow regime, or be immediately followed by a mechanical mixer.

For the open channel there are two general diffuser types: a series of nozzles suspended from flexible hose, or a perforated pipe across the channel.* Open

*The city of Los Angeles has recently developed a novel type of nozzle diffuser for open channel flow[40] with characteristics superior to most open channel diffusers.

channel diffusers require much more care in location selection than do pipeline diffusers because of the poor mixing characteristics of open channel flow. Perforated pipe across the channel is the preferred type. These diffusers should be placed directly in an area of maximum turbulence with a minimum water cover of 6–9 in.

In case a Parshall flume is used as a mixing device, the diffuser must be located upstream of the flume throat and as close as possible to the throat without affecting the head versus flow characteristics. The distance from the diffuser to the turbulent zone should be 5 sec or less at 75 percent of average flow.

An open channel diffuser can be a great nuisance and even a hazard if the perforations do not provide the proper hydraulics for the chlorine solution line. Anytime a negative head exists in the chlorine solution line, molecular chlorine will break out of the chlorine solution and cause serious chlorine gas emission at the diffuser. Therefore, if the sewage level at the diffuser is below the injector throat, the hydraulic gradient from the injector to the diffuser must be calculated based on an assumed back pressure of 5 ft at the discharge of the injector. The friction loss through the diffuser holes plus the line losses minus the difference in elevation between sewage level and injector throat must equal approximately 5 ft of head. This is one of the reasons for having a compound gauge on the solution line at the discharge of each injector. The minimum recommended size for diffuser holes is $3/8$-in. The minimum recommended velocity through the holes is 20–30 ft/sec. All open channel diffusers should be constructed for easy removal. They may become plugged or the hole configuration might have to be changed. This requires a flanged connection so that the entire diffuser can be lifted bodily from the channel. Diffusers are available with projecting pins that fit into slotted wall brackets for easy removal.

Figure 3-18 illustrates a problem in diffuser design for an open channel. Assume a 200 GPM flow through a 3-in. injector and 8000 lb/day chlorine so as to limit the chlorine solution strength to 3500 mg/liter chlorine. Use 4-in. chlorine solution piping to the diffuser. Applying Bernoulli's energy equation we get:

$$P_1 + E_1 + \frac{V_1^2}{2g} - \text{losses} = P_2 + E_2 + \frac{V_2^2}{2g}; \qquad (3\text{-}3)$$

ignoring velocity head differences, from Fig. 3-18 we get the following:

$$5 + 240 - \text{losses} = 0 + 220$$

$$- \text{losses} = -25 \text{ ft}$$

The equivalent length of the chlorine line is assumed to be 150 ft; this is made up of 75 ft linear pipe lengths plus 5 elbows at 15 ft each. Therefore, the friction losses in the solution piping will be about 5 ft. So, to prevent a negative head

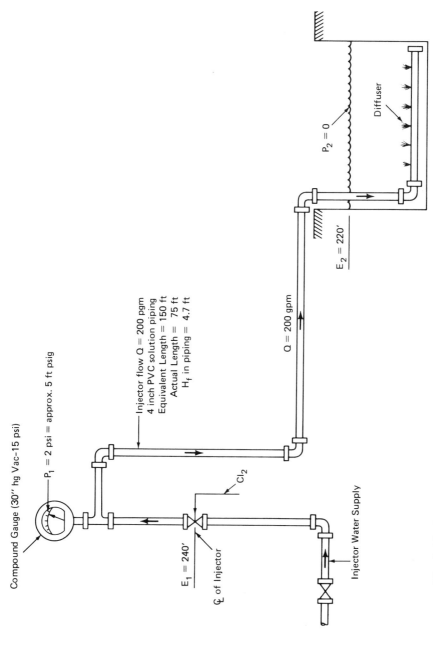

Fig. 3-18   Example of diffuser design for open channel application under negative head conditions.

Compound Gauge (30" hg Vac–15 psi)

$P_1 = 2$ psi = approx. 5 ft psig

Injector flow Q = 200 pgm
4 inch PVC solution piping
Equivalent Length = 150 ft
Actual Length = 75 ft
H_f in piping = 4.7 ft

$P_2 = 0$

Diffuser

$E_2 = 220'$

Q = 200 gpm

Cl_2

$E_1 = 240'$

Ç of Injector

Injector Water Supply

condition at the diffuser by maintaining 5 ft of head at the injector discharge, the diffuser will have to consume 25 - 5 = 20 ft of head loss.

The next step is to choose a diffuser hole size. The minimum acceptable size is $^3/_8$-in. Using the equation

$$Q = CA \sqrt{2\,gh} \qquad (3\text{-}4)$$

where

C = orifice coefficient = 0.75
A = area in ft^2 of orifice
h = head loss in ft
Q = discharge in cfs

Using A = 0.0008 and h = 20, Q calculates to 0.0215 cfs = 9.7 GPM. Therefore, 200/9.7 = 20.6; so use 20 or 21 holes.

Diffusers should always be designed on the basis of *one diffuser* for *each injector*, unless more than one injector is to be operated simultaneously.

Materials of Construction. The materials used for the injector system are those employed in good water works practice for the water system up to the inlet of the injector assembly. From this point forward, a corrosive chlorine solution will be encountered which requires special materials. The chlorine solution lines can be either Sch 80 PVC, Flouroflex-K (Kynar), rubber-lined steel, saran-lined steel, Kynar-lined, or certain types of fiber cast pipe.

Valves on solution lines can be either diaphragm or ball type. Diaphragm types are usually flanged, rubber-lined, or PVC-lined cast iron Saunders-type valves. The ball type PVC valve is preferable up to 2^1/$_2$-in. size, although the ball-type valve is available in much larger sizes. The diaphragm type should be considered for sizes 3-in. and larger.

The injector vacuum line between the chlorinator and the injector is the most controversial of all the piping used for the chlorination facility as to material selection. This line carries moist chlorine gas under a vacuum. It is preferable to use Sch 80 PVC or Kynar pipe, although some types of fiber cast pipe are also suitable. Ball type PVC valves should be used instead of diaphragm type valves on this line. There is no known corrosion resistant metal pipe available for this use. Saran and saran-lined steel pipe have been used for this purpose; glass pipe has been found to be impractical. The only time that this particular chlorine carrying line becomes a consideration is in the case of remote injector installations. It should be noted that when the injector is located in the chlorinator room, the manufacturer usually supplies Sch 80 PVC for the interconnecting piping.

The diffusers and the piping leading to the diffusers are customarily made up of Sch 80 PVC pipe and fittings. If the specifications for underwater piping require steel construction for additional strength, all the underwater piping and diffusers must be made of rubber-lined and rubber-covered steel pipe. The diffuser holes must also be rubber covered. This results in extremely expensive construction costs that are rarely worth the expense.

Nuts and bolts for assembly of the underwater portion should be 316 stainless steel. All other bolts should be galvanized or cadmium plated steel. Pins and slotted wall brackets for diffusers are available in PVC.

As described previously, all gauges on the solution and vacuum lines should be silver diaphragm protected. This is a standard item with the chlorinator manufacturers. Vacuum line gauges are for vacuum only; solution line gauges must be compound gauges as described in the section on Gauges.

Housing:

General. Many important design provisions for chlorination housing relate to the safe use of chlorine and the protection of those working with it. Consequently, many chlorine room design provisions are required elements of State standards.

Chlorinator and sulfonator rooms should be at or above ground level. Container storage should be planned so that it is separate from chlorinators and accessories. It is logical to locate the chlorination room near the point(s) of application to minimize the length of chlorine lines. Other general site considerations include a location which permits ease of access to facilitate container transport and handling, adequate drainage and separation from other work areas.

Separation. Proper design standards require either a completely separate chlorination building or a room completely separate from the remainder of the building with access only through an outside door. There should be no apertures of any type from the chlorination room to other parts of a common building through which chlorine gas could enter other work areas.

Fire Hazard. The building should be designed and constructed to protect all elements of the chlorine system from fire hazards. If flammable materials are stored in the same building, a fire wall should separate the two areas. Fire resistant construction is recommended. Water should be readily available for cooling cylinders in case of fire.

Space Requirements. Modern chlorination equipment is available in modules so that the chlorinators and accessory equipment can be arranged in a panel-like array. There should be about 4 ft between the front of the module and the nearest wall and about 2 ft on the sides and rear. Figure 2-12, page 51, *Handbook of Chlorination*, illustrates space requirements for chlorinator-evaporator installations and Fig. 2-15 on pages 56 and 57 illustrates space requirements for a ton container supply area.

The smallest area for installation of a chlorinator, weighing scales, and a spare cylinder of chlorine should be limited to 6 ft, by 6 ft (150-lb cylinders).

There should be adequate room provided to allow ready access to all equipment for maintenance and repair. There should be sufficient clearance to allow safe handling of the equipment containers. The absolute minimum clearance around and in back of the equipment is 2 ft. Some general minimum space guidelines are as follows.

1. Plants with 1 chlorinator feeding less than 2000 lb/day should have at least 64 ft^2.
2. Plants using 2 chlorinators with a total feed rate of up to 400 lb/day should have at least 160 ft^2.
3. For each chlorinator evaporator unit, 160 ft^2 should be provided.

Ventilation. Adequate forced air ventilation is required for all chlorine equipment rooms. An exception to this would be small chlorinator installations (<100 lb/day) located in separate buildings if the windows and doors can provide the proper cross-circulation. For a small building, windows in opposite walls, a door with a louver near the floor, and a rotating-type vent in the ceiling usually provide the necessary cross-ventilation.

Factors to be considered in the design of a ventilation system are, air turnover rate, exhaust system-type and location, intake location and type, electrical controls, and temperature control.

A forced air system should be capable of providing one complete air change in 2–5 min. Since chlorine gas is 2½ times heavier than air, it is logical to provide air inlet openings for ventilation fans at or near floor level. The usual technique for the exhaust system is to employ an exterior exhaust fan with the intake duct extending to the chlorine room floor. A wall-type exhaust fan is an acceptable alternative. The exhaust system should be completely separate from any other ventilation system. The use of free-moving, gravity operated louvers may be advantageous in colder climates for conserving room heat when the blowers are not in operation; however, venting systems should not have covers.

The discharge should be located where it will not contaminate the air supply of any other room or nearby habitations. It is mandatory that the ventilation discharge be located at a high enough elevation to assure atmospheric dilution (i.e., at the roof of a single-story building).

Air inlets should be so located that they provide cross-ventilation. To prevent a fan from developing a vacuum in the room, thereby making it difficult to open the doors, louvers should be placed above the entrance door and opposite the fan suction. It may be necessary to provide a temperature control on the air supply so that the chlorination system is not adversely affected. A signal light indicating fan operation should be provided at each entrance when the fan can be controlled from more than one point.

Doors. Exit doors from the chlorination room should be equipped with panic hardware and should open outward. Some design guides recommend two means of exit from each room or building in which chlorine is stored, handled, or used; however, this would not appear to be essential in most cases.

Inspection Window. A means of viewing the chlorinator and other equipment in the chlorination room without entering the room should be provided. Usually a clear glass, gas-tight window which is installed in an exterior door or interior wall of the room is recommended. Door windows would appear to be a logical provision even with a separate wall inspection window.

Heating. The chlorinator room should be provided with a means for heating and controlling room air temperatures above 55°F. A minimum room temperature of 65°F has been recommended as a good practice. Ideally, the heating system should be able to reliably maintain a uniform moderate temperature throughout the chlorination room.

Hot water heating is generally preferred because of safety considerations and the uniformity of temperature which is inherent with this type of heating. The problem of temperature extremes, which might be experienced with the failure of a steam heating system, is alleviated. Electric heating is suitable and forced air heating appropriate if an independent system is provided for the chlorination room or building. Central hot air heating is not acceptable since gas could escape through the heating system.

Chlorine vapor leaving a container will condense if the piping temperature is significantly lower than the temperature of the container. Design should provide a higher temperature in the chlorinator room than in the container room. This applies to systems using the gas phase from the containers. Elimination of unnecessary windows may aid in maintaining uniform building temperatures.

If container storage and chlorination equipment are in separate rooms, the temperature of the chlorine containers should not be allowed to drop below 50°F if evaporators are not used.

Drains. It is generally desirable to keep the plant floor-drain system separate from that of the chlorinator. Drainage from a chlorinator relief valve may contain chlorine. Consequently, hose, plastic pipe, or tile drains are recommended. The discharge should be delivered to a point beyond a water sealed trap or disposed of separately where there is ample dilution.

Scale pits are generally designed with floor drains having a water seal trap. In actual practice, most traps probably do not contain ample water to form a seal and it would be preferable to provide a straight pipe drain outside to grade.

Vents. Chlorinators and *external* chlorine pressure reducing and shut-off valves have vents. Chlorinators have a pressure-vacuum relief system which should be carried to the outside atmosphere, without traps, to a safe area, with one vent for each chlorinator. In the case of a malfunction, the operator can easily determine which chlorinator is malfunctioning if the vents are separate. The ends of the vent lines should point down, be covered with copper wire screen to exclude insects, and should not be more than 25 ft above the chlorinator. The line should have a slight downward pitch from the high point (directly above the chlorinator) to drain any condensate away from the chlorinator. It is acceptable to run the vent vertically (but no more than 25 ft) above the chlorinator to the roof, with a 180° return bend at the exit.

External chlorine pressure reducing and shut-off valves should be checked for vents. When supplied, these vents should drain away from the valves. In other words, these valves should be located high enough so that the individual drains will have a continuous downgrade to the outside atmosphere.

Evaporators have a water bath vapor vent which can be manifolded together and discharged to the atmosphere without traps.

Electrical. Controls for fans and lights should operate automatically when the door is opened, and there should be a means of activating these manually from outside the room. Switches for fans and lights should be outside of the room at the entrance. A signal light indicating fan operation should be provided at each entrance when the fan can be controlled from more than one point.

Reliability Provisions. The need for continuous and dependable disinfection has been stressed. The chlorination system can fail due to a number of causes, the system design must either prevent failures or provide for immediate corrective action. Although assured reliability is essential, design provisions for this are often slighted.

Chlorine Supply. As a chemical feed process, one of the most frequent interruptions in treatment is caused by the exhaustion of the chlorine supply. Four features are essential to maintain continuous chlorine feed and are discussed elsewhere in this text. These are: 1) an adequate reserve supply of chlorine sufficient to meet normal needs and to bridge delivery delays and other possible contingencies; 2) a manifolded chlorine-header system; 3) an automatic device for switching to a full chlorine container when the one in use becomes empty; and 4) an alarm system to alert operating personnel of imminent loss of chlorine supply. Without these four features it is not possible to assure uninterrupted chlorine feed even with fulltime operator attendance and no equipment breakdowns.

The chlorine header system is needed both to provide a connected on-line chlorine supply which is adequate to assure uninterrupted flow of feed for whatever period the system may be unattended and to allow switchover to a full cylinder without interruption of feed.

Power Failure. Power outage usually means water supply failure which in turn automatically shuts down the chlorination system. A range of special provisions can be employed to assure reliability of power and water supply depending on the particular situation. As discussed previously, these may be in the form of a standby power source and standby pumps.

Standby Equipment. The design of the chlorine feed system should provide for continued operation in cases of equipment failure. Where both pre- and post-chlorination are to be practiced, separate chlorination systems should be provided plus a standby system. If prechlorination is not to be continuously used, it may be possible to use this system as the standby system for disinfection. The units, piping, and accessories should be designed with this application in mind. If prechlorination must be carried out continuously or if no prechlorination is to be done, a standby system, capable of replacing the postchlorination system during repairs, maintenance or emergencies should be provided. Standby equipment of sufficient capacity should be available to replace the largest unit during shutdowns.

In addition to standby equipment, the equipment manufacturer should be consulted regarding vulnerable components. These components should be a part of the plant's inventory of spare parts.

Water Supply. As mentioned above, during a power failure the injector water system will be shut down unless there is an alternate supply that does not require power, such as an elevated tank. Standby equipment to provide injector water in the event of a power failure would consist of an engine driven injector supply pump. Every injector water supply system should have such a standby pumping unit. There is no way to operate the chlorination system without an adequate water supply.

Chlorine Residual Analyzers. Every system using an analyzer for chlorination control should be backed up by an effluent monitor analyzer that can be switched over to the control function in the event of control analyzer failure.

MIXING

General

The importance of this factor was put in proper perspective by the pilot-plant work of Collins, Selleck, and White.[7] This work revealed that the difference between a plug flow mix in a turbulent regime and that of a back-mixed system (one with short circuiting) amounted to about 2 logs in the reduction of coliform concentration. The plug flow mixing device in this instance was a tubular reactor with a turbulence represented by a Reynolds number of 10^4. Ideally the mixing device should be able to homogenize the chlorine solution and the wastewater in a fraction of a second[8-10]. Mixing Equipment Company of Rochester, New York, claims that it is practical to provide complete mixing within 1-3 sec (re-

Fig. 3-19 Propeller mixer (*courtesy* Mixing Equipment Co. Rochester, N.Y.).

gardless of flow) depending on how much power input is allowable.[10] Such a mixer is illustrated in Fig. 3-19.

This arbitrary mixing time is for either closed conduit or open channel flow using propeller mixers. Based on a comparison of existing installations with various types of mixing, a complete mix in 3 sec would be rated as "excellent." Figure 3-20 illustrates the use of a mechanical mixer in open channel flow. This configuration would be rated as excellent. Sizing procedures for these mechanical mixers are proprietary secrets but for general guidelines Mixing Equipment Co. advises that they size mixers for dispersing chlorine solution in a flowing stream on the basis of 0.3 to 0.6 hp/mgd flow. The mixing basins are sized for theoretical retention times of 5 to 15 sec. This information is helpful for the design of an optimum velocity gradient.[11,12]

Mean Velocity Gradient (G). This is represented as follows.

$$G = \sqrt{\frac{P}{\mu V}} \qquad (3\text{-}5)$$

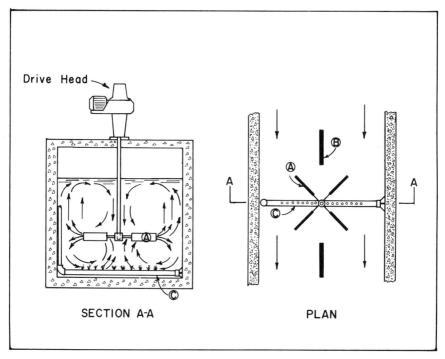

Drive Head

SECTION A-A   PLAN

Fig. 3-20 Propeller mixer in open channel flow showing location of baffles.
A = radial flow impeller, B = baffles, and C = chlorine diffuser (*courtesy* Mixing
Equipment Co. Rochester, N.Y.).

where

G = mean velocity gradient' Sec - 1
P = power requirement; ft lb/sec
V = mixing chamber volume; ft^3
μ = absolute fluid viscosity.

Let us take a condition of 5 mgd flow with 10 sec theoretical detention time in
the mixing chamber, assuming a 3 hp mixer as the design parameter, which is the
upper limit of design quoted by the manufacturers. So G, the mean velocity
gradient, is

$$G = \sqrt{\frac{3 \times 550}{2.35 \times 10^{-5} \times 77.37}} = 952$$

Los Angeles County uses an arbitrary figure of 1 hp/1000 gal. contained in the
mixing compartment. This calculates to a G number of 450. Kennedy Engineers
of San Francisco use a different approach which calculates to a G number of
550. White believes the G number should be closer to 1000 for superior mixing.

This depends entirely on the disinfection requirements. For example; $G = 1000$ should be a minimum for a 2.2/100 ml coliform requirement in a "well oxidized" secondary effluent and nothing lower than $G = 500$ should be used for any other coliform requirement.

An alternate type of mixing device which provides mechanical simplicity and ease of installation is a cluster-jet type mixer illustrated in Fig. 3-20A. This mixer will provide equivalent channel mixing of the mechanical mixer illustrated in Fig. 3-19 and Fig. 3-20. It has the advantage of quickly dispersing the chlorine solution which provides optimum utilization of the disinfectant in a turbulent regime. Further refinements in the concepts of continuous flow chemical mixing utilizing the theories of high-rate mixing, disinfection mechanisms, and chemical dispersion are described below. These new concepts have been explored by Pentech to effect significant improvements in the disinfection process.

A closed conduit can be an excellent mixing device if used properly. In practice the chlorine should be discharged into the center of the pipe (the pipe must be flowing full), and preferably be immediately followed by a mechanical mixer with proper baffling as illustrated in Fig. 3-21. The use of a venturi section of pipe or an orifice plate with the chlorine applied at the throat section of the venturi, or 0.5 D upstream from the orifice plate simulates the mixing device used by Collins, Selleck and White in their pilot-plant study.[7] As of 1976, a scientific evaluation of either of these devices for mixing has not been reported.

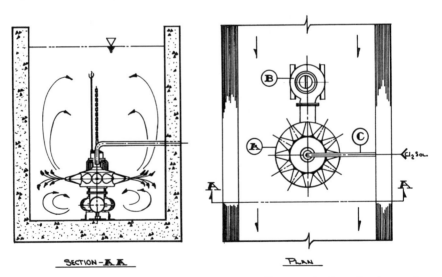

SECTION - A A   PLAN

Fig. 3-20a   Pentech cluster-jet mixer. A = jet diffuser and mixer, B = jet motive pump ∼ (submersible-centrifugal), and C = chlorine solution feed line (*courtesy Pentech-Houidaille*).

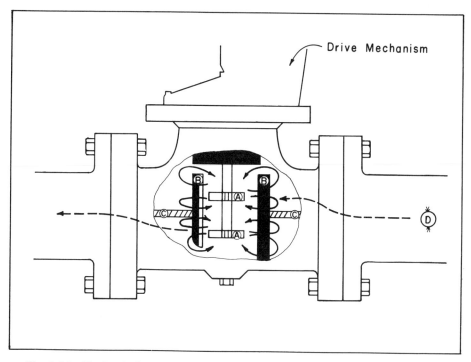

Fig. 3-21 Mechanical mixer in a closed conduit flowing full. A = radial flow impellers, B = baffles, C = separator plate, and D = Cl_2 solution (*courtesy* Mixing Equipment Co.).

Another device available to the engineer for a satisfactory mixing device described above is the hydraulic jump. This was first used as a chemical mixing device in a potable water plant nearly 50 yr. ago[13]. Figure 3-22 illustrates a typical hydraulic jump and the proper location of the chlorine diffuser. One disadvantage of a hydraulic jump is the shifting of the turbulent zone upstream or downstream with the variation in flow. This phenomenon has not been thoroughly examined in a hydraulic jump specifically designed for a chlorine mixing device; therefore, the practical significance of the longitudinal shift of the jump is unknown. Another consideration when using the hydraulic jump is the submergence of the diffuser. This should never be less than about 8–12 in. In spite of these possible drawbacks, a hydraulic jump is an excellent mixing device, whether it is designed as such or whether it is in the turbulent zone of a Parshall flume and/or submerged weir.

The energy required for the propeller mixer versus the hydraulic jump can be easily compared. The hydraulic jump is probably most effective in the range of 1 to 1.5 ft total head loss.

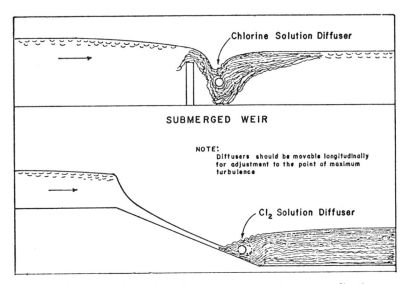

Fig. 3-22   Open channel mixing devices for wastewater application.

The Parshall flume is a type of hydraulic jump which seems to provide a mix rated somewhere between good and excellent. One plant examined by White[14] and rated as having mixing just short of excellent has the chlorine diffuser installed in an open channel immediately upstream from a Parshall flume with an extremely turbulent regime. The location of the diffuser and the velocity at average flow is such that the time required for mixing in this instance is about 4–5 sec. Disinfection as related to mixing at the plant was found to be superior.

A partially closed gate valve immediately downstream from the chlorine diffuser can provide a turbulent regime sufficient to create mixing classified as excellent.

The engineering considerations surrounding the subject of chlorine mixing have been generally overlooked in the past. Since it has been proved that disinfection efficiency is related to adequate mixing, the other aspects of this parameter are being viewed from a strictly chemical engineering approach. One of these aspects is the segregation phenomenon.[15,16]

Segregation Phenomenon. This phenomenon is an extremely complex chemical relationship that exists during the mixing of two fluids with the chemical interaction being the dominant factor. In view of the importance of optimum mixing between a disinfectant and a wastewater stream to provide maximum disinfection efficiency, Selleck[15] has raised the question of the optimum concentration of the disinfectant stream to be applied to the wastewater flow. Simply stated, the segregation phenomenon implies that mixing becomes most efficient between two solutions (a binary mixture) when the solution (chlorine or other chemical) to be mixed with a process flow (wastewater) is of highest practical concentration. This would imply that a hypochlorite solution of 15 percent trade strength (150,000 mg/liter) would mix more efficiently than a chlorine solution discharge from a chlorinator (3500 mg/liter). This is an interesting concept and indeed needs further investigation.

The Pentech System. Perhaps the most efficient chlorine mixing system commercially available is the one developed by Pentech. This system is illustrated in Fig. 3-22A. It utilizes either the direct injection of chlorine gas under a vacuum or chlorine solution from the discharge of a conventional chlorinator. When chlorine gas is injected directly, the pump shown becomes the "injector pump" and the jet nozzle assembly takes the place of the injector.

The entire effluent stream is forced through the reactor tube. This provides plug flow in a highly turbulent regime. This system provides almost instantaneous dispersion of chlorine throughout the mass of effluent resulting in high efficiency of disinfection. Pentech has in effect optimized the system presented by Collins et al.[7] by use of fluid dynamic principles and the theory of high rate mixing. Furthermore, this system prevents any breakout of molecular chlorine which might occur from the improper design of an open channel diffuser in conventional systems.

Another interesting facet of this mixing system is indicated on Fig. 3-22B. This compares the disinfection efficiency of molecular chlorine versus aqueous chlorine (HOCl). The illustrated high rate of virus destruction by molecular chlorine may signify a "kill mechanism" which is different from either free

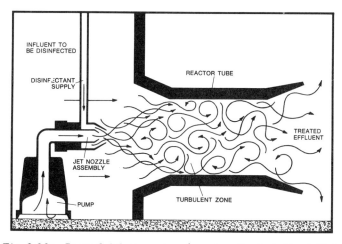

Fig. 3-22a   Pentech injector mixer (*courtesy* Pentech-Houdaille).

chlorine or combined chlorine. It is quite possible that this kill mechanism of viruses by molecular chlorine might be similar to the ozone mechanism. Ozone is known to be a better virucide than either free or combined chlorine. The phenomenon indicated in Fig. 3-22B should be the subject of further investigation.

It is significant to compare the Pentech system with other systems by the use of *the mean velocity gradient* (G numbers, see Eq. 3-5). The G values for jet mixing are enormous. In the case of the example described below the G number exceeds 10,000.

For proper disinfection Pentech[35] recommends a minimum energy dissipation rate be maintained so that the turbulent mixing zone will have a mixing rate t^{-1} (sec^{-1}) of somewhere between 10 sec^{-1} and 5 sec^{-1} depending upon the size of the system and with corresponding mixing residence times of 1.0 sec or less. These limits require mixing in times for less than those described above for other systems. In accordance with fluid dynamic principles, the mixing rate t^{-1} is directly related to the specific turbulent energy dissipation rate in the turbulent mixing zone and is inversely related to the square of the scalar macroscale, L_s, of the turbulence structure of the mixing zone as follows.

$$t^{-1} = K(e/L_s^2)^{1/3} \qquad (3\text{-}6)$$

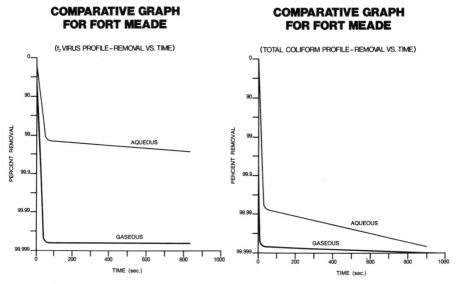

Fig. 3-22b Comparison of disinfection efficiency: molecular chlorine vs. chlorine solution (*courtesy* Pentech-Houdaille).

Where

e = specific turbulent energy dissipation rate
L_s = scalar macroscale
K = constant which is .489 for the cgs system of measurement

The specific turbulent energy dissipation rate e is further defined as:

$$e = \frac{P}{\rho V} \tag{3-7}$$

Where

P = net power lost to fluid
ρ = fluid density
V = fluid volume

The scalar macroscale L_s for a system such as illustrated in Fig. 3-22A may be approximated as about 0.131 D for purposes of calculation for equipment design where D is the mixing parallel diameter (cm).

It is also useful to define a mixing number θt^{-1}, which is the product of the mixing residence time and the mixing rate. This characterizes the product stream inhomogeneity. Mixing efficiency numbers from about 1.5–15 or greater should be applied to achieve superior disinfection results.

Moreover, for a flow through system with continuous mixing, the specific energy requirement (the energy dissipated per unit throughput of product stream or the work done in mixing the product stream) should be at least about 0.2 hp/mgd of treated effluent. For a given level of mixing, the energy requirement will increase with increasing values of L_s, but will generally be in the range of from about 0.2 hp/mgd to 3 hp/mgd of effluent to be treated.

The average residence time θ for the mixer illustrated in Fig. 3-22A, may be readily determined from the volume of the turbulent mixing cone. The effluent and disinfectant may generally be assumed to be mixed to within acceptable disinfectant concentration gradient limits upon reaching a point adjacent to the base of the mixing cone at its intersection with the mixing parallel. The volume V of the mixing cone thus defined may be calculated as follows.

$$V = \frac{\pi D^3}{24 \tan (\alpha/2)} \tag{3-8}$$

Where

D = diameter (cm) of the intercepting conduit at the point of intersection with the mixing cone
α = included angle of the mixing cone.

By way of example, a 5 mgd flow is to be treated utilizing 5 hp input and the jet mixer shown in Figure 3-22A. The mixer parallel diameter is 20 in. $\times$ 2.54 = 50.8 cm and $\alpha = 28°$. Therefore the mixing rate t^{-1}, the residence time θ, and the energy requirement is calculated as follows.

$$V = \frac{\pi D^3}{24 \tan (\alpha/2)} = \frac{\pi (50.8)^3}{24 \tan 14°} = 68,827 \text{ cm}^3 = 2.43 \text{ ft}^3$$

$$\theta = \frac{2.43 \text{ ft}^3}{5 \text{ mgd} \times 1.55 \text{ cfs/mgd}} = 0.31 \text{ sec}$$

$$\epsilon = \frac{P}{\rho V} = \frac{(5)(550)(25.4)(454.5)(980)}{(1)(68,827)} = 452,030 \frac{\text{cm}^2}{\text{sec}^3}$$

$$L_s = (0.131)(20)(2.54) = 6.65 \text{ cm}$$

$$t^{-1} = 0.489 \left[\frac{452,030 \text{ cm}^2}{44.22 \text{ cm}^2 \text{ sec}^3} \right]^{1/3} = 10.61 \text{ sec}^{-1}$$

$$\epsilon = \frac{5 \text{ hp}}{5 \text{ mgd}} = 1 \text{ hp/mgd}$$

It is obvious that the Pentech concept is one which deals with the "ideal" system. This means mixing rates on the order of 0.2 sec with the ability to cope with variable flow rates and to provide a system hydraulic gradient which does not introduce a significant head loss. This concept should be pursued to establish operating reliability.

From a 1976 pilot-plant study at the Fort Meade Sewage Plant No. 2 (a trickling filter plant), Longley[39] investigated the relationship between the mean velocity gradient G, the Prandtl* eddy frequency F, and the Reynolds number N_r as descriptors of rapid mixing in the disinfection process. He found that G and F are highly correlated parameters but that N_r is not a universal descriptor.

CONTACT CHAMBERS (REACTORS)

Function

The chlorine contact chamber must be designed to provide the maximum efficiency of contact between the disinfectant and the microorganisms to be destroyed. There are many considerations required to develop this efficiency. Therefore, the chlorine contact facility should be considered as an integral part of the overall wastewater-treatment process which means it becomes a unit process such as sedimentation, aeration, sludge digestion, etc.

*See Davies, J. T., Turbulence Phenomena, Academic Press, New York, 1972.

In addition to performing the chore of providing chlorine with the necessary time to accomplish microorganism destruction, there will also accrue in this process a significant deposition of suspended solids (at least 50 percent of the remaining SS in the effluent) and some BOD reduction. Scum and oil removal in these chambers is worth considering for some secondary effluents. Thus enhancing the disinfection performance. Therefore, the contact chamber should be arranged for easy cleaning by flushing to remove the inherent accumulation of solids and skimming devices for the removal of scum and oil.

A chlorine contact chamber is not a mixing chamber, and should not be used for that purpose. The applied chlorine should be thoroughly mixed with the wastewater prior to entry into the contact chamber.

Chamber Configuration

The primary configuration of these chambers is to produce a hydraulic flow pattern with a minimum of short circuiting (defined elsewhere as backmixing). The ideal situation is defined as 100 percent plug flow conditions. This situation can be accomplished in outfall conduits or specially designed channels. The basis of design for any contact chamber must be related to the distribution of residence time between the chlorine and the wastewater in the contact chamber. This phenomenon is described below.

Distribution of Residence Time

General Theory. The distribution of residence times may differ appreciably in reactors of different geometrical configuration, even if the reactor volumes and flow rates are identical.[7] The distribution of residence times for ideal plug flow, completely backmixed flow (gross short circuiting), and for an arbitrary flow are presented in Fig. 3-23 from Levenspiel.[17] With ideal plug flow, the flow through the reactor is uniform with no longitudinal mixing along the flow path. The residence time in the reactor is the same for all elements of fluid. In completely backmixed flow, the contents of the reactor are well-mixed and uniform in composition throughout. The exit stream from this reactor has the same composition as the fluid within the reactor. A backmixed reactor should not be confused with a batch reactor.* In a batch reactor, all the reactants added to the vessel are mixed and held for a predetermined time so that chemical reaction can proceed. An example of a batch reactor is a laboratory beaker used for performing chlorine requirement tests.

*Since mixing is not a function of a contact chamber, this type of reactor is ignored in this discussion, except to demonstrate the gross effect on disinfection by short circuiting in a contact chamber.

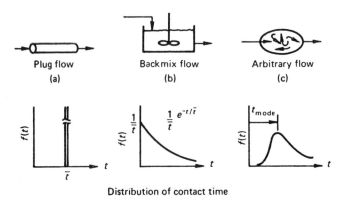

Distribution of contact time

Fig. 3-23 Continuous flow reactors.[17]

Examination of the residence time distribution functions $f(t)$ presented in Fig. 3-23 reveals that a large portion of the flow passes through the backmixed tank in a very short time.[17] It follows that chlorine contact chambers should be designed to approach plug flow reactors and that backmixed flow should be avoided.

In practice, the residence time distribution functions of continuous flow basins may vary between the extreme limits of ideal plug flow and completely backmixed conditions. Ideal plug flow can never be simulated exactly in a real contact basin; it is always accompanied by some backmixing and some dead spaces. Consequently, the distribution of residence times in real basins will be of an intermediate character as indicated in Fig. 3-23c.

Typical results from tracer tests conducted on real chlorine contact chambers are shown in a series of dye concentration curves as a function of time (Fig. 3-24).

Examination of Fig. 3-24 illustrates that plug flow conditions in a contact chamber are ideal. The dye passes through as a slug indicating that all the molecules of liquid passing any given point will have the same residence time for any given length of conduit. Plug-flow conditions are readily achieved in closed conduits flowing full or partly full. Fig. 3-24a illustrates the plug flow characteristics of a 4000-ft outfall line discharging into San Francisco Bay. This represents a dye tracer study by Kennedy Engineers for the City of Richmond, California.

Turbulence. There is no necessity to provide turbulence such as is required in the mixing phase. Turbulence in a contact chamber will not improve coliform destruction; it may in fact be detrimental to the disinfection process. Turbulence during the contact period may reduce the monochloramine residual by aeration and turbulence may cause backmixing. The monochloramine residual is the most potent disinfecting compound of any combined residual occurring in wastewater

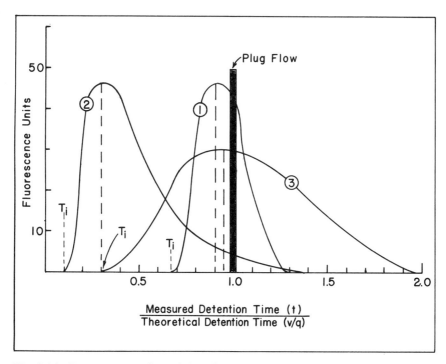

Fig. 3-24 Dye tracer studies of distribution residence times from various chlorine contact chamber configurations.

discharges so it *should not be subjected to any loss by aeration in the contact chamber*.

Characteristics of Rectangular Chambers. Short circuiting (backmixing) characteristics of rectangular chambers vary according to design. Referring to Fig. 3-24, a longitudinally baffled rectangular chamber with proper baffling can achieve the characteristics of curve No. 1 with a modal time of 0.9 and about 95 percent plug flow conditions. Curve No. 2 with a modal time of 0.3 is what may be expected of circular tanks and rectangular tanks without baffles or with a poor baffling arrangement such as cross baffling. Curve No. 3 is an actual dye curve from a seemingly well-baffled tank but which exhibits excessive backmixing which amounts to short circuiting of flow. This tank is considered to be well-baffled since the peak dye concentration occurs at the theoretical contact time (V/q). However, the mere achievement of unit modal time does not satisfy all the requirements of flow in a contact chamber. In this case, the tank is cross-baffled which should be avoided if possible as this design tends to produce short circuiting which in turn reduces the percentage of plug flow conditions.

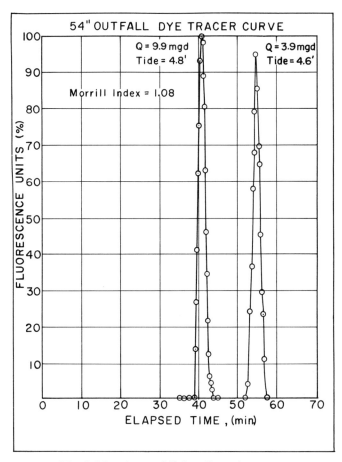

Fig. 3-24a   An outfall dye dispersion curve.

Conventional Analysis

The conventional method of hydraulic analysis of contact chambers is to use the dye tracer dispersion technique. This consists of injecting Rhodamine B dye at the entrance of the chamber and measuring the dye concentration with a fluorometer at the exit of the chamber. The shape of this curve is all important in the analysis of these chambers. The conventional parameters used to describe the performance of a contact basin are presented in Table 3-1.

Each parameter shown in Table 3-1 can be used to predict contact basin performance. Fig. 3-24b illustrates a typical dye dispersion curve. This curve represents a long narrow and shallow gunite lined channel with an L to W ratio of 21:1, an unbaffled pipe entrance and a sharp-crested weir at the outlet.[18] The

TABLE 3-1. Contact chamber performance characteristics

Parameter	Definition
T	Theoretical detention time
t_i	Time interval for the initial indication of the tracer in the effluent
t_p	Time to reach peak concentration
t_g	Time to reach centroid of effluent curve
t_{10}, t_{50}, t_{90}	Time for 10, 50, and 90 percent of the tracer to pass at the effluent end
t_i/T	Index of short-circuiting ⎫ Ideally, all these param-
t_p/T	Index of modal detention time ⎬ eters will approach 1.0
t_g/T	Index of average detention time ⎪ under perfect plug flow
t_{50}/T	Index of mean detention time ⎭ conditions.
t_{90}/t_{10}	Morril Dispersion Index; indicates degree of mixing; as t_{90}/t_{10} increases the degree of mixing increases

importance of the shape of the dye dispersion curve is discussed below (see contact chamber evaluation).

Effects of Short-Circuiting

Ignoring initial mixing effects, the well-mixed batch reactor produces results similar to the plug-flow reactor. The effects of the residence time distribution on process efficiency are illustrated in Fig. 3-25, which compares the reduction of coliform bacteria provided by a stirred batch reactor with that provided by a continuous flow (short-circuiting) backmixed reactor. The amperometric chlorine residuals are approximately equal in the two reactors. In the batch reactor study, a baffled tank was stirred and the chlorine solution was applied instantaneously; replicate samples were then withdrawn at selected intervals of time for bacteriological analysis. In the continuous flow backmixed reactor study, the chlorine was applied continuously to the reactor. Replicate samples were collected for bacteriological analysis when the reactor reached steady state. From the practical viewpoint, a backmixed reactor is a contact chamber with gross short-circuiting while a stirred batch reactor is in reality a plug flow system. Figure 3-25 clearly demonstrates the gross effect of residence time distribution provided by chlorine contact basins (with and without short-circuiting) on coliform reduction, and also demonstrates that a gross design criterion of volume/flow rate (V/Q) is meaningless. For example, there exists a difference of four orders of magnitude in the kill effected in the two reactors for a contact time, $\bar{t}$, of 37 min.

The importance of a contact chamber design that will provide a distribution of residence times approaching plug flow cannot be overemphasized. The best con-

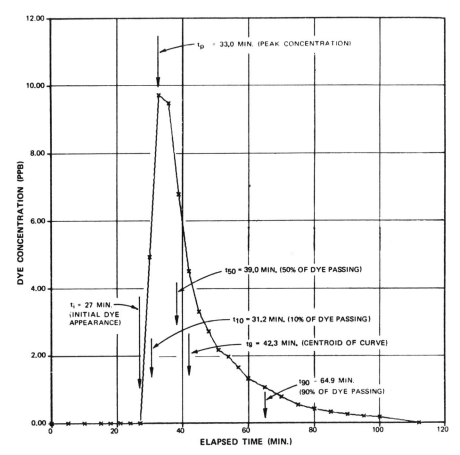

Fig. 3-24b   Typical dye dispersion curve.
Theoretical detention time (V/Q) = 50 min.
Modal time = 0.7–0.8
t_i/T = 0.36–0.54 (depending on wind direction)
Morrill index = 2.1
Chemical engineering dispersion index = 0.08
Percent plug flow = 60–70 percent (depending upon wind direction).
(From Marske and Boyle.[18])

figuration to provide this plug-flow regime is a pipeline; therefore, it is logical that a well-designed chamber should approach pipeline conditions.

Other Physical Characteristics Affecting Residence Time

Marske and Boyle[18] also studied the following physical characteristics used in the design of a contact basin: 1) the depth of the basin as it relates to the effect

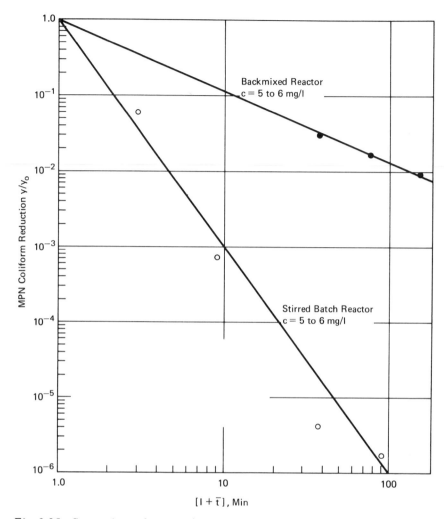

Fig. 3-25   Comparison of reactor (contact chamber) performance.

T = contact time
C = amperometric chlorine residual (total) at the end of the contact time T, mg/liter
Y = final coliform bacteria MPN/100 ml
Y_0 = initial coliform bacteria MPN/100 ml.

of surface wind; 2) the effect of outlet weir configuration; 3) the effect of serpentine baffling configuration; and 4) the effect of length-to-width ratio. The
effects of these physical characteristics are summarized below.

Two tracer tests conducted on a very long, narrow contact channel approximately 3 ft deep, indicated that wind may cause surface currents resulting in

some short-circuiting. In the tests, a down stream wind resulted in a relative time ratio, t_i/T, of 0.36 whereas an upstream wind resulted in a t_i/T of 0.54.*

Consequently, due to wind effects, a unidirectional shallow basin or trench may not provide the plug-flow distribution of residence times that would be expected from the geometrical configuration. In this case, the designer would be well-advised to replace this type of chamber with a closed conduit.

Tracer tests, performed on a 3.6-ft wide Cipolleti weir and on an 18-ft wide sharp-crested weir, indicated that all hydraulic performance parameters were improved when the sharp-crested weir was employed. The value of t_i/T increased from 0.19 to 0.27 and the percentage of plug flow in the basin increased from 38 to 58 percent. Consequently, it is recommended that contact chamber overflow weirs extend across the entire width of the final channel of the contact chamber.

A basin designed with a very high length-to-width ratio would, of course, begin to approach a pipeline. Site limitations will often make a long pipeline for a chlorine contact basin impractical. Consequently, design engineers frequently use rectangular basins with special inlet configurations and serpentine baffles to prevent short-circuiting.

It has been common practice to install basins with length-to-width ratios of 2:1 or less. Such basins are frequently constructed with only 2 or 3 cross baffles which may only extend to the longitudinal centerline. Such a design will not markedly reduce short-circuiting, nor provide plug flow conditions. A poorly baffled basin of this type is depicted in Fig. 3-26. Increasing the number of cross baffles and extending their length would be an improvement. However, longitudinal baffles give much better results.

Longitudinal baffling of a rectangular tank gives the best results. The field investigation by Marske and Boyle[18] evaluated seven different chamber configurations. The one with the most practical value is the one shown in Fig. 3-27. This is a longitudinally baffled chamber with a flow length-to-width ratio of 72:1. This chamber provides 95 percent plug-flow conditions and exhibits a modal time of 0.7. While it would be difficult to improve upon the percentage of plug-flow conditions, the modal time of this tank could be increased to 0.9, as has been demonstrated by others, with some slight changes such as elimination of the square corners in the tank. This work by Marske and Boyle substantiates the claims that long narrow channels and/or conduits make the best chlorine contact chambers.

Cross Section Configuration.   Little mention is made in the literature about the cross-sectional area configuration as a parameter. Some standard should be ap-

---

*The term t_i is the time interval between the injection of the tracer and its first appearance in the effluent, and T is the calculated contact time from V/Q, where V is the basin volume and Q is the flow rate.

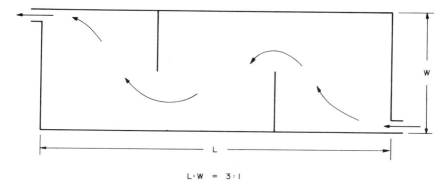

L:W = 3:1

Fig. 3-26   A poorly baffled contact chamber.

plicable to basin depth in order to limit short-circuiting. This is particularly true where stratification might occur due to thermal gradients or skin friction in the chamber. The contact chamber analysis by Trussell and Chao[21] shows that depth can have an effect on the dispersion index, but not nearly to the same extent as the length-to-width ratio. A rectangular cross section with width-to-depth = 2:1 provides a section with minimum wetted perimeter, but requires a length-to-depth ratio of twice the length-to-width ratio because width = twice the depth. Moreover, such a section results in a shallow sidewater depth. This increases the amount of walkways, handrails, and slabs on grade to the extent where costs

FLOW LENGTH:WIDTH RATIO 72:1

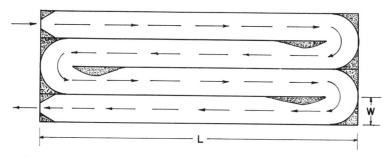

Fig. 3-27   Chlorine contact chamber with longitudinal baffling and optimum plug flow characteristics.

Flow length: Width = 72:1
L:W = 18:1
Modal time = 0.70
% plug flow = 95

(From Marske and Boyle.)

might be prohibitive. On the other hand a deep channel will reduce costs but will affect the length-to-width-to-depth ratio adversely. A good compromise might be to have a square cross section at peak flow (maximum water surface) and a slightly rectangular section (width-to-depth) at lower flows.[38]

Circular Chambers. Circular chambers, least of all circular clarifiers are not acceptable as chlorine contact chambers unless they are specially designed with an outer annular ring. The poor performance of a conventional circular clarifier versus a longitudinally baffled serpentine flow and the special annular ring clarifier is clearly shown in the Marske and Boyle investigation.[18] Any other circular chambers would be expected to perform just as poorly as a conventional circular clarifier.

Theoretical Contact Time. Regardless of the contact chamber configuration the contact time to be selected must be related to the parameters t_i (see Fig. 3-24 and 3-24b) and the modal time. Where severe coliform destruction requirements are required the designer must allow some factor of safety based on the shape of the contact chamber dye concentration curve. For example, conservatism would dictate that t_i shown in Fig. 3-24 should never occur prior to the minimum desired contact time at peak flow. This time interval is a function of the germicidal efficiency of the chlorine residual which is the fundamental basis of the design of a chlorine contact chamber.

In addition to the t_i concept, the modal time must be considered. For example, suppose that a 30-min. detention has been determined for adequate disinfection and the contact chamber is one with longitudinal baffles so as to approach plug-flow characteristics. A well-designed chamber of this configuration will display a modal time of 0.75–0.85 of the theoretical detention time (V/Q). This means that a chamber with a theoretical detention time of 30/0.75 = 40 min. at peak flow will provide the required contact time.

Special Structural Design Situations

Sometimes it is impossible to design a contact chamber with longitudinal baffling due to seismic design considerations. In these cases the horizontal baffling is usually integrated in the design as structural members. In these instances all of the corners should be eliminated as shown in Fig. 3-28 to minimize dead spaces and short-circuiting.

Whenever the effluent MPN coliform requirement is 2.2/100 ml the specifications for the concrete in the contact basins should call for a porcelain like smooth finish which is free from any kind of pits or recesses, because these are breeding areas for the bacteria. Moreover, such situations call for special attention to frequent cleaning. Actually contact chambers designed with either longitudinal or horizontal baffling for effluents which are to achieve a coliform MPN of 2.2/100 ml must be kept as free from slime and algae deposits as a well-kept swimming pool.

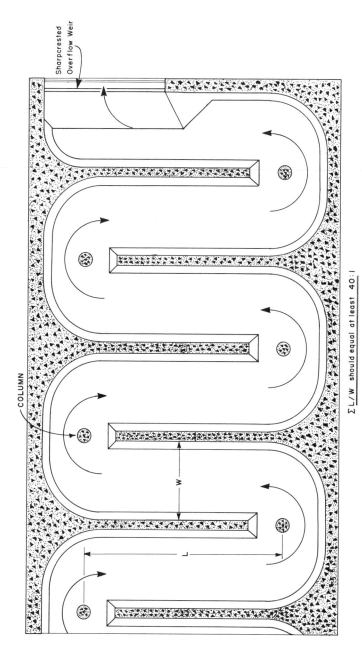

Fig. 3-28   A well baffled contact chamber with horizontal baffles.

Contact Chamber Evaluation

General. All contact chambers should be evaluated *in-situ* as soon after being put into operation as is possible. The information developed must be based on the statistical analysis of the dye dispersion curve of the basin at various flow rates. This describes the hydraulic performance of the contact chamber as related to disinfection and considers the shape of the entire curve rather than the central tendency values.

Chemical Engineering Dispersion Index. There are two acceptable statistical methods of evaluation. One is the chemical engineering dispersion index, d, introduced by Thirumurthi.[36] This dispersion index is calculated from the variance of the dye dispersion curve which relates dye concentration with time. As conditions approach ideal plug flow conditions the value of d approaches zero. The algebraic calculations are delineated in Ref. 17 and 18. The dispersion index d calculated by this method includes all points on the dye dispersion curve illustrated in Fig. 3-24b. The index described by this mathematical approach demonstrates a strong statistical probability of correctly describing the contact chamber efficiency which is related to the dispersion index.

The dispersion index

$$d = \frac{D}{uL}$$

Where

    D = the longitudinal dispersion coefficient (ft^2/sec)

    $u = L/T$ ave chamber velocity (ft/sec)

    L = reactor length (ft)

    σ^2 = Variance of the tracer curve which is a measure of the spread of the curve.* It is the square of the standard deviation.[17]

In computing the dispersion index, boundary conditions must be observed (i.e., whether the chamber is a closed or open vessel).[17] Outfalls would be characterized as open vessels. Most chlorine contact chambers fall into the closed vessel category where the velocity changes abruptly at both the inlet and the exit of the chamber. It is well known that the open vessel category has the characteristics nearest to those of plug flow so computation of the dispersion index for these will be ignored.

For closed vessels the dispersion index is computed from the following expression:

$$\sigma^2 = 2d - 2d^2 (1 - \epsilon^{-1}/d)$$

*Tracer curves often exhibit a long tail caused by recycling of the tracer from dead spaces and backmixing.

When there is a small amount of reduced variance such that d calculates to less than 0.05 the second term of the above equation can be neglected.* However, the complete equation can be easily solved with a Hewlett Packard 21 calculator. The key sequence is shown in the Appendix.

The dispersion index method of contact chamber evaluation is thought to be the best method by many investigators. It has two distinct advantages over other methods: 1) it considers *all points* on the dye tracer curve and 2) it is easy to calculate.

Morrill Index. Another method of analysis which is considered practically equal in statistical confidence as the dispersion index is the Morrill Index.[37] This is a mathematical representation of dispersion such that ideal or 100 percent plug flow produces a Morrill Index of 1.00. This analysis utilizes only two points in the dye dispersion curve shown in Fig. 3-24b. The Morrill Index is characterized by the nomenclature: t_{90}/t_{10}. This means that the time in minutes required for the passing of 90 percent of the dye (65 min.) divided by the time in minutes required for the passing of 10 percent of the dye (31 min.) equals the Morrill dispersion index; or $65/31 = 2.10$. This compares with the longitudinal baffled chamber with the ideal length-to-width ratio of 72:1 and a Morrill Index of 1.48.**

Therefore, it would be safe to declare that any Morrill Index from minimum to maximum flow rates on the order of 1.5–2.5 respectively could be considered as acceptable conditions for disinfection purposes.

Continuing Research. The California State Dept. of Health is engaged (1976–1978) in a project of disinfection facility evaluation. Part of this project is a comprehensive field study of contact chambers by Endel Sepp. The conclusions of this study are expected to be published in 1978. It is expected that the chemical engineering dispersion index d will be the method of choice for evaluating closed vessels as contact chambers. Good graphical correlation has been found to exist between d and other hydraulic efficiency parameters.

It has also been observed that the d values for the same chamber may vary as much as seven times, reflecting different flow conditions, wind velocity, and wind direction. Therefore, it is necessary to make several tracer tests for any given contact chamber in order to get reasonably accurate data.

To date this field research reveals that when the average d values are plotted against the length-to-width ratio the resulting curve appears to form a straight line on log-log graph paper with an ever decreasing d value for an increasing

*See the Appendix for an example of how to compute the dispersion index from a dye tracer study.
**A sample calculation of the Morrill Index is shown in the Appendix.

length-to-width ratio. This indicates a definite correlation between length-to-width ratio and plug-flow conditions of open channel-type contact chambers.

Further contact chamber performance studies are planned to evaluate the effects of mixers at the point of chlorine application and various flow conditions.

Optimum Design Considerations

The optimum design of a contact chamber can be achieved if sufficient information is available. The factors which govern the design are as follows: 1) initial coliform concentration prior to disinfection, (Y_0); 2) final coliform concentration in the disinfected effluent (Y); 3) the 2-3 min. chlorine demand; or the chlorine dosage required to produce the desired residual; and 4) the residual-contact time envelope (ct) to achieve the final coliform concentration (Y). When this information is known for a given effluent, the contact chamber can be designed for minimum annual cost and chlorine contact time.[19]

A differential equation based on the annual cost (capital amortization plus operation and maintenance) can be developed. To find the minimal cost this equation is differentiated with respect to t and the result is set equal to zero. The answer represents the point where annual cost with respect to a change in contact time is a minimum. This equation takes the form of:

$$\frac{\partial AC}{\partial t} = a - b \frac{K}{t^2} = 0 \qquad (3\text{-}9)$$

$$\therefore t = \sqrt{\frac{b}{a} K} \qquad (3\text{-}10)$$

where

AC = annual cost
$K = ct$
c = chlorine residual, mg/liter at time t
t = contact time, min.

The AC (annual cost) consists of capital amortization plus operation and maintenance. This cost is made up of the following factors:

1. CRF (capital recovery factory). This consists of the cost of the chlorine contact chamber, the dechlorination station, the chlorine, and the sulfur dioxide, and the cost of labor. This is usually taken at i (interest) = 7 percent and n = 20 yr.
2. Chlorine Contact Tanks = y/gal. X z gpm X t(min.)
3. Cost of dechlorination station
4. Cost of chlorine
5. Cost of sulfur dioxide

From this a differential equation can be formulated which will be as follows.

AC = CRF (dollars for chlorine contact tank cost) + dechlorination station cost + annual chlorine cost (based on dosage required) + annual sulfur dioxide cost (based on maximum residual required).

When the above term AC is factored out it will declare numerically the relations between contact time and chlorine residual at the end of this contact time which is related to the values of a and b. These are a function of items 1–5 described above. The solution to this equation provides the minimum contact time for a given flow of wastewater to produce the coliform concentration limit (Y) for a given situation.

Referring to Eq. 3-10; $K = ct$; this is the chlorine residual contact time envelope. For a given situation the upper and lower limits for this factor should be selected. From this, contact time t should be tabulated together with chlorine residual. Let us assume that the lower limit is 500 at a given peak flow to accomplish the desired disinfection, then the upper limit based on an average flow would be

$$\frac{Q_1 = \text{peak flow}}{Q_2 = \text{ave flow}} \times 500 = K$$

for contact chamber design conditions. Therefore, c and t for this value of K can be computed and these will be the design parameters for contact time and chlorine residual.

A slightly different approach has been developed by Kennedy Engineers.[38] This method requires knowledge of the wastewater quality the same as for the Los Angeles County approach[19] as follows.

1. The ct envelope versus the log kill of coliform organism (Y/Y_0) is drawn. If there are known seasonal variations of Y_0 an upper and lower ct envelope may be constructed, one for the lower limit of Y_0 and one for the upper limit. The term Y is the coliform concentration at the end of time t and is the disinfection requirement prescribed for the effluent.

2. The next step is to conduct some laboratory chlorine demand studies in order to establish the chlorine dose required to produce the desired residual at several different contact times.

3. The information developed in steps 1 and 2 will provide the necessary information to find the cost of chlorine and sulfur dioxide.

4. The next step is to determine the contact chamber construction cost for various selected contact times.

5. All of the above information is then tabulated against contact times. Then the cost of each function for a given time t is totaled and converted to a "present worth figure."

6. The last step is to draw a curve or series of curves using the present worth figure as the ordinates and contact time as the abcissa. The series of curves may be one each from the two ct envelopes, one dry weather flow, and one for wet

weather flow and another for the present worth figure of the chemicals. The lowest point on the curve to be used will represent the optimum contact time for a given known wastewater quality and disinfection requirement.

DECHLORINATION

Introduction

Recent studies of effluents being discharged into San Francisco Bay have shown that chlorinated effluents are more toxic to aquatic life than unchlorinated effluents. These studies also showed that a dechlorinated effluent is less toxic than either the chlorinated or the unchlorinated effluent.[22] These studies used continuous-flow on-line bioassays and golden shiners (*Notemigonus chrysoleucas*). In these studies the fish were captive in tanks of various chlorine residuals for a minimum of 96 hr. No work has yet been done on fish entering and exiting a sewage plume with a moderate chlorine residual. Unfortunately, the researchers have erred in assuming laboratory conditions are the same as actual environmental conditions. This is not to say, however, that we should not optimize wastewater disinfection systems.

While it is generally agreed by the research biologists that total chlorine residuals of 0.05 mg/liter are harmful to the biota and fish in the receiving waters, it would be more practical to accept chlorine residuals of 0.5 mg/liter, because optimization of the chlorination-dechlorination method can be best controlled by a feedback control system in the range of 0.3 mg/liter total chlorine residual, ±0.2 mg/liter. Very little is known about the chlorine residual die-away in a wastewater plume discharging into fresh and saltwater receiving waters. Only two factors are known for certain: 1) The die-away of a combined chlorine residual in saline waters is a function of dilution only. Saline waters do not exert a chlorine demand on a combined residual; however, the chlorine demand of seawater is about 0.75 mg/liter of free chlorine after 30 min. contact; 2) Freshwater receiving systems do exert a chlorine demand but very little is known about the specific uptake of chlorine residuals from wastewater effluents.

Much more information is needed to optimize the dechlorination process so that the uptake of chlorine residuals by freshwater systems and the maximum stress that may be applied to saline waters via dilution can be evaluated and utilized. Until such time, the alternative is to apply a slight excess of sulfite ion to insure a 0.00 chlorine residual.

Dechlorination Practices

Sulfur Dioxide. Dechlorination of heavily chlorinated potable water by sulfur dioxide has been practiced since 1922 when it was first used in the Toronto water supply. The use of free residual chlorination followed by dechlorination with sulfur dioxide to control the residual of potable water entering a distribution system is an accepted modern water treatment process.[23]

Sulfur dioxide is the most popular dechlorinating agent because chlorination equipment can be used without modification to handle this gas. It is relatively inexpensive ($0.07/lb in 17-ton tank trucks), easy to control, and reacts quickly to remove completely free or combined chlorine residual. It is arithmetically convenient because in practice, one part by wt of SO_2 will remove one part by wt of chlorine residual.

Sulfur dioxide is a colorless gas with a characteristic pungent odor. It may be cooled and compressed to a colorless liquid. When liquid sulfur dioxide in a closed container is in equilibrium with sulfur dioxide gas, the pressure within the container bears a definite relation to the temperature as shown in Fig. 3-29. This relationship is similar to the characteristics of chlorine, but the magnitude of vapor pressure is different. At $70°F$ the vapor pressure of SO_2 is 35 psi as compared to 90 psi for chlorine; however, the density of the two gases is nearly the same. To convert lb/day on a chlorine rotameter to sulfur dioxide, multiply by 0.95.

Sulfur dioxide is soluble in water up to 1 lb/gal. at $60°F$, which amounts to an 11 percent solution. Therefore, sulfur dioxide is considerably more soluble in water than chlorine gas.

Since it is not practical to either record or control a very low chlorine residual, current practice is to add an excess of sulfur dioxide so there will be a surplus of sulfite ions ($SO_3^=$). Instrumentation is not currently available (1976) to continuously monitor the sulfite ion. Moreover, monitoring a zero chlorine residual with a chlorine residual analyzer has not been entirely successful. Sulfur dioxide control and monitoring systems are described below.

Activated Carbon. Dechlorination with activated carbon is extensively used in the brewing and soft drink industries. Little information is available on the dechlorination of wastewater by carbon because it is rarely used for this purpose.

The dechlorination reaction with carbon proceeds concurrently with the adsorption of organic and other contaminants. However, since little information is available on wastewater treatment, and most available data are based on dechlorination of potable water, the effect of organics requires further study. Long-chain organic molecules, such as those of detergents, seem to reduce the dechlorination efficiency, but many common substances such as phenols have little effect. Colloidal impurities can markedly shorten the life of the carbon even with frequent backwashing; therefore, filtration of the influent is desirable.

It is conceivable that for water reuse purposes the additional benefits of the adsorption and absorption characteristics of granular activated carbon might make it the method of choice over the sulfite ion processes. Complete dechlorination by granular activated carbon in a filter bed or column configuration requires

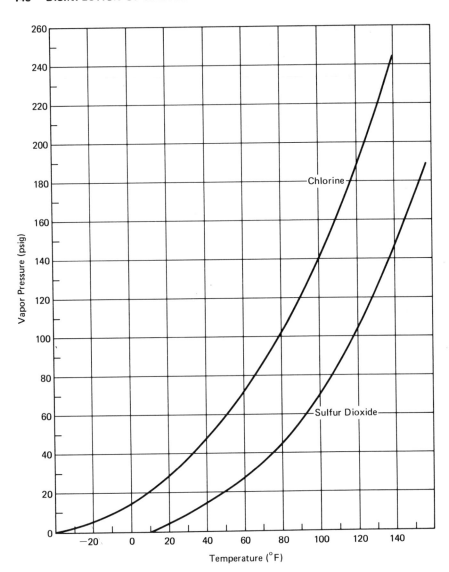

Fig. 3-29 Sulfur dioxide vapor pressure curve compared to chlorine.

10 min. One ft^3 of carbon at 2 gal./ft^2/min. has been found to dechlorinate 5 mg/liter free chlorine residual in 550,000 gal. of water.[24] There is very little information available on the dechlorination capability of activated carbon on wastewater effluents. However, Stasuik et al.,[25] Bauer and Snoeyink,[26] and Barnes et al., [27] while studying the catalytic action of activated carbon in speed-

ing up the breakpoint reaction for ammonia nitrogen removal indicate that activated carbon is equally effective in the removal of chloramines as compared to free chlorine. This is at variance with potable water experience. This apparent variance could be the result of the wide difference of concentrations of the ammonia nitrogen and lack of proper attention to the optimum pH required for maximum efficiency of the chlorine-chloramine reaction with activated carbon.

Another unknown factor is the estimation of the effective life of activated carbon. The various studies on the dechlorination capability of activated carbon show a variation in dechlorination capacity as a function of the hydraulic application rate and particle size. Those data suggest the formation of a dechlorination intermediary compound, nascent oxygen, which collects on the surface of the carbon and causes a gradual loss of dechlorination efficiency. Since the cost of new carbon is extremely high ($800/ton) and since the effective life of the carbon is not precisely predictable, every activated carbon facility would have to include a regeneration system at $400°C$. Regeneration should be based on a conservative estimate of the carbon capacity far in excess of what is required for removal of chlorine. This might be 1 ft^3 of carbon for each 500,000 gal. of water treated assuming 5 mg/liter total chlorine residual to be removed.

It is not likely that activated carbon will ever be a cost-effective competitor of the sulfite ion processes for the chlorination-dechlorination process unless there are circumstances requiring the adsorption-absorption capabilities of activated carbon in addition to its capacity for dechlorination.

Ion Exchange Resins. The synthetic ion exchange resins have been used as dechlorination agents also, but usually on a very small scale. Only certain types of resins have a marked degree of dechlorination efficiency. The use of resins for wastewater dechlorination is not presently practical.

Aeration. Aeration of chlorinated water by submerged or spray aerators will release some combined residual. Reduction of any free residual chlorine, however, is negligible because the calcium or sodium hypochlorite formed by the reactions of HOCl with natural alkalinity of the water is nonvolatile. If any nitrogen trichloride is formed it is readily volatilized by any type of aeration. At pH levels of 7 and above the maximum amount of combined chlorine residual loss which can be expected is no more than about 15 percent.

Storage. Prolonged storage, especially in the presence of sunlight, will result in the gradual disappearance of residual chlorine. Dechlorination by storage may find application in the form of fill-and-draw ponds. Little information is available on the time required to dechlorinate wastewater by this method. Chlorine residual tends to disappear in accord with a zero order reaction (i.e., linearly with time). Under field conditions, however, environmental factors such as wind velocity, temperature, and sunlight may change the overall reaction.

Other Dechlorination Agents. These are classified as compounds which produce the sulfite ion, and are described in Chapter 2.

Factors Affecting Efficiency of Sulfonation Process

Mixing. The application of sulfur dioxide to wastewater differs considerably from the application of chlorine. The reaction is inorganic in nature. The $SO_3^=$ ion reacts with the chlorine residual to convert the active chlorine to chloride and the sulfite ion ($SO_3^=$) to sulfate ion ($SO_4^=$). This reaction takes precedence over any side reaction that might be encountered in wastewater. The reaction of the surplus sulfite ion (after the reaction with chlorine) with dissolved oxygen is very slow at the pH and temperatures usually encountered. Mixing of the sulfur dioxide solution is not as critical as with the application of chlorine. What is needed is simple but rapid dispersion of the sulfur dioxide. This should occur within 45–60 sec.

Contact Time. The chemical reaction between the sulfur dioxide solution (H_2SO_3) and the chlorine residual is instantaneous at the pH and temperature usually encountered in wastewater; therefore, *contact chambers are not required*. A mixing chamber which provides complete dispersion of the sulfur dioxide solution is all that is necessary.*

Control System. Due to the lack of appropriate analytical instrumentation for monitoring sulfur dioxide, the control system must be based on chlorine residual measurement prior to dechlorination. The most logical system is feed-forward. This system requires an effluent flow signal and a chlorine residual signal measured at the end of the contact chamber. The two signals are combined to generate a control signal for the sulfur dioxide metering equipment. This scheme is illustrated by Fig. 3-30.

Another system which has been operating successfully is the one shown in Fig. 3-31.[28] This is a scheme whereby the dechlorinated sample of the sulfonator control analyzer is biased with a constant artificial chlorine dosage. This results in a total residual of approximately 2 mg/liter in the biased sample when there is zero residual in the dechlorinated sample. The residual analyzer-recorder receiving this biased sample is arranged as shown to record chlorine residuals from 0 to +5 mg/liter and 0 to -5 mg/liter. Whenever the chlorine residual

---

*To be conservative dispersion time of the sulfur dioxide solution should be complete on the order of 1–2 min. or 10 pipe diameters in a closed conduit if dispersion is by properly designed diffusers. The maximum reaction time for the complete dechlorination reaction in wastewater at pH 7–8 will probably be less than 200 sec at 68°F.

≤ 4 min.

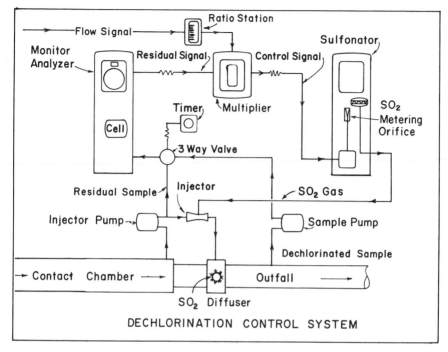

DECHLORINATION CONTROL SYSTEM

Fig. 3-30   Feed forward control of a sulfur dioxide dechlorination system.

appears in the minus area of the chart, this signifies that there is a chlorine residual in the dechlorinated effluent. Therefore, the operators are immediately alerted to this situation by an alarm signal from the residual recorder. This system is the most preferable because it utilizes compound loop feedback control for a zero chlorine residual.

These two systems as described point up the desirability of being able to control the sulfonation process by compound loop control. The hardware and calibration techniques for continuous amperometric analyzers dictate that it is within the realm of practicality to control the sulfonation dechlorination process to a minimum total chlorine residual of 0.3 mg/liter, ± 0.2 mg/liter accuracy. Such a system is exceedingly more desirable than a system which is predicated on the addition of an excess of sulfite ion.

Other systems have been suggested in addition to the one in Fig. 3-30 which relies on a slight excess of dechlorinating agent, and the one in Fig. 3-31 which depends on the biasing of the dechlorinated sample. One of these was originally suggested to White by Carl Petersen systems control engineer CTA, Chicago. His thought was to use the standby sulfonation equipment to provide two step dechlorination. Several of these systems have been designed and are just now

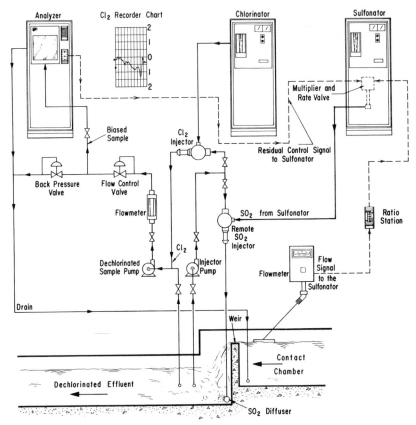

Fig. 3-31 Compound-loop control system for dechlorination using a chlorine residual biased control sample.

(1976) being placed in operation. The idea is a good one so it is hoped that operational experience will bear this out. The system is shown in Fig. 3-32.

Sulfonator No. 1 controls the chlorine residual exiting from the chlorine contact basin to 0.5 mg/liter or thereabouts. Sulfonator No. 2 destroys the remaining residual and is usually set to provide a 0.2-0.3 mg/liter excess sulfur dioxide.

The operation is briefly as follows. The lead sulfonator is controlled by a compound loop control thusly: the chlorine residual analyzer sends a dosage control signal to the sulfonator differential regulating valve and an effluent flow meter sends via a ratio control station a flow signal to the sulfonator metering orifice positioner. The ratio station is imperative because it allows adjustment of wastewater flow and chlorine demand characteristics.

The No. 2 sulfonator is controlled by feed forward control. The same flow signal utilized by the lead sulfonator is sent to another ratio station and thence to a signal multiplier where it combines with the chlorine residual from the effluent chlorine residual analyzer. The signal from the multiplier controls the sulfonator metering orifice positioner.

Proper operation of this system shown in Fig. 3-32 requires proper selection of metering orifices for the two sulfonators. The following example will illustrate this point.

Assume the maximum wastewater flow to be 10 mgd and maximum expected chlorine residual to be dechlorinated is 6 mg/liter. The lead sulfonator will have to provide 1.0 mg/liter SO_2 X 5.5 mg/liter Cl_2 X 10 mgd = 459 lb/day so choose a 500 lb/day rotameter−orifice combination for sulfonator No. 1. To dechlorinate the remaining chlorine residual requires 1.0 mg/liter SO_2 X 0.5 mg/liter Cl_2 X 10 mgd = 42 lb/day, so provide sulfonator No. 2 with a 50 lb/day orifice meter. This arrangement will provide the best flexibility of a system based upon the situation described above.

Another method less sophisticated than the one described above has been suggested by Karl Butters who has been in charge of a most successful chlorination-dechlorination system at Palo Alto, California. This particular system controls both the chlorine dosage and sulfur dioxide dosage by compound loop systems. The chlorine control residual is set at 7 mg/liter and the dechlorinated residual is controlled to 0.5 mg/liter chlorine. This residual dies away to zero at some point in the outfall (2000 ft) before reaching the saline waters of San Francisco bay. Butters has suggested that total dechlorination could be achieved with a single sulfonator as follows. The sulfonator injector discharge could be split to provide a dechlorinating solution for application immediately downstream from the primary or main point of the sulfur dioxide application. The sulfonator would be controlled to a 0.5 mg/liter chlorine residual by a compound loop control system. The second addition of sulfur dioxide would be a small portion (about 10 percent) of the total sulfonator injector discharge sufficient to dechlorinate the remaining chlorine residual. To control this system properly requires two flow measuring devices; one upstream from the sulfonator injector and one in the split to the second point of application. The line to the second point would also have to be equipped with a corrosion resistant control valve possibly supplemented by an orifice-type restrictor. The solution lines to both points of application would have to be carefully analyzed so as to provide the necessary hydraulic conditions to allow precise control of the injector discharge split.

Monitoring System. As can be seen from the discussion above, monitoring a dechlorinated residual can only be done inferentially at present. The following methods are recommended where discharge requirements specify a maximum chlorine residual of 0.10 mg/liter or less in the effluent.

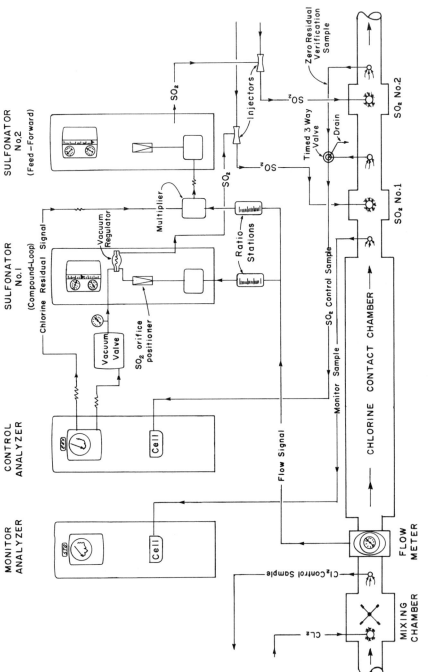

Fig. 3-32 Two step dechlorination system utilizing compound loop control.

Intermittent. This is the simplest, least expensive, and most practical method and employs the dechlorination system shown in Fig. 3-30. Monitoring of the dechlorinated sample is accomplished by the control analyzer on an intermittent basis (i.e., for a small, adjustable, period of time each hour, say from 5-7 min.). During that period, a continuous dechlorinated sample that is being discharged to waste is diverted to the control analyzer which normally receives the chlorinated sample at the end of the contact chamber. This sample, if properly dechlorinated, will show a zero reading on the analyzer. Considering the reliability of chlorination and dechlorination systems that are properly attended and cared for, a zero residual for 5-7 min./hr around the clock is convincing evidence of a continuous zero chlorine residual.

Sulfonator control during the intermittent sampling period of the dechlorinated sample is accomplished as follows:

During the brief verification interval to determine zero residual; a "sample and hold" instrument would continue to generate a signal equal to the last known value of chlorine residual. This allows the system to respond to flow changes. At the end of the verification, the interval timer would again select a sample from the wastewater before dechlorination. However, another timer would delay resumption of normal control (release of sample and hold timer) until the analyzer had been purged of the dechlorinated sample.

Continuous. The only instrumentation available for direct continuous monitoring of a dechlorinated sample is an amperometric chlorine residual analyzer. However, continuous operation of such an instrument on a sample with no chlorine residual presents problems. Two undesirable situations occur: 1) the electrodes display a tendency to drift out of calibration when subjected to long periods of zero chlorine residual; 2) the sample lines and electrodes can easily become fouled with organic debris and slime growths due to the lack of chlorine residual. To overcome this, two analyzers are used for measuring the chlorine residual which provide the residual signal to the sulfonation equipment. These analyzers are so arranged that while one is recording the chlorine residual at the terminus of the chlorine control chamber, the other is recording the dechlorinated residual.* Based on some sort of a definitive calibration drift (with time) program, the function of these two analyzers are reversed. The zero monitor becomes the sulfonator control and the sulfonator control becomes the zero residual monitor. The unknown factor in this situation is the time interval for the switching of the functions of the two analyzers. This arrangement provides complete chlorine residual standby service for the control of the dechlorination process, which is most desirable.

It should be restated that the ideal system would be one that could continuously monitor the concentration of sulfite ions ($SO_3^=$). Since it is not prac-

---

*This is supposedly a zero chlorine residual.

tical to control to a 0.10 mg/liter chlorine residual, current practice is to set up the dechlorination control system to provide a small excess of sulfur dioxide solution (sulfite ion). Although a specific sulfite ion electrode is available, it is only practical for laboratory use and not for continuous field measurement.

Design of Sulfur Dioxide Dechlorination System

Mixing. The mixing required for the most efficient use of sulfur dioxide is to provide complete dispersion of the solution in the effluent before it reaches the receiving water. If, for example, the effluent meter is a Parshall flume, adequate mixing will be provided by placing the SO_2 diffuser just upstream from the flume. If the effluent discharges into a closed but surcharged conduit, the sulfur dioxide solution will be completely dispersed within a distance of ten diameters of the conduit provided the SO_2 solution is discharged into the center of the conduit.[23]

If none of the above situations are available for the mixing of the sulfur dioxide solution, mechanical mixing must be used. The mechanical mixer does not have to achieve complete mixing in the times required for chlorine. Complete mixing in 45–60 sec is adequate for dechlorination by SO_2.

Contact Time. If there is complete dispersion of the sulfur dioxide solution in 45–60 sec the additional time required to achieve complete dechlorination will probably not exceed 3 min.[41]

Sulfur Dioxide Supply System. Sulfur dioxide is available in either ton containers or tank trucks. In California, the maximum amount available in a single tank truck delivery is 17 tons based on allowable axle loading of these vehicles. Tank truck operations require that the user provide storage tank facilities for unloading. The trucks are equipped with a compressor for unloading and a 17-ton delivery can be transferred in 2–3 hr. Storage facilities for SO_2 in bulk should follow the design used for chlorine tank cars or chlorine storage tanks.

Sulfur dioxide can be withdrawn from a storage tank and ton cylinders either as a gas or a liquid. Owing to its low vapor pressure, SO_2 gas reliquefies quite easily. This causes operating problems with the control equipment. This low vapor pressure is also a limiting factor in the movement of the sulfur dioxide from the supply system to the control equipment.

Withdrawing from the gas phase usually requires the application of heat to the cylinders. Gas-phase systems operate best at around 90°–100°F, and it is acceptable practice to apply heat directly to SO_2 ton cylinders provided there is control to limit the heating to 100°F. The maximum gas withdrawal rate at 70°F from a ton container is about 250 lb/day without reliquefaction.

Whenever heat is applied to SO_2 cylinders, a pressure reducing valve must be installed immediately downstream from the cylinders to prevent reliquefaction in the SO_2 header system. Otherwise the benefits of heat application are lost. If the sulfur dioxide is to be withdrawn as a liquid, it is highly desirable to have an air- or nitrogen-padding system available. This can be an automatic air compressor with an electric dryer similar to air-padding systems for chlorine tank cars, or it can be a manually operated system using a nitrogen tank with a pressure regulating valve. Air padding provides the necessary extra pressure required to insure proper movement of the sulfur dioxide. It serves also to insure complete removal of SO_2 from the cylinders or storage tank. The air or nitrogen padding for chlorine and sulfur dioxide must be separate and independent from each other. The sulfur dioxide supply system should be designed similarly to a chlorine system. The materials of construction are the same. The accessories and header system are made of identical components and materials as for chlorine.

While it is desirable to provide a continuously monitored weighing system (load cells or beam scales) to inventory the sulfur dioxide supply, this is not necessary for reliability of the SO_2 supply. Use of the automatic switchover concept provides this reliability.

A nitrogen- or air-purge system for a sulfur dioxide system is as desirable for a sulfur dioxide system as it is for a chlorinator system.

An evaporator enhances the flexibility of any sulfur dioxide supply system. It allows pressure padding to the supply system containers. Since there is always a pressure reducing valve downstream from the evaporator, the problem of reliquefaction in the sulfonator is eliminated. Furthermore, the use of the evaporator diminishes the fallout of impurities, inherent in SO_2, in the supply system piping. Expansion tanks are necessary for SO_2 liquid lines exposed to summer temperatures or to any other conditions where the expansion of the liquid sulfur dioxide exceeds the volumetric expansion of the piping or vessel containers in question. Otherwise rupture of pipe and containers might occur.

Sulfur Dioxide Control System. The first step in designing the SO_2 control system is the determination of equipment capacity and this is based on the amount of chlorine residual that must be removed to produce a zero residual in the effluent. This figure can vary from a minimum of 1 mg/liter to a maximum of 10–15 mg/liter, depending on the local conditions and the efficiency of the chlorination system. When the level of dechlorination is determined, the size of the control equipment is calculated exactly the same as it is for chlorine.

EXAMPLE: The treatment plant is fitted with a 0–100 mgd effluent meter and the expected maximum chlorine residual is 5 mg/liter. The sulfonator should have a capacity of $5 \times 8.34 \times 100 = 4165$ lb/day of SO_2.

The sulfur dioxide control system should be designed in the same fashion as a conventional chlorination system except that it is imperative to install a "ratio station" on the flow signal to the sulfonator. This is a most useful device for operators of any type of feed forward control system. The significance of the ratio station in these situations is described fully in Chapter 4.

The ratio station provides the operator with instant information to make a dosage setting on the sulfonator for any given wastewater flow and chlorine residual necessary to achieve complete dechlorination.

Injector System. The same rules apply for the handling of sulfur dioxide as apply to chlorine with one exception. Chlorine solution discharging from an injector is limited to 3500 mg/liter to prevent breakout of molecular chlorine. Since sulfur dioxide is considerably more soluble in water, the allowable concentration could be raised to 5000 mg/liter or more for a sulfur dioxide solution, thereby utilizing less injector water. However, since mixing efficiency requirements for sulfur dioxide are less than for chlorine, it is recommended that the same 3500 mg/liter solution strength requirements be used. In this way, diffuser design for maximum dispersion will be consistent with that for chlorine. Therefore, the engineer should design the sulfur dioxide injector system in the same fashion and with the same materials as for chlorine. This includes pumps, flow meters, gauges, alarms, and diffusers.

Remote injectors are imperative because of control problems that might arise from the lag time in long solution lines. Usually the point of application of SO_2 is at a considerable distance from the control apparatus; therefore, special attention must be given to the long vacuum lines from the control equipment to the injector. The designer is advised to design the vacuum line on the same basis as for a chlorinator system (i.e., the worst conditions). Regardless of vacuum line length, the total pressure drop at maximum SO_2 flow rate should be limited to 1.5 in. Hg at a vacuum level of 23 in. Hg.

Diffusers. Owing to the much higher solubility of sulfur dioxide than chlorine, SO_2 diffusers can be placed upstream of aeration situations such as overflow weirs, hydraulic jumps, etc., without fear of any molecular SO_2 breakout at the point of application. Similarly, negative head conditions in the sulfur dioxide solution lines are not nearly as critical as those for chlorine solution. However, for good practice and consistency, the sulfur dioxide solution line-diffuser system should be designed hydraulically the same as for chlorine solution diffusers.

Safety Equipment. The safety equipment is similar to that used for chlorine. The breathing apparatus and the emergency container kits used for chlorine are suitable for sulfur dioxide.

A continuous sulfur dioxide leak detector should be supplied with every sulfonation system regardless of SO_2 container size. The SO_2 leak detector is just as necessary as a chlorine leak detector and for the same reasons. The commercially available continuous leak detectors were designed primarily for air pollution monitoring and are generally much more sophisticated than those required for monitoring a leak in an SO_2 supply system. Therefore, those units presently available are expensive compared to chlorine leak detectors, by a factor of 7 to 10. The SO_2 detectors which have been used most often in recent installations (1973-1976) in California are those manufactured by Celesco Industries, a subsidiary of International Biophysics Corp., Irvine, California. Other detectors that have been specified are those manufactured by Thermo Eelectron Corp., Waltham, Massachusetts and Interscan Corp., Chatsworth, California. The designer is cautioned to investigate whether or not the instrument being specified is for continuous monitoring as opposed to the portable type detector which requires on-site operator control for leak detection.

The use of expansion tanks has been described in the section "SO_2 Supply System."

Alarms. Every sulfonation facility should be equipped with an alarm system which adequately alerts the operators in malfunctions or hazardous situations related to sulfur dioxide supply, sulfur dioxide metering equipment, sulfur dioxide leaks, and excess of chlorine residual.

The sulfur dioxide supply system should be monitored for leaks and low pressure. The latter could be an indication of imminent supply loss. High SO_2 pressure alarms are not necessary. Sulfur dioxide leaks in the supply and control area should be monitored by a permanent SO_2 leak detector which meets OSHA requirements. The detector should be capable of sensing atmospheric SO_2 concentrations as low as 1.0 ppm (by volume).

The sulfonator should be equipped with high- and low-vacuum sensing devices. Low vacuum signifies failure of the injector system and high vacuum signifies loss of sulfur dioxide supply.

Installations using evaporators require alarms for warning of high- and low-water bath temperature and low-water bath level.

Chlorine residual analyzers arranged for zero residual monitoring of the dechlorinated effluent should be equipped with adjustable high- and low-residual alarms.

Systems using air padding by a compressor with an air dryer should have a high humidity alarm on the dried compressed air.

Monitoring System. The subject of monitoring has already been thoroughly discussed; therefore, only certain design features are pointed out here.

The continuous monitoring system shown on Fig. 3-31 utilizes the chlorination of a minor flow of dechlorinated effluent at a constant dosage. The following example illustrates the use of this system.

A constant regulated flow of 75 GPM of dechlorinated effluent is discharged back into the chlorine contact chamber. A manually-controlled chlorinator equipped with a 10 lb/day rotameter is used to chlorinate the 75 GPM of dechlorinated effluent. Assuming a 2 mg/liter dosage, this would require a chlorine feed rate of approximately 1.8 lb/day assuming zero chlorine demand other than excess sulfite ion. A sample of the rechlorinated effluent is sent to a monitor analyzer. The amount of chlorine or sulfur dioxide residual can then be calculated for any moment of operation by comparing the residual analyzer reading with the effluent flow and chlorinator feed rate. If more reliability is required, a chlorinator can be furnished which can be fitted with a chlorine flow recorder. Then from the chlorine flow, dechlorinated effluent flow and analyzer reading, the precise chlorine residual or excess of sulfur dioxide can be accurately determined. The system illustrated in Fig. 3-31 has been operating successfully for more than one year.

Sample Lines. All sample lines must be provided with means for flushing to remove slime growths. Sample lines carrying dechlorinated effluent will be subject to a rapid development of organic debris and slimes throughout the transport system. Any such biofouling introduces errors in the analyzer system, whether it be for control or for monitoring.

Sample lines should be designed to provide velocities up to 10 ft/sec. This acts to decrease dead time in the control loop and produces some scouring action.

Reliability. The dechlorination system can fail due to the same reasons which cause failure of the chlorination system. Therefore, the reliability provisions discussed above are generally applicable to sulfonation systems.

PROCESS MONITORING AND CONTROL

Chlorine Residual Analytical Methods

General. Precise measurement and control of the magnitude of the chlorine residual in wastewater is critical in preventing adverse effects associated with over or under chlorination. Process control however, is complicated by the improper use of the available standard methods of chlorine analyses. The concern over the presence of measurable toxic chlorine concentrations of 0.1 mg/liter or less in wastewater, demands the use of the most precise analytical method of measurement that is practicable. *It is meaningless to specify a maximum residual such as 0.1 mg/liter without specifying the method of analysis.* Toxic residuals several times this magnitude may go undetected using the orthotolidine method of analysis in current use.

Iodometric Method. The iodometric method of chlorine residual analysis is much more precise than the orthotolidine procedure. With the iodometric method, two ways of detecting the end point are possible: starch iodide and amperometric. The amperometric method is inherently more accurate. The *end point* means the stage in a titration at which equivalency is attained and revealed by a change that can be observed or measured, such as color development or reaching a specified pH. Standard practice is to use the indirect procedure (back titration) in analyzing for chlorine in wastewater to avoid any contact between the liberated iodine and the wastewater.[4] This procedure is considered necessary regardless of the method of end point detection.

The back titration procedure of the iodometric method, using an amperometric titrator for end point detection, should always be used for precise analyses where trace but toxic residuals are of concern. This method will detect chlorine residuals less than 0.05 mg/liter.*

DPD Method. The DPD method is the only colorimetric method considered reliable enough for measuring chlorine residuals in wastewater. Actually this method which uses the *FAS titrimetric procedure* is the one favored by such eminent researchers as J. C. Morris of Harvard, V. L. Snoeyink of Univ. of Illinois, and Bernard Saunier of Rennes, France. This method is described in detail under "Palin's methods" in *Handbook of Chlorination*[1], Chapter 5.[1]

Orthotolidine Procedure. Orthotolidine analyses may indicate chlorine residuals that are a variable percentage of the iodometric residuals in sewages of different composition. The method is unreliable and erroneous[2] and should not be used to control the wastewater chlorination process.

Chapter 3—Summary

A chlorination-dechlorination facility is necessarily divided into its component parts. These parts are usually classified as follows:

- chemical storage and supply system
- metering and control apparatus
- injector system
- chemical solution lines
- diffusers
- monitoring apparatus

---

*The amperometric titrator can be modified in the field to measure residuals as low as 0.001 mg/liter (see p. 182, Chapter 4).

These are the major pieces of the hardware required to accomplish the injection of chlorine and sulfur dioxide into the wastewater stream.

Other requirements concern the alarms and monitoring devices as follows.

- high- and low-supply pressure
- high- and low-vacuum condition within the metering equipment
- low pressure in the injector water supply system
- high pressure in the chemical solution discharge piping downstream from the injectors
- chemical flow recording apparatus
- leak detection equipment
- chlorine residual analyzers with high and low alarms
- safety equipment for the operators

The chorine supply system should have a monitoring device to provide the operator with chemical inventory information. This can be weighing apparatus or automatic switchover systems.

Where ton containers are involved, lifting and other special handling equipment is required. This includes container positioning trunnions and constraining devices, lifting bar, and lifting hoist. Pressure gages, switches, and alarms are recommended in the locations described in the text.

Special attention should be given to the use of evaporators in lieu of gas withdrawal from ton containers. When evaporators are used, a strict requirement for an automatic pressure relief device is required by new ASME pressure vessel codes.

Absorption tanks are highly desirable to take care of liquid chlorine in the piping system between the supply containers and the metering apparatus. The ability to clear this system quickly of liquid chlorine and confine the chlorine to the supply container is of utmost importance. The absorption chamber should be designed to absorb this amount of liquid chlorine.

The reserve tank concept should be considered as a combination back-up device for reserve capacity and automatic switchover system of supply containers. This concept eliminates the dependency upon weighing devices for this purpose.

All on-site storage tanks must conform to the latest practices and design of tank cars as recommended by the Chlorine Institute.

All chlorine and sulfur dioxide supply piping should be Sch. 80 black steel. Unions should be lead gasketed, two-bolt ammonia type. Fittings should be 3000 lb forged steel. Socket weld is preferable to threaded fittings. *Bushings should never be used.* Use reducing fittings instead.

Line valves and expansion tanks should be in accordance with the Chlorine Institute recommendations.

Safety equipment should include continuous leak monitors, oxygen or air breathing apparatus and container emergency kits.

The metering and control equipment commonly referred to as chlorinators and sulfonators should be the vacuum solution type. The flow of chlorine should be controlled by the flow of wastewater to be treated as well as the variation in chemical demand (chlorine residual).

The injector system should utilize treated effluent as the injector operating water supply. The control equipment manufacturer should be consulted for these requirements. In all cases it is desirable to locate the injectors as close as is possible to the point of application.

Diffusers should be designed to prevent break-out (off-gassing) of molecular chlorine at the point of application. This is not a serious concern with sulfur dioxide. All diffusers should be easily removable for periodic inspection.

Equipment capacity should be related to the maximum reading of the waste-water-flow measuring device.

Residual control equipment should consist of a chlorine residual analyzer to control the dosage rate immediately downstream from the point of application and a second analyzer monitoring the residual at the end of the contact chamber. This equipment should be supplemented with standby analyzers.

Analyzer sample lines should be arranged for periodic purging with caustic solution to remove any zoological slimes.

Housing of the metering and control equipment should permit easy access and egress in case of an emergency. The temperature should not be allowed to go below 50°F and ventilation should be on the order of one complete air change in 2-5 min. Fire resistant construction is recommended. If flammable materials are stored in the same building, a fire wall should separate the two areas. Water should be readily available to cool chlorine containers or storage tanks in case of fire.

Special attention must be given to chlorine and/or sulfur dioxide vent lines from the control apparatus.

Standby power should be available to operate the injector system in the event of a power failure. This should also include the power necessary to operate vaporizing equipment.

Standby equipment should be provided to allow for equipment failure. In situations where both pre- and postchlorination are to be practiced, separate chlorination systems should be provided plus a standby system for disinfection. However, if prechlorination is not to be considered for continuous use, the equipment set aside for this purpose can be in most cases used as standby for the disinfection equipment.

In addition to the details required for proper control of the chemical application, it is of equal importance that the chlorine solution undergoes rapid mixing at the point of application. Ideally the chlorine solution should be completely mixed with the wastewater within a span of *2-5 sec*.

Assuming the chlorine has been rapidly mixed into the wastewater stream the next most important factor is the contact time and the configuration of the contact chamber. The contact chamber should have a minimum of short-circuiting and should demonstrate a chemical engineering dispersion index (d) (based on a dye study) of about 0.02-0.03.

Mixing requirements are not so easily or readily defined but it is believed that if the mean velocity gradient (G) is on the order of 800-1000 this could be comparable to complete mixing in about 5 sec. This is considered good mixing. The Pentech system achieves superior mixing in less time. The definition of adequate, superior, or sufficient mixing in the case of chlorine or hypochlorite dosing has not been quantified and still remains an unknown variable in the disinfection process.

Chlorine chamber contact time should never be less than 15 min. nor more than 2 hr. The latter appears to be nominal for virus destruction. Fifteen to thirty min. is usually sufficient to achieve the various coliform destruction requirements.

The geometrical configuration together with the hydraulic properties of the inlet and outlet of any contact chamber contribute significantly to its efficiency.

All contact chambers should be evaluated and documented as to either the dispersion index (d) or the Morrill Index by a dye test. This test should encompass periods of low, average, and peak dry weather flow.

Additionally the operator should know the minimum time it takes for the first appearance of the dye at these different flow rates. This alerts the operator to the absolute minimum contact time regardless of the contact chamber configuration.

Dechlorination of the chlorination process is most readily accomplished by the application of sulfur dioxide. It is the most simple, accurate, and least expensive of all dechlorinating methods. The same equipment is used for dechlorination by sulfur dioxide as is used for chlorination.

Mixing and detention times are not important factors in the dechlorination process because the chemical reaction between sulfur dioxide and chlorine residuals is practically instantaneous.* Moreover, this reaction takes precedence over any other probable side reactions. Diffusers can be placed in aeration situations (upstream from a weir) owing to the high solubility of the sulfur dioxide solution.

Control systems to achieve zero chlorine residual must necessarily depend upon the chlorine residual in the wastewater discharge. Since it is not possible to calibrate a chlorine residual analyzer to a zero chlorine residual, because of calibration drift, other control schemes must be employed. These include feed forward dosage control, absolute dosage control using a chlorine biased sample (one having a specified and controlled chlorine residual); or a two-stage dechlorination application whereby the first stage of dechlorination reduces the chlorine residual somewhere between 0.5 and 1.0 mg/liter by feedback control. This is followed

---

*Probably on the order of 2-3 min at pH 7.5 and 68°F.

by a second stage of dechlorination which is accomplished by feed forward control of a second sulfonator which reduces the residual to zero. This second stage is controlled by both a flow signal and a residual signal from the monitor analyzer.

A sulfur dioxide system design generally conforms to its chlorine counterpart. The supply system should be artificially pressurized to 50 or 60 psi in order to eliminate reliquefaction. This requires the use of liquid withdrawal which means the installation of evaporators.

A ratio station with a range of 0.4–4.0 on the flow signal to the sulfonators is imperative.

The analyzer measuring the chlorine residual at the end of the contact chamber should be in duplicate. Sample lines to all analyzers should be arranged for occasional purging with caustic.

Two methods are available for process monitoring and detection of trace residuals. These are the amperometric titrator procedure and the Palin DPD, FAS titrimetric method. The amperometric titrator can be modified to measure chlorine residuals as small as 0.001 mg/liter.

Another residual measuring method is described in Standard Methods as the iodometric with the starch-iodide endpoint. It is not as precise as the other two methods described above.

The Optimum chlorination system should consist of the following elements.

1. A reliable chlorine control system with appropriate malfunction alarms.
2. A reliable control and monitoring system for sulfur dioxide dechlorination.
3. Rapid mixing at the point of chlorine application.
4. A chlorine contact chamber with plug-flow characteristics.
5. Dedicated operating personnel.
6. Safety equipment conveniently available to operating personnel.
7. Chlorine and sulfur dioxide leak detecting monitors with alarms.

REFERENCES

1. "Load Hugger 5000 lb Capacity Bulletin." Lift All Products, Manheim, Pa. (1974), and Liftex Slings Inc. Bulletin A76–TD, Libertyville, Ill. (Dec. 1976).

2. White, G. C. *Handbook of Chlorination* Van Nostrand Reinhold Co., N.Y. (1972).

3. Wallace and Tiernan Cat. File No. 25052 "Series V–800 Remote Vacuum Arrangement". (June 1976).

4. Shera, B. L. *Modular Design and Construction of Chemical Preparation Plant.* Pulp and Paper p. 60 (Dec. 1972).

5. Fisher, H. S. and Shera, B. L. "Chlorine Tank Car Depletion Alarm and Reserve Supply." Pennwalt Corp. Bulletin C–4015 (1970).

6. White, G. C. and Stone, R. W. "Factors Affecting the Feed Rate Response Time in a Long Injector Vacuum Line for Chorinators and Sulfonators". Unpublished in-house report for Brown and Caldwell consulting engineers, Walnut Creek, Calif. (Mar. 1974).

7. Collins, H. F., Selleck, R. E., and White, G. C. "Problems in Obtaining Adequate Sewage Disinfection." *ASCE San. Engr. Div.* 97, SA5, Proc. #8430 (Oct. 1971).

8. Collins, H. F. and Deaner, D. G. "Sewage Chlorination Versus Toxicity–a Dilemma? *ASCE J. Env. Engr. Div.* 99: 761 (Dec. 1975).

9. Vrale, L. and Jordan, R. M. "Rapid Mixing in Water Treatment." *J. AWWA* 63: 52 (Jan. 1971).

10. Private Communication. Mixing Equipment Co., Rochester, N.Y. (1973).

11. Metcalf and Eddy Inc. "Wastewater Engineering". McGraw-Hill Inc., N.Y., p. 281 (1972).

12. Camp, T. R. and Stein, P. C. "Velocity Gradients and Internal Work in Fluid Motion," p. 203 in Civil Engineering Classics: Outstanding Papers, by T. R. Camp ASCE, N.Y. (1973).

13. Levy, A. G. and Ellms, J. W. "Hydraulic Jump as a Mixing Device." *J. AWWA* 17: 1 (Jan. 1927).

14. White, G. C. "Disinfection Practices in the San Francisco Bay Area." *J. WPCF* **46**: 84 (Jan. 1974).

15. Selleck, R. E. Private Communication Univ. of Calif., Berkeley, Calif. (1976).

16. Rietema, K. "Heterogeneous Reactions in the Liquid–Liquid Phase: Influence of Residence Time Distribution and Interaction in the Dispersed Phase." *Chem. Eng. Sci.* 8: 103 (1958).

17. Levenspiel, O. *Chemical Reaction Engineering*, John Wiley & Sons, Inc., N.Y., p. 250 (1962), 2nd. Ed., p. 260 (1972).

18. Marske, D. M. and Boyle, V. D. "Chlorine Contact Chamber Design–a Field Evaluation." *Water & Sewage Works* **120**: 70 (Jan. 1973).

19. Stahl, J. F. "Chlorine Contact Tank Design, Pomona Water Renovation Plant Filters." Unpublished report for Sanitary Districts of Los Angeles County, Calif., Whittier, Calif. (Circa 1971–1972).

20. Collins, H. F., Selleck, R. E., and Saunier, B. M. "Optimization of the Wastewater Chlorination Process." A Paper Presented at the National Conference Env. Engring. Research Development and Design, Env. Eng. Div. ASCE Seattle, Wash. (July 12, 1976).

21. Trussell, R. R. and Chao, J. L. "Rational Design of Chlorine Contact Facilities." Paper Presented at 48th Ann. Conf., WPCF, Miami, Fla. (Oct. 1975).

22. Esvelt, L. A., Kaufman, W. J., and Selleck, R. E. "Toxicity Removal from Municipal Wastewater." SERL Report No. 71-7. Sanitary Engineering Research Lab., Univ. of Calif., Berkeley, Calif. (Oct. 1971).

23. White, G. C. "Chlorination and Dechlorination: A Scientific and Practical Approach." *J. AWWA* **60**: 540 (May 1968).

24. "Process Design Manual for Nitrogen Control". U.S.A. Environmental Protection Agency Technology Transfer, Chapter 6 p. 21 (Oct. 1975).

25. Stasiuk, W. N., Hetling, L. J., and Shuster, W. W. "Removal of Ammonia Nitrogen by Breakpoint Chlorination Using an Activated Carbon Catalyst." New York State Environmental Conservation Tech. Paper No. 26 (April 1973).

26. Bauer, R. C. and Snoeyink, V. H. "Ammonia Removal by Chlorination, Followed by Activated Carbon". Paper presented at the Student Symposium, Central States Water Pollution Control Assoc. Meeting, Milwaukee, Wis. (June 14-16, 1972).

27. Barnes, R. A., Atkins, P. F. Jr., and Scherger, D. A. "Ammonia Removal in a Physical-Chemical Wastewater Treatment Process". EPA-R2-72-123, Wash., D.C. (Nov. 1972).

28. White, G. C. Unpublished investigative notes on the City of Santa Rosa, Calif., Laguna Water Pollution Control Plant (1976).

29. Albers, R. G. "Manufacturers' Data Reports". Private Communication issued by Pressed Steel Tank Co., Milwaukee, Wis. (Dec. 22, 1975).

30. "Chlorine Vaporizing Equipment". Chlorine Inst. Pamphlet No. 9, 2nd Ed., N.Y. (1970).

31. "Technical Information for Handling Chlorine, Sulfur Dioxide, and Ammonia from Supply to Point of Application." Fischer and Porter Instr. Bull. 70-9001 Revision 1, Pub. No. 22155, Warminster, Pa. (1977).

32. "Evaporator Series 50.202." Wallace and Tiernan Div. Pennwalt Corp. Belleville, N.J. (Rev. Jan. 1977).

33. Walker, T. B. Wallace and Tiernan Division, Pennwalt Corp., Personal Communication (June 1977).

34. Nagel, Wm. Fischer and Porter Co. Private communication (July 1977).

35. Mandt, M G. Private communication. Pentech Div., Houdaille Ind. Inc., Cedar Falls, Iowa (March 1977).

36. Thirumurthi, D. "A Breakthrough in the Tracer Studies of Sedimentation Tanks." *JWPCF* **41**, Part 2, pp. R405 (Nov. 1969).

37. Morrill, A. B. "Sedimentation Basin Research and Design". *J. AWWA* **24**: 1442 (1932).

38. Calmer, J. C and Adams, R. M. "Design Guide Chlorination/Dechlorination Contact Facilities." In House Report Kennedy Engineers, San Francisco, Calif. (July 1977).

39. Longley, K. E. "Mixing and Clorine Disinfection". A paper presented at the Annual WPCF Conference Minneapolis, Minn. (Oct. 4, 1976).

40. Egan, J. T., "Chlorine Solution Diffusers and Mixers in Effluent Channels," City of Los Angeles Hyperion Plant Work Order No. 31232 Private communication April 1978 and p. 38 CWPCA Bulletin (Apr. 1978).

41. Lister, M. W. and Rosenblum, P. "Rates of Reaction of Hypochlorite Ions with Sulphite and Iodide Ions". *Can. Jour. of Chemistry* **41**: 3013 (1963).

4

Operation and maintenance of a chlorination - dechlorination system

---

LABORATORY PROCEDURES

Chlorine Demand

The chlorine demand test is a valuable tool that can be used to estimate the chlorine residual and contact time required to reach a specified coliform bacteria density.[1] The test is conducted using batch reactors and rapid initial mixing of the chlorine and wastewater. Results of the tests indicate the degree of disinfection that may be expected in a full-scale treatment plant under ideal conditions of: 1) a clean chlorine contact chamber; 2) no short-circuiting; and 3) adequate initial mixing. The statistically significant differences between test results and actual plant results indicate the shortcomings in existing treatment plant disinfection facilities and their operation. Therefore, this is one of the most important tests in evaluating disinfection efficiency.

In conducting chlorine demand tests in the laboratory, samples of wastewater are collected and chlorinated using varying dosages to give a range of chlorine residual concentrations. Aliquots of the chlorinated samples are then withdrawn with time for chlorine residual and bacteriological analyses. The results of these tests indicate the chlorine residual and contact time requirements necessary to obtain specific coliform reduction densities under ideal conditions. There is no other way to prove if a given effluent can in fact be disinfected to the level of the requirements set forth.

Correlation of Laboratory Results with Full-Scale Facilities

Results obtained in laboratory studies can be correlated with plantscale results, provided that the initial mixing is comparable, and the plant contact basin provides a distribution of residence times approaching that of a plug flow reactor. Therefore, the designer should utilize the laboratory results as a guide and then design the chlorination facilities to approach ideal conditions. The designer should also make available to his client a dye study of the distribution of residence times in the chlorine contact chamber.

Mathematical Model

Equation 2-11, describes the relationship between the wastewater quality and the required chlorine residual–contact time envelope, yielding a given coliform concentration after disinfection. This serves as a valuable tool. This equation can be used to estimate the required chlorine residual to obtain a given coliform concentration under plant-scale conditions. For example, if the MPN of coliform organisms is $5 \times 10^7/100$ ml in an unchlorinated primary effluent and the final effluent must meet a bacteriological density of 1000 MPN/100 ml, then for a chlorine contact time (batch or plug flow conditions) of 30 min., the chlorine residual–contact time envelope would calculate to be 156 as follows.

$$y/y_o = (1 + 0.23 \, ct)^{-3}$$

and

$$0.23 \, ct = \sqrt[3]{\frac{y_o}{y}} - 1$$

$$0.23 \, ct = \sqrt[3]{\frac{5 \times 10^7}{1000}} - 1$$

$$ct = 156$$

Therefore, if the contact time is 30 min, then the required residual at that contact time would have to be on the order of 5.2 mg/liter. For effluents of higher than primary quality, y_o will be considerably less; therefore, as the quality of the effluent improves the ct envelope will be reduced accordingly.

Using this mathematical model and knowing the chlorine contact time from dye studies, it is relatively easy to compare the ideal system (laboratory mixing and contact chamber design) with the prototype. *This is the first step in evaluatting an operating system.*

RECORDS AND REPORTS

General

Records will serve to establish a reliable continuing record of proof of performance, justifying decisions, expenditures, and recommendations They also serve as a source of information for plant operation, modification, and maintenance.

Adequate records of disinfection are also important from the standpoint of regulatory agencies. Records will facilitate public health surveillance and enable the regulatory agencies to assess compliance with State regulations.

There are two classes of records which should be maintained at wastewater-treatment plants using chlorination: 1) records of a descriptive, planning, or inventory type, related to the physical plant; and 2) records of performance.

Physical Facilities

The following records, referring to the chlorination-dechlorination unit, should be available for reference at the plant:

1. Design engineer's report, including basis of design, equipment capabilities, population served, design flow, reliability features, and other data.
2. Contract, and "as built," plans and specifications.
3. Shop drawings and operating instructions for all equipment.
4. Costs of each equipment item.
5. Detail plans of all piping and electrical wiring.
6. Complete record of each piece of equipment, including name of manufacturer, identifying number, rated capacity, and dates of purchase and installation.
7. Supply of chlorine and dechlorinating agent, including reserves and availability estimate.

Records of Operation

Daily records should be kept of the following:

1. Chlorine: daily quantities used, both for pre- and postchlorination; chlorine residuals, including method used (preferably continuous recording of residuals); and dosage.
2. Sulfur Dioxide (or other dechlorinating agent): daily quantities used; dosage.

3. Wastewater flow: preferably continuous recording; total daily flow treated; maximum, minimum, and average daily.
4. Results of bacteriological analyses: coliform bacteria; and other required tests.
5. Daily inspection: including record of operational problems: equipment breakdowns; periods of chlorinator outage; diversions to emergency disposal; and all corrective and preventive action taken.

A typical daily record form is shown in Table 4-1.

In addition, the following important records and plans should be kept at each facility:

1. Routine equipment maintenance schedule and record.
2. Annual equipment inspection and maintenance record.
3. Plan for prearranged repair service.
4. Emergency plan for chlorinator failure.
5. Emergency plan for accidental chlorine or sulfur dioxide release.

TABLE 4-1
Operational record of a chlorination-dechlorination plant

Month _____   Operator _____

Day	Total Flow (mg)	Chlorine Use (lb/day) Prechlorin.	Postchlorin.	SO$_2$ Use (lb/day)	Chlorine Residual (mg/liter) Total	Max.	Min.	Interruptions In Treatment
1								
2								
3								
Through								
31								
Sum								
Mean								

Significant operational problems: _____

Reports

Monthly operating reports should be prepared, containing information on chemical usage, wastewater flows, laboratory analyses, and significant operational problems.

STORAGE AND SUPPLY SYSTEM (CHLORINE AND SULFUR DIOXIDE)

System Preparation

Introduction. The critical period in the operation of the supply system is during the first start-up after the contractor has completed the piping and all the equipment has been connected. This period is most critical because on many construction jobs the piping and evaporators are exposed to moisture in the atmosphere for long periods of time. In one instance severe damage was caused to a new system when it was discovered that the evaporators had accumulated about three in. of water in the bottom of the chlorine container vessel. Therefore, precautions must be taken to ensure the entire supply system to be clean, dry, and gas tight.

Piping System:

Welded Pipe. If the supply piping is assembled with welded forged steel fittings, then thread cleaning and lubrication is unnecessary.

Threaded Pipe. All pipe threads must be thoroughly cleaned before being assembled with trichlorethylene, or suitable chlorinated solvent. (Never use hydrocarbons or alcohols. The residual solvent may react with the chlorine.) Following this, some suitable thread lubricant should be carefully applied to each joint. Teflon tape is satisfactory for pipes up to one-inch diameter. Other lubricants can be: John Crane Plastic Lead Seal No. 2; a mixture of linseed oil and graphite; a mixture of linseed oil and white lead; or a mixture of litharge and glycerine. The latter must be prepared as a heavy syruplike consistency from litharge powder and glycerin. Since it hardens or "sets-up" quite rapidly, it must not be made up too far in advance of its use. While this mixture does harden in a short time it should never be relied upon as a joint sealant. There is no substitute for good threads in a chlorine or sulfur dioxide piping system.

System Testing

The most comprehensive procedure is to first wash the system with water to make sure it is open and clean. Then apply 100 psi air to locate obvious leaks.

This is followed by a 300 psi hydrostatic test. The system must then be emptied of the water and dried to a dew point of $-40°F$.

During the hydrostatic test all equipment such as pressure reducing valves, pressure switches, frangible discs, chlorinators, and sulfonators should be isolated to prevent any possible overpressure damage.

Alternate Pickling Procedure

In some quarters it is believed advisable to "pickle" the entire supply system: piping, evaporators, reserve tanks etc. This process removes all mill scale, rust, and other deleterious iron compounds that are likely to find their way into the system.

If pickling is to be carried out it should be done following the hydrostatic test. One way to accomplish this is to empty the system of water, but before drying fill the system with chlorine gas or sulfur dioxide and let stand for several hours. Then relieve the system of any gas pressure and flush with water until the water flows clear. The system is then emptied of water and dried first with steam followed by air drying until the exiting air has a $-40°F$ dew point.

Another method of pickling is by the acid cleaning method. This is done after the hydrostatic test and after the system has been thoroughly dried. This method consists of washing the piping and evaporators with acid and then neutralizing the acid with a caustic or other corrosion inhibitor. This must also be followed by drying the system thoroughly.

Drying the System

After the hydrostatic test the system must be dried thoroughly. This is best accomplished with the application of steam followed by air purging. Compressed nitrogen in cylinders can be used instead of air if desired. When evaporators* are involved, the water bath should be filled and heated to the operating temperature of about $180°F$ during the drying procedure. This will hasten the drying. The dryer the system is before initial start-up, the less corrosion. This translates to fewer operational problems with the metering and control equipment; therefore, less maintenance.

OPERATION–CHLORINATION EQUIPMENT

Chlorine Supply System

100- and 150-lb Cylinders. Operating personnel should know the maximum allowable cylinder withdrawal rates. At $68°F$ the allowable rate for these cylinders

---

*Evaporator manufacturers supply a cleaning kit which includes a water operated aspirator. This allows the chlorine vessel to be subjected to a high vacuum which insures proper drying.

is about 40 lb/day. Higher rates will produce undesirable excessive cooling of the cylinder. The withdrawal rate is less at lower temperatures.

Never connect a "hot" cylinder to a chlorine supply system. Allow cylinders that have been exposed to sunlight to cool off overnight. If a hot cylinder must be connected, it should be artificially cooled by a minor flow of cold water. This can best be done by wrapping the cylinder with water absorbent material such as burlap and then pouring a continuous but small stream of water on the top of the cylinder. At multicylinder installations where the chlorine feed rate is in excess of 40 lb/day, an easy way to cool off a hot cylinder is to make a solo connection to the supply system and withdraw the gas at rates higher than 40 lb/day. As an example, withdrawing chlorine from a 100- or 150-lb cylinder at 100 lb/day will reduce the temperature and consequently the pressure by at least 20°F in the first 30 min.

The flexible connection is the most vulnerable part of a cylinder supply system. To insure the ability to replace these connections without hazard to the operator, connect an auxiliary cylinder valve between the cylinder outlet valve and flexible connection.

Ton Containers.   Ton containers have two outlet valves. One is for liquid withdrawal and one is for gas withdrawal. When used for gas withdrawal, ton containers behave in the same way as do the 100- and 150-lb cylinders. The maximum withdrawal rate at 68°F is about 400 lb/day. However, these containers must be positioned so that the two outlet valves line up in a vertical plane. In this position, the top valve is for gas withdrawal and the bottom valve is for liquid withdrawal. If these two outlet valves are not aligned vertically, serious difficulties will be encountered. A gas-phase withdrawal system may withdraw liquid chlorine, which will severely damage the chlorination equipment. A liquid-withdrawal system will withdraw gas; this will immediately reflect in a severe loss of chlorine supply.

Never manifold a hot ton cylinder with others that are at a cooler temperature. Cool off the hot cylinder by the same procedure described for cooling the 100- and 150-lb cylinders. If this practice is not followed, the possibility arises of filling the partially empty cool ton cylinders by one or more new hot cylinders. Filling a cylinder to liquid capacity could result in the eventual hydrostatic rupture of the filled cylinder(s). Such a situation would obviously cause a disastrous chlorine accident. Therefore, always be sure that the new cylinder being connected is at the same temperature as the cylinders already connected before turning them into the system. This will be revealed by the vapor pressure of the cylinders. The operator is cautioned to connect the new cylinder(s) in such a way that he can verify and compare the vapor pressure between the new cylinders and those already in use. If there is a significant difference in these pressures, the new cylinders will have to be allowed to reach the temperature of the cylinders on the line by remaining in the storage area overnight before being connected. Otherwise, the new cylinders will have to be artificially cooled as described for the 100- and 150-lb cylinders.

The best way to operate a ton container supply system is to use a group of cylinders until empty. When this occurs, secure the empty cylinders and activate a new "full" group of cylinders until empty and so on. It is best never to turn a new cylinder into a supply system of partly full cylinders.

Start-up (Gas System). Probably the most critical period in the operation cycle occurs when a system is put into service for the first time after a prolonged shutdown.

First, check that all union joints on the supply system are properly gasketed and tight and then check that all chlorine supply valves are closed.

Second, check the injector system for proper vacuum at the chlorinator vacuum gage and freedom of the chlorine solution lines of obstruction. This can be done by observing the appropriate gages, or, if such gages do not exist, disconnect the vacuum line to the injector and place the hand over the inlet to the injector. If the injector is performing properly, the injector vacuum will be felt immediately. If the suction is feeble, then the injector and solution line hydraulics are not proper and should be investigated further.

When the injector system is operating properly, the chlorine gas may be turned on, but before doing so the chlorine feed rate should be in a partially open position; about 25 percent of maximum feed rate. Automatic control chlorinators should be placed in the manual mode for this type of start-up.

Third, open one cylinder slightly by cracking the outlet valve. Progressively check all joints for leaks and begin opening the valves leading from the opened cylinder to the chlorinator. If there are no leaks, proceed to open the required number of cylinders. The chlorinator is now ready for further testing—that is, for rangeability, automatic control, and so on.

If there is a leak, the first step is to close the cylinder valve and relieve the entire supply system of chlorine pressure. NEVER ATTEMPT TO REPAIR A LEAK BY TIGHTENING A PACKING GLAND, PIPE FITTING, OR ANYTHING ELSE WHEN GAS OR LIQUID PRESSURE EXISTS. THIS LEADS TO DISASTER. If the cylinder valve is only cracked open, it can be closed quickly, which is the purpose of this procedure. Next open all other valves wide, increase the chlorine feed rate to maximum. This will reduce the pressure to zero very quickly. Repair any leaks which result from a poor fit at a gasketed joint. This might require, in addition to a new gasket, refacing the cylinder valve seat with a flat file. When this is accomplished, the system is ready to be started again.

Fourth, verify that the chlorinator can reach its maximum capacity. This is the most important operative criterion. If this is not possible, there can be no reliance on proper operation from the automatic mode of action. If the hydraulics of the system are proper and there is sufficient injector vacuum at all times, the difficulty lies in the chlorine supply portion of the system. Usually, however, such difficulty is traceable to the injector system and will consist of either of the following: 1) vacuum leak; 2) insufficient water pressure; 3) excessive back pressure in the solution line caused by inadequate pipe size, too much static head in system or an obstruction in the diffuser; or 4) air or gas binding in the solution line. A pressure gage immediately downstream from the injector will verify any of these conditions.

Tank Cars and Storage Tanks. The most important operating consideration for this type of supply system is proper instruction. A qualified representative from the chlorine supplier should be engaged to demonstrate the use of safety equipment and to outline safety precautions and preventive maintenance procedures. The most critical time for these large systems is the original start-up, or start-up after a prolonged shutdown.

Start-up (Liquid System). The procedure for start-up on a system using liquid chlorine is generally similar to that using gas; however, the one big difference is caused by the presence of an evaporator.

The evaporator is an extension of the chlorine container system. Whatever occurs in the container is reflected in the evaporator. The danger existing in a liquid system is the possibility of trapping chlorine liquid in a pipeline. If this occurs and there is a significant temperature rise, the liquid chlorine will expand and may rupture the pipe. For this reason, the liquid line between the evaporator and the chlorine supply system should always remain open while the evaporator is operating.[1]

The first step preparatory to starting up a liquid system is to verify that the system is dry. This can be done by heating the water in the evaporator and passing dry air (- 40°F dew point) through the evaporator cylinder and all the chlorine lines between the containers and the chlorinators. This may take several hours but it is worthwhile, for it can save many days in maintenance time.

When the operator is convinced that the system is dry and if the evaporator water bath is at the appropriate temperature (about 170°F), the system is ready to be started. First, start the injector water and proceed exactly as in the start-up procedure using gas; then follow the same procedure to check for chlorine leaks.

Always start a liquid system first with gas. If the system is started with liquid and a leak occurs many times more chlorine will be released at the leak than if the system was started on gas; one volume of liquid chlorine is equal to 456.8 volumes of gas.

Superheat. The proper operation of an evaporator depends upon the water bath temperature that will provide 20°F superheat to the gas being vaporized. This condition is dependent on the vapor pressure of the liquid. For example, read the liquid vapor pressure of the evaporator instrument panel and determine the corresponding liquid temperature from the chlorine vapor pressure curve. Then read the chlorine gas temperature of the gas exiting the evaporator. This gas temperature should be kept 20°F higher than the corresponding liquid temperature obtained from the chlorine vapor pressure curve. When it is no longer possible to achieve 20°F of superheat, the chlorine container vessel in the evaporator must be cleaned or the immersion heaters must be replaced.

Excessive Chlorine Supply Pressure. A condition that is likely to be encountered in tank cars and sometimes in ton containers is abnormally high-chlorine vapor pressure. In tank cars this is usually caused by excessive air padding. The vapor pressure in a newly arrived tank car should be about 85–90 psi. If the pressure is greater than this, the chlorine liquid temperature has risen too much or the car has been excessively air padded. In these cases, the chlorination system should be used to withdraw from the gas phase of the new tank car directly and for a long enough time to reduce the vapor pressure down to about 70–75 psi. After equilibrium at this pressure and the absence of leaks has been verified, the system can then be switched to the liquid phase using the evaporators to provide gas to the chlorinators.

To Stop or Secure an Installation. First, shut off the chlorine supply system. If the shutdown is to be of short duration, any auxiliary valve near the chlorinator may be used for this purpose. If it is to be a long shutdown or for major repairs, it is best to shutdown at the main container valve. In this way, the chlorine pressure in the entire system can be reduced to zero gage pressure.

When the chlorine pressure gage on the chlorinator reaches zero and if a disassembly of equipment is involved, remove the plastic plug which is usually located in the chlorinator inlet reducing valve assembly;* this evacuates all the chlorine gas within the chlorinator piping.

After the chlorine has been purged to the satisfaction of the operator, the injector system may be shut down, thus securing the entire installation. At this point, replace the plastic plug.

Chlorine Control and Metering System

Injector Systems. The control and metering system will not function without an adequate injector system. The injector system provides the power to pull gas

*This is shown as the pressure-vacuum regulating valve in Fig. 3-10, Chapter 3.

from the containers through the chlorinator and then dissolves this gas into the injector water supply to provide a chlorine solution at the point of application. All injector systems should be operated with sufficient water to provide a solution concentration not to exceed 3500 mg/liter of chlorine. This usually figures to be 40 gal./day/lb of chlorine.

The injector system should be evaluated as follows. Turn on the water supply and allow the hydraulics of the system to stabilize. With the chlorine supply secured, the injector system should be indicating a vacuum in excess of 20 in. Hg. Otherwise, something is wrong. Assuming the vacuum valve is operating properly, open the chlorine supply to the chlorinator and adjust the chlorine feed rate to about 25 percent of full scale. If the system is adequate, the admission of the chlorine to the injector may only lower the vacuum to about 15 in. Hg or higher. As long as this vacuum does not deteriorate rapidly, it can be assumed the injector system is in order. Let the system remain at this condition momentarily then change the chlorine feed rate to 100 percent of full scale. If the injector system is adequate, the chlorine feed rate response will be instantaneous and the injector vacuum gauge will not deteriorate much below 15 in. Hg. Large installations where 100 percent of full scale is 2000 lb/day or more, and where the injector back pressure is greater than 3-5 psi, the injector vacuum might deteriorate to 10 in. Hg or less upon feed rate change from 25-100 percent of full scale on the rotameter feed rate indicator. Chlorinators as a general rule will not operate properly at injector vacuum values much less than 10 in. Hg. This figure should be checked with the manufacturer.

Booster Pumps. Chlorinators used for wastewater treatment commonly utilize the effluent as the injector water. Therefore, these installations require injector water pumps that are designed solely to provide the injector water supply for the chlorinators. These pumps are either the centrifugal or turbine type. If a turbine-type pump is involved, the operator should be aware of the special attention that is required. Whenever a turbine-type pump is used for an injector water supply, it comes equipped with an adjustable bypass assembly that is connected between the discharge and the suction of the pump. The relief valve is on the discharge side of the bypass assembly and is arranged to blow off to the atmosphere or a floor drain system. The adjustable needle valve in the bypass is used to adjust the discharge pressure of the pump.

Upon start-up, the needle valve is adjusted so that the chlorinator will just barely pull the maximum amount of chlorine. The discharge pressure at this condition is noted. The needle valve is then adjusted to raise this pressure by 5 psi to provide some safety factor. As the impeller of the pump wears, due to usage, the needle valve may be closed further to maintain the proper pressure. Centrifugal pumps do not require an adjustable pressure by-pass assembly.

Chlorinators. If the chlorinator feed rate adjustment is manual, there are no critical adjustments to be made. The unit is adjusted manually to give the desired residual. No further adjustment or calibration is necessary. It is important however, to choose a rotameter of the proper and most useful range. For example, if the desired feed rate is 15 lb/day, a 30 lb/day rotameter would be a wiser selection than a 100 lb/day size.

When the chlorine feed rate is paced by a flow signal, the chlorinator must be adjusted for zero and span. These adjustments are carefully outlined in the manufacturer's instruction book. This adjustment is usually a combination of signal input and mechanical linkage, depending on the type of signal—pneumatic or electric.

The important point when attempting adjustments to the chlorinator control mechanism is that these adjustments can only be made properly when there is a flow of chlorine through the machine. Furthermore, a chlorinator is not designed for zero flow conditions while the injector is in operation. Thus, the zero adjustment is contingent upon a mechanical linkage. As long as the injector is operating and even though the chlorine metering orifice is adjusted to read zero on the rotameter scale, some chlorine will be passing through the chlorinator. This is easily verified. Shut off the inlet gas valve to the chlorinator and watch the chlorine gas gauge pressure on the chlorinator instrument panel gradually creep towards zero.

The most important aid to chlorinator adjustment and calibration is the use of signal simulators to simulate input signals from flow and dosage measurement devices to the chlorinator. The signal simulators used should have the capability of providing input signals over the entire range of the chlorinator.

Special adjustments are required for chlorinators operating from a variable vacuum signal as the sole means of controlling the chlorine feed rate. This mode is commonly used for straight residual control. The special adjusting procedure is outlined in the manufacturer's instruction manual and has to do with the zero adjustment of the chlorine inlet reducing valve. The accuracy of this adjustment, as well as the overall accuracy of this method of control, is greatly improved by the use of an external pressure reducing valve installed in the supply system upstream from the chlorinator inlet. This valve takes the burden off the pressure reducing valve in the chlorinator. Without that additional valve, the chlorinator inlet pressure reducing valve would be the primary instrument of control in a variable vacuum system.

Chlorine Residual Analyzer. This unit has zero, span, and temperature adjustment. If a thermistor is used, it automatically compensates for temperature changes in the sample.

The zero adjustment is a simple one. The sample flow to the analyzer is shut off, or one lead to the cell is disconnected (depending on the manufacturer). Under these conditions the indicator or pen on the recorder should go to zero. If not, the potentiometer marked zero is rotated until the pen reads zero. The span adjustment depends upon the calibration.

The analyzer may be calibrated by amperometric titration, colorimetric (DPD or 0-T), or starch-iodide procedures when the sample is potable water.[2] However, for wastewaters, calibration should be limited to the amperometric back titration procedure for total chlorine residual or the titrimetric DPD-FAS procedure. The exceptions to this would be those advance waste-treatment plants that are chlorinating to a free residual. Then calibration would be limited to the forward titration procedure using either the amperometric titrator or the DPD-FAS titrimetric procedure.[2]

It is desirable to have a signal simulator available whenever extensive calibration of the analyzer is required. This instrument provides check points throughout the entire range of the cell output.

Once the analyzer has been put into operation, it is not necessary to respan the instrument each day. It is more desirable to plot a graph of the indicated versus the titrated sample readings. The curves of each of these groups of reading should indicate a span adjustment either upward or downward. This procedure should be done at first on a weekly basis, inasmuch as most wastewater systems require respanning the analyzer once a week.

Two critical points should be observed in the operation of a wastewater analyzer. 1) The sample pH must be kept at 4.5-5.0, preferably 4.5. Multicolored pH papers are satisfactory for this determination. 2) Sufficient potassium iodide must be added to the pH 4 buffer solution to allow the chemical reaction between the KI and the chloramine residual to go to completion. The chlorine residual reacts with the potassium iodide to release free iodine in proportion to the total residual. The analyzer cell measures the free iodine (or free chlorine). Therefore, the amount of KI to be added to the buffer solution depends upon the sample flow through the analyzer cell and the magnitude of the chlorine residual. The cell flow rates vary from one manufacturer to another (i.e., the Capital Controls cell flow is 100 ml/min. while the Fischer and Porter cell is 150 ml/min. and the Wallace and Tiernan cell is 360 ml/min.). However, the Wallace and Tiernan cell is arranged for fresh water dilution* when measuring residuals in wastewater. A rule of thumb for KI addition is 30 grams/gal. of buffer solution per 150 ml/min. cell flow of undiluted sample per mg/liter of total chlorine residual. The operator is advised to consult the manufacturers instruction book to make sure of the proper calibration and operating procedures with respect to the buffer solution, KI addition etc. *This is extremely important in the operation of a wastewater analyzer.* White has found that some

*The recommended dilution of the sample with fresh water is 1 to 1.

wastewater residual control systems are operating without benefit of KI addition to the buffer solution. This means that the residual cell is responding to combined chlorine residual. This is supposed to be unreliable because the cell response to chloramines is not linear as it is with free iodine or free chlorine. However, these systems not using KI show remarkably consistent control, somewhat superior to those systems using KI for measuring total chlorine residual. This phenomenon requires further investigation.

Owing to the importance of the addition of a buffer solution and KI, it is obvious that the operation of the reagent pumps is critical. Therefore, the operator must check the sample pH daily. If there is any doubt as to whether or not the pH is low enough, add 0.5-1.0 ml of pH 4 buffer to the sample inlet tank. If there is no large change in the cell output, the pH is low enough. The same should be done to check on the amount of KI being added. However, in this case a small change in the cell output response is indicative of insufficient KI addition. Therefore, more KI should be added to the reagent bottle until this condition has been corrected.

Daily inspection of the sample system to the cell is necessary to assure a continuous sample.

Amperometric Titrator. Operating personnel should practice the titration procedure initially one or two per day, to establish their confidence in the reliability of their results. The proper degree of confidence can come only after a great many titrations. Daily titrations after that are in order.*

When standing idle, the electrodes should be immersed in the sample jar containing tap water with a slight concentration of free iodine. This precaution provides a continuous sensitivity of the electrodes to the back titration procedure.

OPERATION—DECHLORINATION EQUIPMENT

Sulfur Dioxide Supply System

General. The vapor characteristics of sulfur dioxide make it prone to reliquefaction—an undesirable situation when attempting to meter a gas. Unless the container is artificially padded with a "propellant," the cylinder pressure at room temperature will be about 35 psi. If the cylinders are stored outdoors in a

---

*The sensitivity of the W and T amperometric titrator can be increased from 0.1 mg/liter to 0.001 mg/liter by the following procedures.
  1. Remove the ammeter from the titrator and substitute a Rochester converter which will deliver a milliamp output signal.
  2. Carefully encase the sample jar in aluminum foil and ground it.
  3. Dilute the phenylarsene-oxide four times.
  This modification can measure 0.001 mg/liter chlorine residual.

"carport-type" structure, the SO_2 gas will surely reliquefy in the header between the container and sulfonator. The only way to relieve this situation is by the application of heat on the cylinders followed by a pressure reducing valve installed immediately downstream from the cylinders. A better alternative is to install an evaporator and artifically raise the vapor pressure of the cylinders by an air-pad system or a bottle(s) of nitrogen. The design considerations to provide an operable system have been described previously.

Ton Containers. The present supply (1977) of sulfur dioxide is generally limited to ton containers. If the gas phase of these containers is to be used, then heat must be applied to the containers. Strip heaters or hot air blower-type heaters that are arranged to automatically shut off at $100°F$ are required. Care must be taken that the heat applied to SO_2 cylinders is not directed to adjacent chlorine cylinders. Sulfur dioxide gas-phase systems do not operate satisfactorily at ambient temperatures much below $85°F$. They operate best at $100°F$ with an external reducing valve in the SO_2 header adjacent to the cylinders. The installation of an evaporator for these systems is the most practical solution.

Although manufacturers of sulfur dioxide may claim that gas withdrawal rates from ton containers can be as high as 500 lb/day, the operator is best served if the withdrawal rate is kept to 225–250 lb/day. The reliquefaction problem is kept to a minimum at this rate.

Storage Tanks. Owing to the fact that sulfur dioxide tankers are available and that the pressure of SO_2 vapor is low, the use of sulfur dioxide storage tanks is popular. The maximum delivery in road tankers is 17 tons. It takes about 2-3 hr to unload such a tanker. After the storage tank has received the tanker load, it should be air padded to about 50 psi. This is an optimum operating pressure. At this figure, the gas leaving the evaporator should be reduced to a sulfonator inlet pressure of about 15 psi. The sulfur dioxide storage tank should be designed so that the suction line on the tanker's compressor can be connected to the gas phase of the storage tank. This speeds up the unloading operation by utilizing the SO_2 tank vapor pressure.

To Start Up a System. The operator is referred to this same heading under "Chlorine Supply System." The same guidelines apply in this chapter. See pages 176 and 177.

Superheat. In the use of an SO_2 evaporator which is a highly desirable piece of ancillary equipment, the same amount of superheat is required as with chlorine to assure satisfactory operation of this unit. Therefore, the operator should review the identical heading under "Chlorine Supply System."

Sulfur Dioxide Supply Pressure. As compared to chlorine, the higher the SO_2 vapor pressure the better it is for the operation of the facility. There is rarely any likelihood that "high" SO_2 vapor pressure would ever be an operating problem. If such a situation would occur, the same procedures outlined for chlorine would be applicable.

To Stop or Secure a System. The operator should follow the same procedure described under this heading "Chlorine Supply System." The identical guidelines apply for sulfur dioxide.

Sulfur Dioxide Control and Metering System

Injector System. The only difference between the use of an injector system for sulfur dioxide and chlorine is that SO_2 gas is much more soluble in water than is chlorine. Therefore, if the operation of a sulfur dioxide injector system is fashioned the same as that for chlorine, the operator will be well on the side of safety. *The use of 3500 mg/liter as the maximum strength of sulfurous* acid solution in the injector discharge will prevent breakout of molecular SO_2 at the point of application.

Booster Pumps. The same applies here as for the chlorination system.

Sulfonators. The same operating procedures are required for the sulfonators as has been previously described for chlorinators. This includes adjustments and calibration. However, since a great many sulfonator control systems are based on feed forward control, a slightly different set of operational conditions faces the operator. This concerns the function of the Ratio Station installed on the flow pacing signal which is ahead of the flow-residual signal multiplier (see Fig. 3-30 Chapter 3).

The Ratio Station is a precise dosage control instrument particularly valuable in "feed forward systems." Figure 4-1 provides the operator with dosage values for a given sulfonator orifice meter. The one shown is a 4000 lb/day meter used with a 200 mgd (maximum) plant flow meter. When the ratio station is set at unity (1.0) the SO_2 dosage will be 2.4 mg/liter. At a ratio setting of 2.0 the dosage will be doubled and so on.

EXAMPLE: Operating with a 1.0 ratio station setting, a 4000 lb/day SO_2 meter controlled from a 0-200 mgd flow meter the SO_2 dosage would be

$$\frac{4000 \text{ lb/day}}{200 \text{ mgd} \times 8.34 \text{ lb/mg}} = 2.4 \text{ mg/liter}$$

for any flow from 0-200 mgd.

Figure 4-2 provides the operator with the necessary information to set the system to dechlorinate a given chlorine residual to zero for any plant flow rate. The set of curves shown on Fig. 4-2 are for a 4000 lb/day SO_2 rotameter. The

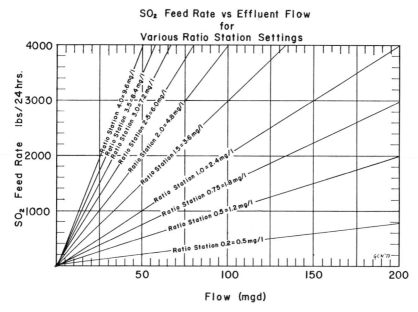

Fig. 4-1 Sulfur dioxide feed rate vs. effluent flow for various ratio station settings.

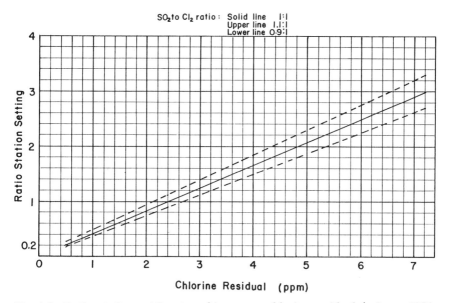

Fig. 4-2 Ratio station setting to achieve zero chlorine residual (using a 4000 lb/day SO_2 orifice meter).

solid line represents a sulfur dioxide to chlorine ratio of $1:1$ (by wt). The lower dotted line represents the stoichiometric ratio of $0.9:1$. The upper dotted line is the probable upper limit of SO_2 to Cl_2 as $1.1:1.0$.

EXAMPLE: If the chlorine residual analyzer is showing 4 mg/liter residual, the *proper setting* of the Ratio Station to feed enough SO_2 to produce a zero residual in the dechlorinated effluent will be about 1.65 using the solid line as a reference. If the selected Ratio Station setting to produce a zero residual falls above or below the dotted lines of Fig. 4-2 for a 4 mg/liter chlorine residual (i.e., is greater than 1.8 or less than 1.5), *something is wrong*. Either the analyzer has drifted out of calibration or the analytical technique of measuring the chlorine residual is in error. Therefore, the laboratory personnel should make further analyses of both chlorinated and dechlorinated effluent samples to verify the calibration of the analyzer. Figure 4-2 gives Ratio Station settings values for chlorine residuals regardless of plant flow between 0-200 mgd. The same approach would be used regardless of the effluent plant flow meter range.

Analyzers.   At present (1977), the chlorine residual analyzer is the only instrument available for the control of the sulfur dioxide system. This is well documented in other sections of this text. If instrumentation ever becomes available to measure the concentration of sulfurous acid (H_2SO_3) or the sulfite ion ($SO_3^=$), the operating requirements are certain to be substantially different from those of the chlorine residual analyzers.

MAINTENANCE—CHLORINATION EQUIPMENT

Chlorine Supply System

General (150-lb and Ton Containers).   The most vulnerable parts of a chlorine supply system are the flexible connections. These are usually made of 2000-lb annealed copper tubing, cadmium plated. Whenever these connections are exposed to the moisture in the atmosphere during a cylinder change, corrosion sets in. Internal corrosion of these flexible connectors is a function of the number of times the tubing is exposed to the atmosphere. A positive way to determine the reliability of a flexible connection is to bend it slightly. If it "screeches," discard it immediately. This phenomenon signifies that internal corrosion is excessive and the tubing is liable to rupture prematurely by crystallization.

The Supply System Should Be Checked Regularly for any Signs of a Leak. The most potentially dangerous leaks are the ones that start out at such small magnitudes that they cannot be detected by the usual application of an aqueous ammonia solution and are undetectable by the average sense of smell. However, there are telltale signs for such minute leaks. One sign is the gathering of moisture near or at the point of leakage. Another is the sign of discoloration of the metal at the leaking joint. This occurs even in the most arid climates. Any chlorine leak will, over a period of time, cause a discoloration of the cadmium plated chlorine header valves and flexible connections. Even if the hardware were not cadmium plated, the discoloration caused by dezincification of the brass and the corrosion of copper by the chlorine would be obvious. This calls for immediate corrective action.

After a thorough cleaning of all the threads by use of a steel wire brush the flexible connection in question should be regasketed. Small and apparent insignificant leaks that can result in devastating leaks often occur at the chlorine cylinder outlet valve. To prevent this situation, the operator should always carry a 1-in. flat file so that he can reface any chlorine cylinder outlet valve. This allows the proper seating surface between the cylinder valve and the auxiliary cylinder valve. This maintenance procedure has proved to be well worth the effort.

Never reuse a gasket that has been in a joint after the joint has been disassembled.

The question of lead gaskets versus asbestos or paper composition gaskets is a continuing controversy. The paper- or asbestos type gaskets tend to cover up gross deficiencies in chlorine cylinder outlet valve maintenance. This gasket is favored by the chlorine cylinder packagers. On the other hand, once a lead gasket has passed the leak test of a joint, it will never result in a chlorine leak at that particular joint providing it is not disturbed by disassembly and reassembly. However, a paper composition gasket can develop a serious leak by failure during operation. This type of gasket should be avoided because a gasket failure during operation can result in a much more serious leak than with a lead gasket. Furthermore, the failure of a lead gasket has never been proved to be the cause of a chlorine leak once that system was connected without leaks. A lead gasket joint will leak from the very beginning if it is not properly seated. This is the reason a lead gasket joint should be regasketed each time it is disassembled.

Repairing Leaks. The most important advice to an operator is: *never ever attempt to repair a leak on the chlorine supply system when it is under the chlorine container pressure.* Tightening packing glands, union joints or flexible connections etc. while the system is under pressure can only lead to disaster.

If the outlet valve of a container, tank car, or storage tank is leaking, the following steps should be taken in order. 1) Call the chlorine supplier to send expert help. 2) Don air or oxygen breathing apparatus (not a canister type mask). 3) Ready the container emergency kit for action. 4) First try "gingerly" to tighten the valve packing; if this does not stop the leak apply the emergency container kit and await the arrival of the chlorine packager representative.

Ton Containers. The same advice for this size cylinder is applicable as the advice for the 100- and 150-lb cylinders. The exception is the more rapid deterioration of the flexible connections. These should be discarded after each 50 tons of chlorine or 12 months service, whichever is sooner.

Tank Cars and Storage Tanks. The maintenance watchword of these systems is the continuous surveillance for leaks. As with any other chlorine supply system, the most critical time is during start-up procedures.

Replacement of diaphragm protected pressure gauges and pressure switches should be considered on a once in 3–5 yr. basis, unless other maintenance information prevails.

Storage tanks should be placed on the same routine service procedures as for tank cars.

Safety Equipment. Emergency kits for all sizes of tanks should be maintained and ready for instant use. Practice using these devices should be a part of the plant-maintenance schedule. The same applies to the emergency breathing apparatus. Special drills at prescribed intervals should be made available for all operating personnel on both the use and handling of the container emergency kits and the breathing apparatus.

Evaporators. This piece of equipment is subject to filling with a gooey mass of sludge that accumulates from inherent impurities in liquid chlorine. The amount of sludge that accumulates in the bottom of the liquid chlorine vessel is primarily a function of the amount of chlorine passing through the evaporator. In general, evaporators should be inspected for sludge once a year or after passing 250 tons of chlorine, whichever comes first. In addition to this, the operator should keep a close watch on the evaporator "superheat." If this cannot be maintained at 20°F, it signifies that the evaporator is losing capacity due to sludge build-up in the liquid container or that there is an immersion heater failure.

Immersion heaters can and do build up a scale deposit from hardness in the water-bath water. This can be so severe as to require annual replacement. Although the water-bath surfaces are cathodically protected from corrosion, the anodes used are sacrificial. Therefore, annual inspection and possible replacement of these anodes is desirable. The ammeter on the evaporator instrument panel does provide an indication of whether or not the cathodic protection system is functioning properly. Regardless of this indication, annual inspection is desirable.

Cleaning the evaporator liquid container consists of dismantling and removing the chlorine vessel and flushing this with cold water until the flushing water is clean. The inside is visually inspected for pitting. If the pitting is severe, the vessel should be replaced.

After all the flushing water has been removed, the evaporator is reassembled and the water-bath is filled and heated to 180°F. Then an aspirator is attached so that a vacuum can be exerted on the inside of the vessel. The vacuum should be about 25 in. Hg and should continue for 24 hr with the water-bath at 180°F to remove all moisture from the inside of the chlorine vessel.

Chlorine Gas Filter. This unit should be inspected every six months. The filter element should be replaced at each inspection and the sediment trap washed in cold water and dried before reassembly. The lead gasket at the flanged joint should also be replaced.

External Chlorine Pressure Reducing Valve. Chlorine impurities are most likely to precipitate out at points of pressure drop. So whatever impurities are not picked up by the filter, described above, will plate out on the spring loaded valve stem and seat of this unit. Deposits that cannot be removed with a soft cloth can be removed with trichlorethylene. The spring opposing the diaphragm action in this valve will eventually suffer from metal fatigue. Depending on local conditions, the life of this spring is from 2–5 yr.

Chlorine Control and Metering System

Injectors. All injectors have removable internal components. These consist of a throat and tailway and an adjustable stem assembly. The one-in. or fixed throat injectors do not have an adjustable stem. Injectors should be disassembled and cleaned every six months or more often if required. Even though the velocities through an injector provide a certain amount of scouring, iron or manganese in the injector water will plate out on the throat, thereby reducing the efficiency of the injector. This deposit is easily removed by washing with muriatic acid. Abrasion from sand in the injector water can damage the throat and tailway. This can be determined from visual inspection.

Booster Pumps. Maintenance of booster pumps is the same as for any other pump, except that an injector water pump must put out a minimum specified flow at a minimum critical pressure. As soon as this is not achieved, the chlorinator injector system will fail even though the pump appears outwardly to be doing a satisfactory job.

Chlorinators. Modern chlorinators consist of a series of spring-loaded diaphragm units that form the basis for control of the chlorine gas through the equipment with certain vacuum values at various points. Therefore, it is essential that all the joints be vacuum-tight for proper operation.

Most of the maintenance problems occur from metal fatigue of the springs in the various diaphragm assemblies, and from improper stem and seat closure in these diaphragms caused by impurities in the chlorine gas. All springs should be replaced every two yr. The stem and seat units should be inspected and cleaned annually.

The rotameter tube and float assembly should be removed and cleaned periodically at least every six months. The chlorine metering orifice should be dismantled for inspection after six months operation, because the impurities deposited here will give a clue as to the condition of the rest of the system.

Automatic control devices are made up of plug-in transistorized components which give long and reliable service provided the circuitry is balanced by proper calibration procedures at the time of start-up. Devices with moving parts such as potentiometers or alarm devices dependent upon mercury switches are subject to wear, and will require eventual replacement. The operator should have these items on hand as spares. Reversible motors used in control devices are also subject to failure, even though they do give long and reliable service. The operator should check the source and availability of these motors. The best advice for the operator is to consult the manufacturer for a reasonable spare parts list to keep on hand, even though the installation consists of standby equipment.

Chlorine Residual Analyzer. When the chlorine residual analyzer is arranged for control of the chlorination system, it is the heart of the installation; so it must be carefully maintained. In wastewater application, the most vulnerable part of the analyzer is the sample flow to the cell, next is the condition of the electrodes, and finally the maintenance of the reagent addition system.

All wastewater analyzers must be equipped with a motorized disc-type filter. This filter should be operated so that it is continuously flushed to waste. The drain valve controlling this waste must be in the wide-open position at all times. The operator should keep a spare outer filter case on hand at all times. This piece is subject to frequent failure due to corrosion.

In some analyzers there is an additional filter of some sort downstream from the motorized unit. These filters should be cleaned on a daily basis. If there are other screens upstream from the motorized filter, these screens should also be cleaned on a regular basis.

The sample flow to the cell should be checked on a daily basis.

All of the sample lines beginning at the suction of the sample pumps should be purged on a regular basis to remove organic deposits, slimes, and grease. These accumulations cause gross inaccuracies in the analyzer system.

The electrodes should be cleaned (flushing with wastewater—not the use of abrasives) on a weekly or biweekly basis depending upon the operation of the analyzer. If the output signal is erratic and the cell refuses to remain in proper calibration, the electrodes should be cleaned with the recommended abrasives.

The operator is advised to follow the manufacturer's recommended procedure for cleaning these electrodes and placing the analyzer back into operation.

The reagent additive system will not require any regular maintenance unless the pumping system fails to perform properly.

Amperometric Titrator. The best maintenance for this unit is frequent use. The electrodes should be soaked when idle in an iodine solution (10 mg/liter concentration) and cleaned periodically.* If a regular routine such as this is followed, the titrator will show sharp responses with clear-cut end points and repeatable results.

MAINTENANCE—DECHLORINATION EQUIPMENT

Sulfur Dioxide Supply System

The characteristics of sulfur dioxide are so similar to chlorine that everything said under this same heading for chlorine is applicable to the sulfur dioxide system. The only exception is possibly the life of the diaphragm protectors on the gas pressure gages and switches. Because SO_2 vapor pressure is less than half that of chlorine, the wear on these units is much less.

As for corrosivity, the active ingredient, sulfurous acid (H_2SO_3), is equally aggressive as its chlorine counterpart hypochlorous acid (HOCl). Therefore, considerations of corrosion are identical.

One of the major maintenance differences between chlorine and sulfur dioxide is that of continuous leak detection equipment. While the spot check use of a solution of NH_4OH to locate leaks is the same for SO_2 as it is for chlorine, the continuous leak detectors are vastly different. The procedure for detecting SO_2 in the air is much more involved and therefore the equipment is more complex than a chlorine leak detector. The best advice is to adhere to the manufacturers procedure for calibration and recommendations for maintenance.

Sulfur Dioxide Control and Metering System

Everything that has been said for chlorine under this section applies to sulfur dioxide. The only precaution to the operator is this: never interchange chlorination equipment, particularly evaporators, for sulfur dioxide duty. While chlorination equipment and sulfonation equipment are, as presently manufactured, identical for all practical purposes, they should never be used alternately for chlorine and then sulfur dioxide.

The reason for this precaution relates to the corrosion products of each gas.

---

*This maintenance procedure is for titrators used primarily for total chlorine residual determinations.

SAFETY IN HANDLING CHLORINE AND SULFUR DIOXIDE

Case Histories of Massive Chlorine Gas Leaks

A One-Ton Cylinder Liquid Leak. This accident resulted in the most serious consequences of any chlorine leak in North America associated with the handling of chlorine containers at the consumer level, because it caused the untimely and quite unnecessary deaths of two people close to but not associated with the handling of the chlorine. This particular accident occurred in May 1969. At the time, the leak was quickly blamed on a "faulty gasket." A team of newspaper reporters made a comprehensive investigation of this accident and arrived at the following conclusions.

1. The accident probably would not have occurred if there had not been a power failure during a brief rain squall.
2. The accident probably would not have occurred if the city had not removed the auxiliary municipal light plant service to the filtration plant one month before the accident.
3. If working gas masks were *conveniently* available the leak could have been halted immediately and no injuries would have resulted.
4. Residents adjacent to the plant would not have been affected if the chlorine container room had been farther than 65 ft away from their homes.

An analysis of this accident and the newspaper reporters conclusions is deserving of the following observations.

The power failure plunged the chlorine container room into practically total darkness (a rain squall at 4 P.M. shuts out most of the light). This situation caused a great delay in the operator's response to all of his duties required during the power failure. It is easy to imagine the panic which struck his heart at smelling a severe chlorine leak in the dark! He could not find the gas masks quickly so he left the area immediately (as he should have) to go for help, as he was the sole operator on duty.

With the power out, the operator was unable to relieve the chlorine piping system of pressure, even though the container valve at the leaking connection might have been closed. During the time of a power failure, all of the liquid chlorine in the piping system between the containers and the metering equipment can exit through any leaking joint unless there are appropriate isolating valves. When power is available the metering equipment is able to withdraw the liquid chlorine from the supply piping system very quickly, thereby averting a massive chlorine leak.

Therefore, auxiliary lighting to show the operator the way and auxiliary power to operate the chlorine withdrawal system are essential for safety in chlorine handling.

Easy access to chlorine gas masks is of top priority. Just as soon as the fire department arrived with the necessary assistance to provide gas masks, and emergency lights, the operator was able to locate the leaking cylinder connection. By this time power had been restored and the operator was able to relieve the system of pressure and repair the leak.

In these situations where an operator faces overwhelming odds he needs all the isolating valves available to him to reduce the length of piping under pressure adjacent to the leak. This is why every container should be connected with an auxiliary cylinder valve attached to the cylinder outlet valve. The outlet of the auxiliary valve is then connected to the inlet of the flexible connection and the outlet to the stationary header valve. Some operators prefer an auxiliary header valve at the outlet of the flexible connection so that for liquid withdrawal the flexible connection can be shut off at each end and removed during each cylinder change without the fear of discharging liquid chlorine. The best way to operate is to use duplicate header systems so that one header can be completely emptied of liquid chlorine during a cylinder change.

The deaths which occurred in this accident could have been avoided had the residents been told to get out of their houses. The children of the deceased did run away from the house which undoubtedly saved their lives. By staying in a house through which the chlorine gas has passed, any occupant is subject to continuous chlorine exposure because the gas becomes absorbed immediately by upholstered furniture, drapes, carpets, and clothing. Body heat and perspiration enhances the chlorine absorption by the clothing. This exposes the victim to continuous chlorine inhalation from both the clothing and house furnishings.

A Four-Ton Liquid Leak. This leak allowed escape of the liquid chlorine contents of four one-ton chlorine cylinders which were "on line" at the time of the leak. Fortunately, there was no loss of life but many residents nearby were "treated" for various degrees of chlorine inhalation.

This massive leak was the result of the following factors, in their order of importance:

1. The operator attempted to stop a leak at the stem of a chlorine header valve while the system was under full pressure from the four cylinders.
2. This leak was caused by the structural failure of a bushing ($3/4'' \times 1''$) in the 1-inch chlorine header into which the leaking header valve was threaded.
3. The failure of the bushing was the result of pernicious corrosion over a long period of time.

The lessons to be learned from this accident are as follows:

1. Never use bushings in the chlorine supply system.
2. Operators should be warned *never* to attempt repair of a chlorine leak while the system is under supply pressure.

3. The first step in repairing a leak is to relieve the system of the chlorine cylinder pressure. Necessary gauges must be included in the design to provide the operator with this vital information.

4. Duplicate header systems should be provided so that these piping systems can be replaced in their entirety on a regular basis (i.e., every five yr. for systems passing two tons or more per day and every ten yr. for all others passing less than two tons per day).

A 1600-lb Gas Leak. This case was a freak accident compounded by mistakes resulting from lack of experience in coping with a massive chlorine leak.

A workman for a natural gas utility company was busy with a cutting torch working at several isolated locations in an industrial area. He was cutting into sections of empty unused natural gas pipe lines preparatory to sealing them off. By mistake he cut into a 6-in. underground chlorine pipeline about 7000 ft long. This pipeline was used to transport chlorine gas from a chemical plant to a nearby plastics plant. The pipe was under tank pressure of about 85 psi at the time of the accident.

This calculates to about 1600 lb of chlorine gas which escaped into the atmosphere. Fortunately, the line was equipped with automatic shut-off valves at both ends. When the pressure dropped due to the hole by the cutters torch these valves automatically closed.

Coincidentally, with the action of the pipe cutter which was about 7.45 A.M. some 300-400 people were waiting to go to work at a nearby refinery. These people were all gathered in a small cluster waiting for the gates to open at 8:00 A.M. They were a short distance downwind from the leak. The morning air was cool and the humidity was high—about 70 percent. The general location is only a few miles from the ocean surf and the wind blowing in their direction was 10-15 mph. Even though this is considered a massive leak only 30-40 of these people were taken to the hospital. It is probable that the high humidity coupled with a strong wind was responsible for rapid dilution and dispersal of the chlorine into the atmosphere. However, the humidity had an adverse effect. The combination of body moisture and high humidity conspired to absorb chlorine into the workers clothing. The people taken to the hospital were herded into a small room and were immediately exposed to more chlorine coming from their clothing. Some hospital attendant realized what was happening and they were ushered outside and then taken in one by one and had their clothing removed. They remained in the hospital long enough for a thorough observation and time to get a change of clothes. None stayed more than one night in the hospital.

The lessons to be learned from this accident are as follows:

1. Pipelines carrying chlorine liquid or gas should never be buried in the ground. The preferred method is a grate covered concrete channel at grade level or in overhead support systems.

2. Whenever a person is exposed to chlorine their clothing should be removed as soon as possible and the body showered thoroughly with warm water if available to avoid further shock.
3. Never allow a person subjected to chlorine exposure to remain in a confined area, particularly a room with rugs, carpets, drapes or upholstered furniture. Get them into the open air as soon as possible.

Additional information on the hazards of chlorine and the treatment for chlorine exposure can be found on pages 28–36, *Handbook of Chlorination* Van Nostrand Reinhold Co., N.Y. 1972.

Chapter 4—Summary

Selection of a chlorine residual measuring procedure which provides the operator with a maximum confidence in its reliability is of the utmost importance. The operator *must* find a method he can *believe in and swear by*. Once this is accomplished the rest is routine chemistry and equipment maintenance.

Of equal importance, but for a different reason, is that of safety as applied to liquid-gas chlorine systems. This means monitoring for leaks, recognizing leaks which are too small to be picked up by leak detection equipment, and observing all the safety rules in the handling of chlorine supply systems.

In order to be conversant and aware of the variables in any treatment plant system the operator is advised to conduct a series of chlorine demand tests over a period of time. This will provide a historical record of the chlorine demand "quality" of the treated effluent so that "upsets" in the plant can be related to chlorine demand. This is a most useful tool.

The operator should limit his choice of *total* chlorine residual measurements to one of the following: amperometric titration procedure, Palin's FAS titrimetric method, or Standard Methods starch-iodide procedure.

The operator should insist upon being able to determine the MPN coliform concentration of the treated effluent before disinfection. This governs the required dosage of disinfectant to produce and effluent meeting the health requirement standards.

The most important part of any chlorination facility is the storage and supply system. It sets the tone of the overall safety of the system. Therefore, the operator must recognize the necessary operational and maintenance procedures to prevent leaks and maintain an "even temperature" operating installation whether it be for 150-lb cylinders or ton containers, or whether it be liquid or gas withdrawal. There are a great many do's and don'ts (see the text).

The operator must acquaint himself with start-up and emergency procedures. This knowledge is imperative. For example, *never ever* try to repair a chlorine gas leak while there is container pressure in the system. Procedure for securing an installation is equally important. It is usually the reverse of a start-up procedure.

Maintenance of a chlorination system revolves mostly around the chlorine storage and supply system. Chlorine as produced by the various manufacturers contains an inherent amount of impurities. This requires that the operator will be faced with a cleaning program related to the amount of chlorine passing through the system (see the text of this for frequency of cleaning).

One of the most important pieces of equipment in any disinfection system whether it be chlorine, chlorine dioxide, bromine, or ozone is the residual analyzer. This is the heart of the control system. The operator must acquaint himself with the manufacturer's instruction manual in order to be able to establish his confidence in the control system.

A dechlorination facility utilizes the same equipment as does the chlorination system so long as the chemical to be used is sulfur dioxide. Therefore, the same rules apply for the operation and maintenance of dechlorination systems. However, there is one exception. *The control of a dechlorination system is entirely dependent on a chlorine residual analyzer installed to monitor the residual at the end of the chlorine contact chamber.*

This instrument provides the signal necessary to control the sulfur dioxide feed rate to achieve zero chlorine residual.

Sulfur dioxide supply systems are extremely sensitive to the reliquefaction phenomenon which interferes with the operation of the metering equipment. Sulfur dioxide containers and/or storage tanks should be artificially pressurized by either air or nitrogen. This presupposes the use of liquid phase for SO_2 containers.

Sulfur dioxide leak detection equipment is much more complicated than comparable detection equipment for chlorine. This means that SO_2 leak detectors are 5 to 6 times more expensive than their chlorine counterpart.

Very little attention is required for the mixing and detention time of sulfur dioxide. It is more soluble in water than is chlorine so that off-gassing does not occur as a result of turbulence at the point of application. The chemical reaction between SO_2 and the chlorine residual is practically instantaneous.

Every installation of the chlorination-dechlorination process should be equipped with the following safety equipment: chlorine and sulfur dioxide leak detectors; air or oxygen breathing apparatus as distinguished from canister-type gas masks; and container emergency repair kits.

Finally, chlorine and sulfur dioxide accidents are avoidable provided every operator is alert to the hazards and knows the safety procedures. Part of an operator's responsibility is to organize and perform safety drills using the oxygen or air breathing apparatus. The local Fire Department should be consulted for their advice on the use of this equipment. Additionally, the plant supervisor should see to it that once every year or two the personnel responsible for the chlorination system should be exposed to the current PPG safety film for chlorine handling.

REFERENCES

1. White, G. C. *Handbook of Chlorination*, Van Nostrand Reinhold Co., N.Y. (1972).
2. "Standard Methods for the Examination of Water and Wastewater." by APHA, AWWA, and WPCF, 14th Ed. (1975).

5

Hypochlorination

INTRODUCTION

One of the first known uses of chlorine for disinfection was in the form of hypochlorite, known as chloride of lime. Snow used it in 1850 in an attempt to disinfect the Broad Street Pump water supply in London after an outbreak of cholera caused by sewage contamination. Sims Woodhead used "bleach solution" as a temporary measure to sterilize potable water distribution mains at Maidstone, Kent (England) following a typhoid outbreak in 1897.[1]

The first United States-produced dry calcium hypochlorite appeared on the market in 1928. This bleaching compound contains about 70 percent available chlorine, and is sold under the trade names of HTH, Perchloron, Pittchlor, and others. This product has largely replaced Tennant's bleaching powder in the United States.

Liquid bleach (sodium hypochlorite) came into widespread use about 1930 for laundry, household, and general disinfecting uses. Today it is the most widely used of all the chlorinated bleaches. More than 150 tons/day are now used in the United States alone. Its preparation is a modification of Labarraque's method, using a lower-residual alkali, which simplifies purification and sedimentation while maintaining a pH of about 11 for stability.

The application of hypochlorite in wastewater treatment achieves the same results as does that of chlorine gas: sodium hypochlorite

$$NaOCl + H_2O \longrightarrow HOCl + Na^+(OH)^- \qquad (5\text{-}1)$$

or calcium hypochlorite

$$Ca(OCl)_2 + 2H_2O \longrightarrow 2HOCl + Ca^{++}(OH^-)_2 \qquad (5\text{-}2)$$

The active ingredient is the hypochlorite ion (OCl^-), which hydrolyzes to form hypochlorous acid. The only difference between the reactions of the hypochlorites and chlorine gas is the side reaction of the end products. The reaction with the hypochlorites increases the hydroxyl ions by the formation of sodium hydroxide (Eq. 5-1) or calcium hydroxide (Eq. 5-2); the reaction with chlorine gas and water increases the H^+ ion concentration by the formation of hydrochloric acid.

The choice between the use of hypochlorite and liquid chlorine would always favor the latter in wastewater disinfection if the sole criterion were chemical cost. Depending upon the size and complexity of the installation, capital cost could in some cases favor hypochlorite. Accurate and dependable metering systems are available for both large and small installations for either chemical. The choice of hypochlorite instead of chlorine gas is the factor of safety. For on-site generation of hypochlorite a second factor is always presented: reliability of a continuous supply of chemical.

CURRENT PRACTICES

In recent years, the stress on safety has caused large metropolitan areas to consider the use of hypochlorite rather than chlorine gas where large amounts of the gas are stored in either storage tanks, multiple-ton containers, and/or tank cars.

Since the first use of chlorine gas the injuries to persons ingesting it, in the United States, have been minimal. However, authorities are concerned over the possible necessity of mass evacuation. Despite the considerable additional cost of hypochlorite over chlorine gas (two to four times) and its inherent unwieldy and cumbersome handling problems, the city of New York in 1967 changed from gas to hypochlorite at certain of their wastewater treatment plants.

The first to experiment with the use of large amounts of hypochlorite was the power industry. Electric generator stations use considerable quantities of chlorine for slime control of the condenser cooling water. In one such case, a hurricane along the northeast coast of the United States ripped loose a ton cylinder of chlorine at one of these power stations and carried it a considerable distance. No damage was done, but the management decided it was time to eliminate any possible hazard from chlorine gas.

Sometime in 1955, the Narragansett Electric Company of Providence, Rhode Island, effected the switchover to hypochlorite.[2] Other stations in Chicago and New York followed suit.

A notable experiment in the application of hypochlorite for wastewater disinfection has been done by the Metropolitan Sanitary District of Greater Chicago (MSDGC). Installation of disinfection facilities using sodium hypochlorite at 15 percent "trade" strength were begun in 1969 at the 330 mgd North Side Plant, the 900 mgd West-Southwest Plant and the 220 mgd Calumet Plant.[3,4,5,18,19] Since July 1972, the District has been continuously chlorinating all the effluents from these plants.[6] The newest facility of MSDGC is the John E. Egan plant at Schaumburg, Illinois which was first used in 1976.

Currently the city of Houston, Texas uses sodium hypochlorite at its largest plants and liquid-gaseous chlorine at its smallest plants.[19]

Soi le of the large hypochlorite users have switched back to chlorine gas simply because of greater monetary savings. One of these is the city of Cleveland which in 1975 decided to make the change from hypochlorite to gas at the Easterly Wastewater Treatment Plant.

Also, the interest shown in the use of hypochlorites (because of the potential danger in the handling of compressed chlorine gas) has created serious consideration of on-site chlorine generating apparatus. These systems are described below.

CHOICE OF CHEMICAL

When considering the use of hypochlorites there are three choices:

1. Liquid sodium hypochlorite.
2. Sodium hypochlorite flakes.
3. Granular calcium hypochlorite.

The best choice by far is liquid sodium hypochlorite (also called soda bleach), whether the installation is large or small, provided of course, the availability is reliable. Sodium hypochlorite flakes and granular calcium hypochlorites present an array of difficult handling and storage problems. Many fires of spontaneous origin have been caused by improperly stored calcium hypochlorite. Sodium hypochlorite flakes deteriorate rapidly in the presence of any moisture.

Sodium Hypochlorite

The strength of sodium hypochlorite solution is commonly expressed in terms of its available chlorine content as "trade percent" or "percent by volume."

A more accurate expression is the actual weight percent of the available chlorine or sodium hypochlorite. The relationship between these values is as follows.

$$\text{Trade percent (percent by volume)} = \frac{\text{Grams/liter available Cl}_2}{10} \qquad (5\text{-}3)$$

$$\text{Weight percent available Cl}_2 = \frac{\text{Trade percent}}{\text{Specific gravity of solution}} \qquad (5\text{-}4)$$

$$\text{Weight percent sodium hypochlorite} = \frac{\text{Grams/liter sodium hypochlorite}}{\text{Specific gravity of solution} \times 10} \qquad (5\text{-}5)$$

$$\text{Weight percent sodium hypochlorite} = \frac{\text{Grams/liter available Cl}_2 \times 1.05}{\text{Specific gravity of solution} \times 10} \qquad (5\text{-}6)$$

The weight in terms of lb/gal. for a particular strength is not a constant. It varies, depending on the amount of excess sodium hydroxide the manufacturer uses to promote stability.

Table 5-1 gives approximate values of weights suitable for calculating dosages of sodium hypochlorite.

The stability of hypochlorite solutions is greatly affected by heat, light, pH, and *the presence of heavy metal cations*. These solutions will deteriorate at various rates depending upon the following factors:

1. The higher the temperature, the faster the rate of deterioration.
2. The higher the concentration, the more rapid the deterioration.
3. The presence of iron, copper, nickel, and cobalt catalyzes the rate of deterioration of hypochlorite.

Iron as low as 1 mg/liter (a common maximum for industrial grades of hypochlorites) is sufficient to cause this effect.[7] Source of the iron is usually from the caustic used in the manufacture of sodium hypochlorite.

TABLE 5-1

Trade % Avail. Cl_2 gm/liter / 10	Approx wt of 1 gal. (lb)	Available Cl_2 (lb/gal.)	GPD required to dose 1 mgd to 1.0 ppm	GPM required to dose 1 mgd to 1.0 ppm
1.00	8.45	0.083	100	0.0694
5.00	8.92	0.42	20	0.0138
5.25	8.95	0.44	19	0.0132
10.0	9.50	0.83	10	0.0069
15.0	10.1	1.25	6.6	0.0046

Copper content should be kept as low as possible; not in excess of 1 mg/liter in the finished solution. It is generally present because of the flexible copper connections and brass body line valves used in the chlorine supply system. Great care must be taken to prevent active corrosion of these parts by keeping them internally free of moisture. This is difficult to accomplish.

The most stable solutions are those of low-hypochlorite concentration, with a pH of 11 and low iron, copper, and nickel content, stored in the dark at low temperature.

Caustic (NaOH) is used purely as a stabilizing factor. A large excess of alkalinity, however, does not stabilize a hypochlorite solution any more than the slight excess given in Table 5-1. However, *if the pH drops below 11*, decomposition becomes more rapid.

Table 5-2 shows the half-life of various concentrations of sodium hypochlorite solutions at different temperatures.[8] Half-life is the best way of expressing "shelf life" when the rate of decay is nonlinear. Hypochlorite does not lose its strength at a constant rate per day, but at a decreasing rate as it loses strength. It is estimated that the rate of decomposition of 10 and 15 percent solutions nearly doubles with every 10°F temperature rise.

It is important to note that a 167 g/liter solution (16.7 trade percent) stored at 80°F will decay in strength 10 percent in 10 days, 20 percent in 25 days, and 30 percent in 43 days.[9]

The influence of light on the solutions of sodium hypochlorite is easy to demonstrate by putting one portion in a clear container and the other in an amber container and exposing both to sunlight. The half-life of 10–15 percent available chlorine solution will be reduced about three or four times by sunlight. For stronger solutions up to 20 percent, the result is a reduction of half-life of about six times.[8]

These solutions will freeze, *but at temperatures considerably lower than the freezing point of water*. (See Fig. 5-1.)

All bulk sodium hypochlorite shipments for large installations should be purchased on the basis of specifications delineating the available chlorine, iron,

TABLE 5-2

Percent Available Chlorine	Half-life Days			
	212°F	140°F	77°F	59°F
10.0	0.079	3.5	220	800
5.0	0.25	13.0	790	5000
2.5	0.63	28.	1800	
0.5	2.5	100.	6000	

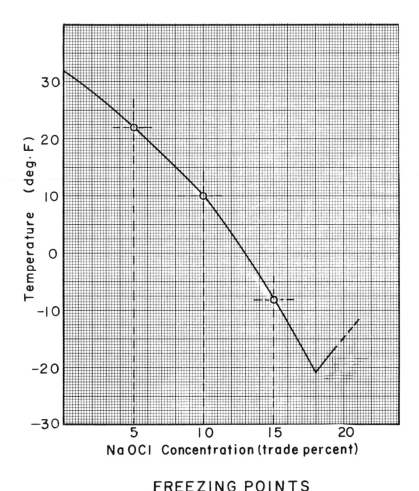

FREEZING POINTS

Fig. 5-1 Freezing temperatures of hypochlorite solutions (*courtesy* Dow Chemical USA).

and copper content and excess caustic. Copper should be limited to a maximum of 1 mg/liter and iron to 2 mg/liter. Excess caustic should be limited by a pH not to exceed 11.2. If these limitations cannot be met by the suppliers, then compromises with price and chlorine strength will have to be made. Furthermore, it should be specified that the material be free of sediment and suspended solids. All shipments of bulk sodium hypochlorite should be analyzed upon receipt for available chlorine, iron, copper, and excess caustic.

Calcium Hypochlorite

Details on the use and manufacture of this chemical can be found in Chapter 12 *Handbook of Chlorination*, Van Nostrand Reinhold Co., 1972. There will be no further discussion of this chemical in this text as its use is not recommended for permanent installations.

CHEMISTRY OF HYPOCHLORITES

The use of hypochlorite in the disinfection of wastewaters immediately raises the question of its cost and efficiency compared to that of chlorine gas. As shown by Eq. 5-1 and Eq. 5-2 the reactions of both the hypochlorites and aqueous solution of chlorine gas produce the same active ingredient, hypochlorous acid (HOCl). The only chemical difference between these two solutions is in the pH. This is extremely important since pH materially affects the efficiency of the combined chlorine species.

The aqueous solution of chlorine gas is predominantly undissociated HOCl with some molecular chlorine, and arrives at the point of application at about pH 2.

Hypochlorite solutions are buffered to pH 11, and at this level the active ingredient is the hypochlorite ion (OCl⁻), which is about 1/100 as efficient as the undissociated HOCl. However, as soon as the hypochlorite solution becomes diluted with the wastewater, it approaches the pH of the wastewater and hydrolysis occurs resulting in the formation of HOCl:

$$Na^+ + OCl^- + H_2O \longrightarrow HOCl + NaOH \qquad (5\text{-}7)$$

There is reason to speculate that a chlorine gas solution at pH 2–3 will always be somewhat more effective than a solution of hypochlorite at pH 11–12 at the immediate area of the point of application. This is because there is more of the active ingredient, HOCl, and possibly some extremely active molecular chlorine due to the low pH of the chlorine gas solution. It is a well-known fact that at pH 11 or 12 the HOCl is almost completely dissociated to the ineffective hypochlorite ion as follows.

$$HOCl \longrightarrow H^+ + OCl^- \qquad (5\text{-}8)$$

This high pH condition will exist only momentarily at the interfaces of the hypochlorite solution and the wastewater to be treated. There is also reason to speculate in the opposite direction.

Sawyer[10] reported in 1957 greater efficiency from hypochlorite solutions in the disinfection of wastewater than with chlorine gas solutions. He theorizes there may be undesirable side reactions occurring in the lower pH environment of the chlorine gas solutions, robbing it of some of its disinfecting powers. These side reactions may be in the nature of the formation of organic chloramines, with little or no germicidal efficiency, and yet will appear as part of the total chlorine residual. Sawyer's laboratory work indicated that 1 lb available chlorine in the form of hypochlorite was equivalent in disinfecting power to about 1.5 lb of aqueous chlorine gas solution. This fact was not substantiated in treatment-plant application. However, the most impressive findings of Sawyer's work showed significantly higher amounts of residual chlorine—as measured by amperometric titration—after 15 min. of contact in the case of hypochlorite treated samples for all dosages.

EXAMPLE: at a chlorine dosage of 10 mg/liter (Providence sewage) the ampero-metric residual for aqueous chlorine gas solution at the end of 15 min. contact was 0.80 mg/liter versus 1.40 mg/liter for sodium hypochlorite solution. Similarly, sewage from Worcester, Massachusetts, showed, for the same dosage and contact time as above, 1.75 mg/liter for aqueous chlorine gas solution versus 2.75 for sodium hypochlorite solution.

Unfortunately these observations when applied to field conditions were not of the same magnitude or consistency, but they indicate that any given wastewater is likely to demonstrate a higher chlorine demand when aqueous chlorine gas solution is used than when sodium hypochlorite is used. This becomes an economic consideration for wastewaters of poor quality and high chlorine demand. These situations call for highly efficient mixing to eliminate the possibility of undesirable side reactions. Such reactions are more prone to occur in waters of high chlorine demand in the low pH area surrounding the point of application of aqueous chlorine gas solutions. This theory has yet to be proved but should be investigated.

The next consideration is the total effect on the pH of the effluent due to chlorination. This depends upon three factors: 1) the buffering capacity of the wastewater; 2) the chlorine dosage; and 3) whether chlorine is in the form of hypochlorite or aqueous gas solution. Sawyer concluded that the variation in pH of the final sewage mixture is not a significant factor. However, as the pH drops below 7.5 the ratio of monochloramine to dichloramine shifts toward the formation of more dichloramine; therefore, the disinfection efficiency should improve because dichloramine has at least twice the germicidal efficiency of monochloramine.[11] The pH of the effluent in the 36 plants investigated by White[12] was usually between 6.9 and 7.3. Only a few plants noted any appreciable downward shift because of chlorination. One plant using a very high dose of chlorine (25-30 mg/liter), which lowers the pH to about 6.3 from 7.0, reported considerable improvement in disinfection efficiency. Such increased efficiency has been reported by others.[13] In another case (primary effluent), experiments were carried out to determine the difference between the efficiency of hypochlorite and aqueous gas solution. A hypochlorite dose of 15 mg/liter raised the pH from 7.1 to 7.4 but was lowered to 6.85 with the same dose of an aqueous chlorine gas solution. Likewise, another primary effluent in the same city, but with a high chlorine demand and a pH of 7.4 prior to chlorination, will reached a pH of approximately 8.0 after a hypochlorite dose of 40 mg/liter. After a similar dose of aqueous chlorine gas solution it reached a final pH of approximately 6.5. In both of these instances the disinfection efficiency was significantly increased as the pH was lowered due to the use of the aqueous gas solution.[14]

However, of the 36 plants White observed in 1972 and an additional 20 from 1973 to 1976 the final pH was not in itself a major factor contributing to disinfection efficiency with the two exceptions mentioned above.[12]

An additional consideration must be made comparing hypochlorite versus aqueous chlorine gas solution, and that is the concentration of available chlorine in the solution. By design the chlorine concentration in the injector discharge of a conventional chlorinator is limited to 3500 mg/liter as this is the upper limit of containment of the molecular chlorine (off-gassing) inherent in these solutions. However, a hypochlorite solution of 15 percent available chlorine has an available chlorine concentration of 150,000 mg/liter or about 43 times the concentration of a conventional chlorinator discharge. This may be of great significance in the practical application of chlorine because it relates to the segregation phenomenon.[15] This phenomenon briefly states that to mix two liquids, one a chemical to be intimately mixed with the process flow, the mixing is most efficient when the chemical added is in the smallest possible amount as compared to the process flow. In other words, a chemical solution of the highest possible concentration is the optimum procedure for better mixing. This phenomenon is explained in further detail in Chapter 2. To date (1976) this phenomenon has not been investigated for disinfection efficiency relative to the concentration of the chlorine solution applied.

HYPOCHLORITE FACILITY DESIGN

Imported Hypochlorite Solution (10–15 percent)

Pumped System. Feeding and control systems should be categorized by the quantity of hypochlorite solution to be applied. Up to rates of approximately 200 gal./hr the best choice would be a positive displacement diaphragm pump. An appropriate example would be the Wallace and Tiernan 44 Series metering pumps; single or dual heads.[16] These pumps can be automatically controlled by either a variable speed drive (SCR control) or by a stroke length positioner (electric or pneumatic) or both.[16] Therefore, a compound loop control system is practical. In practice the flow signal is sent to the SCR controller which provides a 20:1 metering range and the chlorine residual signal is sent to the stroke positioner which as a 10:1 range. This is more than ample for any wastewater disinfection application.

BIF of Providence, Rhode Island, makes a line of diaphragm metering pumps capable of handling hypochlorite solutions. One group utilizes a mechanical diaphragm, and is limited to 48 gal./hr in the dual head version. Automatic control is limited to variable speed drive. Their hydraulic diaphragm series metering pump has a maximum capacity of 830 gal./hr in the dual head version. These pumps can be arranged for compound loop control (i.e., flow signal to variable speed drive and dosage signal to automatic stroke positioner).[34]

Pulsafeeder of Rochester, New York, also makes a suitable diaphragm pump with an automatic feed rate control for hypochlorite solutions.

The Pulsafeeder approach is to resolve the entire variation of solution discharge as a function of stroke length, maintaining the stroke speed constant at all times. This means that a system using the Pulsafeeder would have to have a multiplier to combine the flow and residual signals into a single signal. Therefore, the concept of compound loop control is not yet available with a Pulsafeeder system.[35]

Diaphragm metering pumps suitable for this service can tolerate back pressures as high as 150 psi. This is far in excess of any need that could occur in a wastewater system.

Systems requiring pumping rates in excess of the capacities available in diaphragm pumps can use a centrifugal pump. Feed rate control in these situations is by the use of a modulating rate valve downstream from the pump discharge. The Duriron Co. of Dayton, Ohio, makes a pure titanium pump which is excellent for this service. This pump has a long and satisfactory record of pumping chlorinator injector discharge solutions into transmission lines where pressures are too high for the usual injector booster pump upstream from the injector. This pump can also be used as a transfer pump or a low-lift pump for any hypochlorite installation. The "Durco Titanium" line of pumps comes in a variety of sizes for either low- or high-head purposes.

Pumps other than those mentioned above should be carefully investigated for their suitability for hypochlorite service. Recently a chlorine manufacturer put out a bulletin on the advantages of hypochlorite which recommended the use of graphite pumps as suitable for use with high-strength (15 percent) hypochlorite solutions. Upon investigation the manufacturer of such a pump (Union Carbide Corp.) advised that their Karbate line of graphite pumps is *not recommended* for this purpose.[16a]

Plunger- or gear-type pumps are not recommended owing to their high maintenance costs and poor reliability. There is insufficient operating experience available on these types of pumps.

Gravity Systems. The preferred system is to meter the hypochlorite through a modulated rate valve and allow the solution to flow by gravity to the point of application. If the hydraulic gradient from the storage tanks to the point of application cannot provide gravity flow, the hypochlorite could be pumped to an intermediate head tank. This can be done with a sonic-type level control switch allowing automatic intermittent operation of the supply pumps which would always pump at a constant rate. The modulating rate valve would be somewhere on the downstream side from the intermediate head tank.

The Metropolitan Sanitary District of Greater Chicago, has used all three systems of hypochlorite delivery. They have found the gravity system to be the most reliable and the one that requires the least amount of maintenance.

Eductor System. Eductors can be used effectively up to about 10-15 psi total back pressure. These are normally powered by pumps using the plant effluent. The principle of operation is similar to that of a chlorinator injector system. Of the three different systems described, this one is the least reliable for controlling flows. Another disadvantage is the loss of available chlorine caused by the presence of ammonia nitrogen in the plant effluent. The breakpoint reaction when using a mixture of hypochlorite solution of about pH 10 takes place almost instantaneously. Therefore, in the eductor system there will be a loss of chlorine to complete this reaction by a factor of 10 parts (by wt) chlorine for each part of ammonia nitrogen. As an example, the eductor system at the Calumet treatment plant of the MSDGC requires about 65 GPM of water to provide the necessary minor flow of sodium hypochlorite (about 2-4 GPM). If the treated effluent is used for powering the eductor system and if the effluent is not nitrified, the ammonia nitrogen content may be as high as 20 mg/liter. Therefore, the chlorine consumption in the eductor water would be as follows.

$$65 \text{ GPM} \times 1440 = 0.0936 \text{ mgd}$$
$$20 \text{ mg/liter} \times 8.34 = 167 \text{ lb/mg}$$
$$167 \text{ lb/mg} \times 0.094 \text{ mgd} \times 10 \text{ mg/liter Cl}_2 = 157 \text{ lb Cl}_2/\text{day}$$

So regardless of the chlorine dosage 157 lb of available chlorine will be used in the reaction with the effluent eductor water. Assuming a 15 percent trade strength hypochlorite solution this would amount to about 126 gal. of hypochlorite.*

Control System. The elements of an adequate control system are the utilization of both flow pacing and chlorine residual information plus the chlorine residual monitoring of the final effluent. The flow and residual signals can be cascaded or they can be a compound loop. Figure 5-2 illustrates an electronically controlled

---

*One gal. of 15 percent hypochlorite contains 1.25 lb available chlorine.

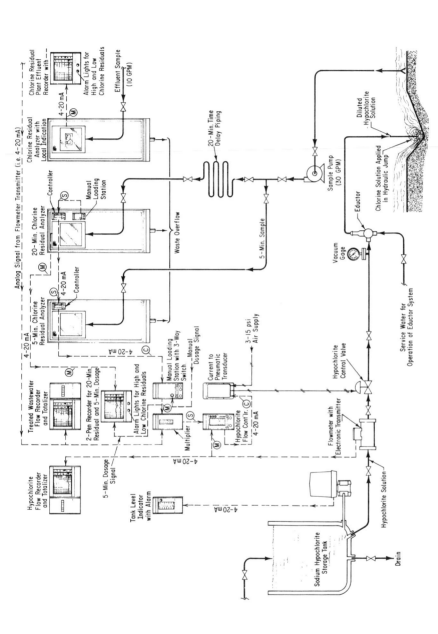

Fig. 5-2 Layout of hypochlorite dosing system using electronic controls. Key: S = set point; M = measured output; C = control signal; [twenty-minute chlorine residual analyzer output signal (measured) is used to control the output of the five-minute analyzer. This output becomes the chlorine dosage signal]. (*courtesy* of Fischer & Porter Co., Warminster, Pa.).

system utilizing the cascade principle where the flow signal is combined with the dosage signal from the chlorine residual analyzer. An alternate system would be to substitute diaphragm pumps for the eductors and change the control system to a compound loop. If either centrifugal-type pumps or a gravity system were used the control principle (cascade) as shown in Fig. 5-2 would be used.

The system illustrated has the added control feature of a second residual analyzer operating on a 20-min. sample detention time to "set" the control point on the 5 min. sample time analyzer. The 5-min. analyzer provides the dosage control signal to the system. A third analyzer is used to monitor the plant effluent and control the dechlorination system (sulfonator) if required. To provide the operator with all the elements necessary for proper operation, the system should include the following functions: continuous recording of hypochlorite flow; tank-level indicator; continuous recording of the chlorine control residual and the 20- or 30-min. residuals; and continuous recording of residual in the final effluent. Alarms should be provided for low-hypochlorite supply level, low- and high-chlorine residual on all analyzers, loss of hypochlorite flow, and loss of water supply to eductor.

It is of interest to note that in 1977 Fischer and Porter announced the marketing of their "Feedrator II Liquid Chemical Feed Dispenser"[21] which is a packaged system for the continuous feeding (automatic or manual) of chemicals normally used in potable water and wastewater. This includes high-strength hypochlorite. Referring to Fig. 5-2 the Fisher and Porter package mounts in a single cabinet the magnetic flowmeter with transmitter, the hypochlorite control valve, the eductor, the hypochlorite flow controller, the current to pneumatic transducer, and the signal multiplier. One of the main features of this system is the incorporation of their new Chloromatic-type valve constructed of PVC and teflon. This unit is used up to 5 GPM of hypochlorite. Above 5 GPM a Saunders-type diaphragm valve is used. In this system the Chloromatic operator along with the magmeter becomes the flow controller. The plant flow meter and residual signal are fed into this operator which multiplies these two signals to provide the control signal to the chloromatic valve. This eliminates the need for the external controller as shown in Fig. 5-2. This system is also equipped with a potentiometer to manually set the hypochlorite flow rate. In this mode of operation (without chlorine residual signal) the control action of the chloromatic valve is adequate (when the magmeter combination is still operable). The hypochlorite flow will remain proportional to the flow signal regardless of changing vacuum conditions in the eductor. This results in a subsequent change of pressure across the rate valve.

These systems can be either electronic or pneumatic. The advantages of each are:

Electronic System
- Has a rapid response.
- Can operate over long distances without a time lag.

- Is not subject to cold weather difficulties.
- Is compatible with telemetering and data processing systems.
- The medium is always clean.

Pneumatic System

- Low first cost of equipment.
- Easy to service and understand.
- No explosion or fire hazard.
- Valve positioners operate directly from the control medium without the need for conversion.

Storage Tanks. The first large storage tanks built for the Chicago projects were of the filament-wound fiberglass type. These tanks proved unsatisfactory having a life of only about four yr.[20] About 1975 these tanks were converted to hand lay-up fabricated fiberglass tanks utilizing a vinyl resin binder. This change was made based upon positive results obtained from another operating installation. The performance of these new tanks appears to be acceptable.[20]

The experience of the Metropolitan Sanitary District of Greater Chicago with underground concrete tanks has indicated the fiberglass lining of concrete is undesirable. Laminate failures have caused clogging of valves, pumps, and diffusers, and the use of these tanks has been discontinued. For the newest MSDGC at the John E. Egan wastewater plant, which was put into operation in 1976, the hypochlorite storage tanks were put underground because of their size (12 ft in diameter and 35 ft long—approximately 30,000 gal.). These are plastic lined, continuous weld, full weight carbon steel tanks. The lining which was applied on-site, consisted of two coats of fiberglass reinforced polyester material applied at a nominal thickness of 35 mil.

Rubber lined steel tanks would be equally as satisfactory as any type of PVC lining. This method has been used in various aspects of chlorination for many years.[1]

Storage tanks should be equipped with a level guage and transmitter for continuous readout of its contents. They should also have vents and some method of manually sampling both the incoming delivery and the contents of the tank.

The fill piping system should discharge through the top of the tank. A pressure relief and overflow pipe should also be provided. The fill pipe connection to the delivery vehicle should be a Hastelloy C or a titanium nipple securely braced to the tank. The fill pipe itself can be PVC, R.L. steel or saran lined steel, or Resistoflex (Kynar®). A high capacity drain or sewer with a proper drainage gradient adjacent to the storage tanks should be provided in the event of a storage tank rupture. Underground tanks should be equipped with equivalent facilities consisting of a completely corrosion resistant sump pump.

Piping. Above ground piping can be Sch 80 PVC, Kynar, hard rubber lined steel, or saran lined steel. The steel lined pipe presents some installation difficulties. The preferred material is Kynar for both pipe and fittings. This pipe is also known as Fluoroflex®-K. The latter is a trademark of Resistoflex Co.,[17] and Kynar is a Pennwalt Corp. trademark.

Underground piping should be some type of lined steel pipe. Chicago favors either a PVC or polypropylene lining. Hard rubber or saran lining is also acceptable and is preferred by others.

Valves. Plug valves are preferable to ball valves. Ball valves are subject to stem breakage. This limits a ball valve operating life to about six months.[20] Plug valves made of steel and lined with PVC or polypropylene are preferable.

Diffusers. The distribution of the hypochlorite solution at the point of application needs special attention because of the very high concentration of the solution (15 percent = 150,000 mg/liter). This compares to the chlorine concentration of 3,500 mg/liter (maximum) in a conventional chlorinator discharge. Although the segregation phenomenon favors better mixing with a higher concentration of the process chemical, efficient dispersion is the goal for a hypochlorite diffuser. The diffusers should be designed with perforations for across-the-channel installations using about a 25–30 ft/sec velocity at the perforations.

The diffusers should be made of either PVC, Kynar or rubber lined and rubber covered steel.

Hypochlorite Flow Meters. These can be either teflon lined magnetic flow meters manufactured by Fischer and Porter or PVC Straight-Through Vareameters (with transmitter) electric or pneumatic manufactured by Wallace and Tiernan Div. Pennwalt Corp.

Rate Control Valves. The most common control valve (diaphragm) in use is the Saunders-type using a rubber diaphragm with teflon facing. The valve body can be all PVC; or a steel body lined with either PVC, Kynar, hard rubber, or saran. These valves can be equipped with electric or pneumatic actuators. Plug-type valves are also available for this service. PVC ball-type construction are not recomended for this service. Maintenance is too high and reliability is low.

For flows up to 5 GPM Fischer and Porter uses their Chloromatic type PVC and teflon valve; above 5 GPM a Saunders-type diaphragm valve is used.

Eductors. These must be all PVC construction because of the corrosivity of the hypochlorite solution. The eductor used at the MSDGC Calumet wastewater-treatment plant is a Penberthy Model No. 168P (PVC). This unit operates at about 35 psi. and uses approximately 70 GPM wastewater for a maximum back pressure of 10 ft.[18] It can deliver about 3 GPM of 15 percent hypochlorite.

Operating Costs

General. The operating cost of any imported hypochlorite system will depend entirely upon the amount of chemical to be delivered at one time and the total amount to be consumed over a contract period. The following are examples of hypochlorite and chlorine liquid-gas prices in various metropolitan areas.

CHICAGO, ILLINOIS

The Metropolitan Sanitary District of Greater Chicago (MSDGC) purchases 15 percent (trade) sodium hypochlorite under contract agreements with two suppliers: K.A. Steele and Barton Chemical. The unit price paid for sodium hypochlorite is adjusted each month based upon quantity requirements of a facility, published prices of certain raw materials, transportation, fixed costs, and the Consumer Price Index. Utilizing this rather involved calculation, one of the potential suppliers in the Chicago area supplied 15 percent hypochlorite on the basis of one tank truck (4000 gal.) minimum delivery at a cost of 37¢/gal. for the month of October 1976. The other supplier delivered hypochlorite to the smaller plants at a cost of 59¢/gal. The 37¢/gal. figure calculates to 30¢/lb available chlorine and 59¢/gal. calculates to 47¢/lb available chlorine.

Chlorine gas in the Chicago area is grossly dependent upon the size of the containers, because of a peculiar safety ruling of long standing. The various water-treatment facilities for the city of Chicago are compelled to receive their chlorine supply by the truck-load (14-ton containers). By law, chlorine tank cars are not allowed to have access to the Lake Michigan water treatment plants. Therefore, the cost per ton of chlorine gas at these plants is $204/ton in 14-ton lots (10¢/lb).[22] If 90-ton tank cars were allowed access to these plants the chlorine cost would be about $145/ton = 7¢/lb.[23]

HOUSTON, TEXAS

Sodium hypochlorite supplied by Dixie Chemicals is quoted in 1977 at 25¢/lb. available chlorine in tank truck quantities. This supplier is quoting chlorine liquid-gas in ton containers at 13¢/lb in one-ton lots.[24]

Diamond Shamrock is currently (1977) quoting chlorine in tank cars of any size at $135/ton (7¢/lb).[24]

SAN FRANCISCO BAY AREA

Sodium hypochlorite (14 percent) is available at 40¢/gal. FOB Richmond, California, in 4000 gal. tank truck quantities. This is roughly equivalent to 33¢/lb available chlorine.

Liquid-gas chlorine in lots of 5–10 one-ton cylinders varies from $190–210/ton (10–11¢/lb). Delivered in 55-ton tank cars the cost is about $152/ton delivered anywhere in the San Francisco bay area (8¢/lb).[25]

As can be seen from the above tabulation, the cost of chlorine gas and hypochlorite varies widely depending upon the locality, demand, and availability. The price spread between hypochlorite and chlorine gas increases significantly as the distance from the source of chlorine gas manufacture and the user increases.

The most optimistic estimate is that imported hypochlorite will cost at least three, or more likely, four times that of liquid-gas chlorine.

Cost comparisons of the chlorination facility between liquid chlorine and hypochlorite should include storage and supply facilities, metering equipment, instrumentation, and monitoring equipment.

Generally speaking, the metering and feeding equipment for chlorine gas is more expensive than that for hypochlorite, but the expense of storage facilities for hypochlorite are far greater and more than offset the equipment difference. Maintenance of a hypochlorite system requires more man hours than the gas system.

Reliability

For maximum reliability the hypochlorite flow control system should consist of two separate and independent systems which accurately control the flow of hypochlorite to provide a predetermined chlorine residual in the plant effluent.

If pumps are required to deliver the hypochlorite or operate eductors, both standby equipment and standby power should be available to prevent interruption of hypochlorite flow to the point of application.

ON-SITE GENERATION OF HYPOCHLORITE

Historical Background

On-site generation of hypochlorite in the United States dates back to the early part of the 20th century. This was largely inspired by the use of hypochlorite solutions during World War I which came to be known as the Carrell-Dakin solution.[28] The success of this antiseptic treatment of open wounds led to the on-site generation of this solution in hospitals. One of the first electrolytic cells for this purpose was developed by Van Peursem et al.[29] in 1929. This cell was designed to produce the equivalent of the Carrel-Dakin solution.[30]

The concept of on-site chlorine generation inspired Wallace and Tiernan, manufacturers of chlorine gas metering equipment, to develop an electrolytic cell for the production of chlorine for swimming pools located in buildings where people slept. As early as 1940, they established a policy that chlorine gas equipment must not be installed in such buildings. For this purpose they developed an electrolytic chlorinator.[31] This unit aroused the interest of Pan American Airways who were developing stopover (refueling) sites on their San Francisco to Sydney and Orient flights. The electrolytic chlorinator for their water supply at these way stations was ideal. Of course, World War II changed all that.

The enthusiasm for on-site generation disappeared until the potential hazard of chlorine gas in containers was evaluated owing to the proliferation of chlorine gas installations at wastewater and potable water treatment plants. At the same time small units began appearing on the market (ca. 1950) for use in backyard swimming pools. However, the cost of these units and the manufacturer's inability to provide satisfactory service discouraged the popularity of this equipment.

Now (1970's) the popularity of on-site generation is on the rise again, largely because of the potential hazards of liquid-gas systems using chlorine stored in containers, and the availability of Federal funds for the necessary research and development of reliable equipment.

Importance of Raw Material

Brine Systems. Quality of the raw material is of great concern in the operation of any electrolytic process. Chlorine manufacturers have long realized that successful cell operation is dependent upon the use of a pure brine. The best product, which needs very little pretreatment except for the brine makeup water, is food-grade salt which is mostly refined solar salt extracted from seawater by evaporation. Solar salt which is not refined must be treated by the on-site system. This is known as "stack" salt.

Then there is mined salt, or brines naturally occurring in the earth's crust. This material, unless it is of exceptional quality also has to be treated at the site. Any underground brines should be checked for ammonia nitrogen content as this is most undesirable in the electrolytic production of chlorine.

Salt or brine impurities seriously affect the operation and maintenance of any type of membrane cell. The allowable brine hardness for membrane cells is on the order of 3 mg/l as $CaCO_3$.*

Seawater Systems. These systems are designed primarily for seawater and less saline waters. They are supposed to use these waters without any treatment except for some form of screening or microstraining if required.

The most significant factor when using these raw materials is the concern of the total dissolved solids concentration introduced into the process stream. Seawater which is not subject to any immediate dilution from surface run-off usually contains about 30,000 mg/liter TDS. All of this is returned to the process stream from the generating cells. In some cases this may be a critical factor in

*This is in accordance with the recommendations of Diamond-Shemrock, one of the developers of the membrane cell.

the choice of systems (i.e., purified brine, or seawater). Water reuse would almost certainly require a purified brine system. The amount of TDS contributed by a seawater electrolytic system is considerable because only a small fraction of the chloride ion content is converted to chlorine (hypochlorite). This is an inherent requirement in these systems because lots of water is required to perform the function of sweeping out the inevitable deposits of calcium and magnesium which occur in this process. Seawater systems are usually identified as those capable of producing a maximum of 0.8 percent hypochlorite solution (8000 mg/liter). This varies in accordance with chloride content of the product and whether or not recycling is involved. Maximum strength from seawater without recycling is on the order of a 1000 mg/liter solution of NaOCl. This translates to a very high dilution ratio of product to process stream.

Equipment Development

Membrane Cell. Circa 1968–1970, the U.S. Environmental Protection Agency became actively concerned over the environmental impact of storm water overflows (from large combined sewer systems) discharging into confined receiving waters such as the Charles River, San Francisco Bay, Chesapeake Bay, etc. These massive overflows have long been considered hazardous to health (water contact sports and shellfish growing areas) in the absence of disinfection. Since the application of chlorine would have to be at the point of storm water overflow, the chlorine containers would surely have to be transported through congested areas, thus creating the potential hazard of a chlorine spill.

The EPA, considering these factors, funded a study for on-site generation of chlorine carried out by Ionics, Inc., of Watertown, Massachusetts.[26,27] This project resulted in the development of an extremely efficient electrolytic cell similar to, but more advanced than, some of the older designs currently used by manufacturers of chlorine gas. Details covering the development and features of this cell are well documented.[26]

The two-compartment membrane cell with expanded electrodes as developed by Ionics, Inc. is illustrated in Fig. 5-3. The most important feature of this cell is the membrane which separates the anode compartment from the cathode compartment. This membrane separator concept is not new. Hooker Chemicals and Plastics Corp. began a program in 1950 which recently culminated in the introduction of the MX chlor-alkali cell for commercial production of liquid-gas chlorine.[32] The concept reached economic feasibility with the recent availability from DuPont of Nafion membrane material. The Nafion-family membranes consist of a 2–10 mil-thick film of perfluorosulfonic acid resin; a copolymer of tetra fluoroethylene; and another monomer to which negative sulfonic groups are attached.

The world's first *commercial membrane cell* chlor-alkali plant has been op-

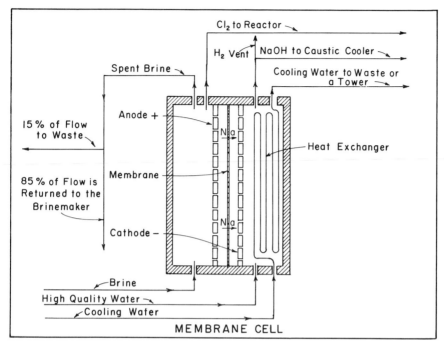

Fig. 5-3 Chloromat membrane cell with expanded electrodes.

erating successfully since April 1975 for Asahi Chemical Co. at Nobeoka, Japan using Nafion 315 membrane cells.[33] Recently, however, a perfluorocarboxylic acid membrane developed by Asahi Chemical is reported to give a higher current efficiency and is being used to replace the Nafion membranes.[33]

In a membrane cell (Fig. 5-3) the anode and cathode compartments are separated by the cation-exchange membrane. This membrane inhibits negative ions (anions) from moving through the membrane but allows the positive ions (Na^+, K^+, H_3O^+ cations) to move freely. This effect is known as the Donnan exclusion.

There is no direct hydraulic flow from the anode compartment to the cathode compartment. The only water that passes through the membrane is endosmotic water, which is associated with the ions being transferred. The anolyte liquor (spent brine) can be sent to waste or partially recycled. Chlorine from the anode chamber is sent to a water cooled reactor where it is mixed with the caustic solution.

Brine is fed to the anode compartment. Water is fed to the cathode compartment to sweep out the sodium hydroxide (caustic) that is produced. The cathode compartment is cooled by water flowing through a heat exchanger.

The caustic and hydrogen produced in the cathode compartment discharges from a common port. The hydrogen is vented to the atmosphere.

The cathode compartment cooling water can be tertiary effluent of high quality but should not be scale-forming.

The electrodes are described as "expanded." This refers to their shape. They are rectangular pieces of metal perforated so that they look like a grating. The anode is dimensionally stable (DSA), this means that it is nonsacrificial. It is made of pure titanium with a platinized coating. The cathode has the same configuration as the anode except that it is sacrificial as it is made of iron.

One of the major advantages of a membrane cell is that the anode and cathode can be placed close to the membrane. This increases the current efficiency and reduces the space required in stacking the cells side by side.

The Cloromat System (Ionics, Inc.). Ionics, Inc. use the membrane cell described above. They are assembled in modular form as in a filter press. The cells within a module are connected in series electrically, but connected in parallel hydraulically. Manifolds molded into the cell frames provide the connections for parallel fluid flow to and from the cells. The module and individual cells are designed for easy dismantling and reassembly required for annual maintenance. To increase the production of a unit, cells are added to the cell module in increments of about 165 lb/day chlorine per cell. A 2500 lb unit consists of 15 cells. The system is designed to produce an 8 percent hypochlorite solution. For this strength solution a 2500 lb/day unit requires about 2 GPM of high-quality water to sweep out the caustic from the catholyte compartment of the 15 cells involved. The flow diagram of a 2500 lb/day unit is shown on Fig. 5-4. The brine maker requires about 1 GPM so that the total amount of water to be produced by the water-treatment unit is on the order of 3–4 GPM.

A "nonscale forming" water supply of 25 GPM is required for cooling the cells, the caustic cooler, and the chlorine-caustic reactor (see Fig. 5-4).

The brine solution pumped from the brinemaker to the cells is about 1 GPM containing 3 lb/gal. NaCl (4400 lb salt/day). This calculates to 1.75 lb salt/lb chlorine.

The spent brine should be discharged to waste unless recycling is planned as part of the system. Recycled spent brine must be subjected to special treatment before it is allowed to discharge back into the system. The spent brine flow is about 0.6 GPM for a 2500 lb/day unit (see Fig. 5-4).

All of the mass-balance figures shown on Fig. 5-4 are based upon food-grade salt. This represents 1.75 lb NaCl/lb chlorine, at 2.0 kWh/lb of chlorine. In the San Francisco area electrical energy is expected to cost about 2¢/kWh.

When considering this system the designer must be careful to specify the grade of salt to be supplied. According to Ionics, Inc.,[38] food-grade salt, which is refined, is so free from the usual impurities normally associated with sodium chloride that the hypochlorite solution thus produced is characterized as "Rayon

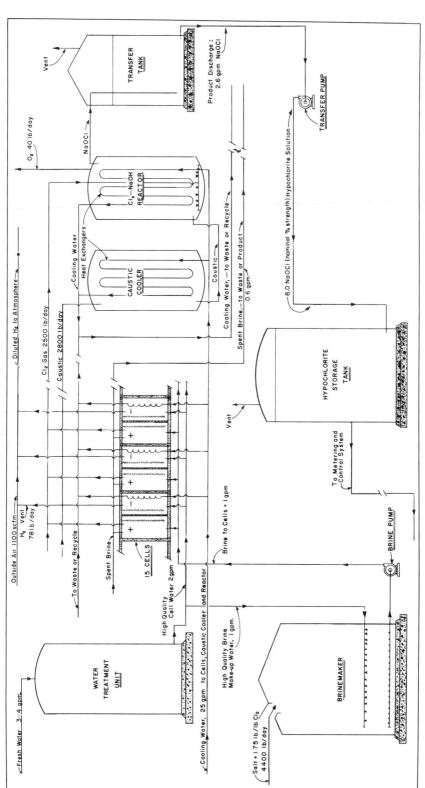

Fig. 5-4 Ionics, Inc. Chloromat hypochlorite generating system 2500 lb/day chlorine capacity.

Grade Bleach." This means that this hypochlorite does not contain the usual impurities (heavy metal ions) which contribute to the instability of the hypochlorite solution strength during average storage conditions. For example, the usual 10 percent commercial bleach (NaOCl) deteriorates from 10 to 8.5 percent in 40 days as compared to the Cloromat 8.43 percent solution which deteriorates to 7.5 percent in the same length of time. However, it becomes a question of economics whether or not food-grade or stack-grade (unrefined) solar salt should be used.

Stack-grade salt requires a pretreatment system which, in terms of power, is estimated at 0.32 kWh/lb. chlorine.

Food-grade salt in the San Francisco Bay Area at the harvester costs (1977) 2.0¢/lb. while stack grade is 1.0¢/lb.

The Cloromat system is of course a proprietary manufacturing process. When it is compared to other on-site processes there will always be claims and counter-claims for the various processes.* The designer is therefore warned to consider in detail all of the suggested pretreatment requirements of raw materials presented and recommended by the various manufacturers. These will include, but will not be limited to the treatment of cell water, cooling water (if any), brine, and re-cycled brine. Other factors are the variations required in the materials of construction. These can be dependent upon the degree of pretreatment and the use of reclaimed raw material.

For example, there will always be a small amount of molecular chlorine in re-cycled brine, so the brinemaker tank should be made of filament wound fiber-glass if brine recycling is involved.

There are not enough of these systems in operation at wastewater or potable water treatment plants to comment on the reliability of the process. Since it is a manufacturing process it presents a new dimension for operating personnel. They must incorporate manufacturing techniques and attitudes into the control of the disinfection process. The majority of installations (ca. 1977) of this process are for the manufacture of liquid sodium hypochlorite for household bleach, hypochlorite for pools, etc.

Engelhard Industries

Introduction. Early in the 1970's, the United States Department of Interior Office of Saline Water awarded a contract to construct, install, and operate a seawater hypochlorite generator at their desalting test facility, Wrightsville Beach, North Carolina. The contract for this project was awarded to Engelhard Industries following an 18 month test of a prototype unit at the OSW facility.[39]

---

*One of the most important considerations of any on-site generation system is the strength of hypochlorite solution produced. This involves size of storage tanks, pumps and piping systems.

The unit developed by Engelhard for desalting plants is also applicable to other seawater situations, the most important being the control of marine fouling organisms in seawater-piping systems. Originally the Engelhard Chloropac system was designed specifically for shipboard use to combat slime and marine growths of all kinds which normally thrive in the seachests and the ship's seawater systems. Protection from the proliferation of these growths is needed wherever seawater is used (i.e., condenser cooling, general engine room services, circulating water in the ship's air conditioning system, fire system, and other seawater piping throughout the vessel). Moreover, chlorination is now mandatory for the potable-water supply aboard ship even though the water is produced by the ship's distillation process.

One other important application which gives validity to this system and thereby enhances its reliability is the widespread use of the Engelhard Chloropac on seawater drilling platforms and seawater supertanker storage platforms. These storage systems require reliable prevention of marine growth in the pump passages and storage-system piping.

Projecting this concept of making hypochlorite from saline water will, therefore, surely include its use for wastewater disinfection whenever a reasonable supply of seawater is available.

Raw Material. All of the Engelhard systems are designed to handle an electrolyte equivalent in strength to a normal seawater (i.e., 19,000 mg/liter chlorides and 30,000 mg/liter TDS). This holds true for all of their systems whether it be recycled seawater or brine. The brine is always diluted to the optimum seawater chloride content before it is sent to the cells.

This presents a significant parameter, namely, the amount of raw product required to provide the chlorine generating capacities. This amounts to a minimum of 20 gpm of electrolyte (seawater) for a module containing a series grouping of 2 to 10 cells. Each ten-cell module can produce up to 240 lb/day chlorine as hypochlorite. The modules are connected hydraulically in parallel so a 480 lb/day unit consisting of two 10-cell modules requires 40 gpm seawater and so on. The maximum concentration of chlorine in the hypochlorite solution generated in the once through seawater system is about 1200 mg/liter and the TDS will be on the order of 30,000 mg/liter. If refined brine is used instead of seawater, the TDS will be about 20,000 mg/liter.

Cell Configuration. The Engelhard cell is a flow-through or "open-type" cell where the saline water is subjected to electrolytic decomposition on an incremental basis as the salt water flows through a series of these cells. The cell is designed so that the salt water flows through an annular opening at a velocity of 5–7 ft/sec. This velocity is supposed to continuously flush out precipitates of insoluble anions normally found in seawater or other brackish waters. The patented Chloropac® cell is illustrated in Fig. 5-5. It is constructed of three titanium cylinders. Two cylinders are placed axially in line and connected by their flanges with an insulating cylindrical spacer to form a smooth-bore pipe. The third cylinder is small in diameter and longer than the first two pipes. This third cylinder is sealed at each end and placed inside the pipe formed by the first two (see Fig. 5-5.) The salt water electrolyte flows in the annular space between the inside of the outer cylinders and the outside of the inner cylinder.

The inside surface of the outer cylinder on the left is coated with a proprietary platinum alloy which allows it to respond as though the entire cylinder were a solid platinum alloy. This cylinder connected to the positive terminal generates molecular chlorine on its inner surface. The outside of the inner cylinder adjacent (to the right) receives the electric current as a cathode and releases the cathode products (i.e., sodium hydroxide and hydrogen). The right hand half of the cell operates in the same fashion except the roles of the anode and cathode surfaces are reversed. Here the outside of the inner titanium surface is coated with platinum alloy and the current passes from the outside of the inner cylinder to the inside of the second outer cylinder. The flowing stream of electrolyte mixes the products produced at the anodes and cathodes which produces a weak sodium hypochlorite solution and minute bubbles of hydrogen gas. This chemical reaction is summarized by the equation shown on Fig. 5-5.

It should be noted that both the inner and outer pipes of this cell are made of titanium, whether platinized or not. Titanium is resistant to salt water corrosion. Moreover, titanium possesses the unique chemical characteristic of being able to form a protective oxide coating so that it will receive but not emit a current in the 8–12 V dc range. The platinized anodes are consumed at the rate of 6 mg/A year. This calculates to an expected anode life of about 6 years.

As long as the system stays within the power design specified by the manufacturers, the titanium cathode will not be consumed in the electrolytic process. It is therefore labeled as an "infinite life" electrode by the manufacturer.

Chloropac Generator.   The Engelhard system is usually arranged for 10–12 cells in series. This is identified as a module, and the arrangement of these modules is further identified with a model number as to equivalent chlorine capacity in lb/day. Figure 5-6 illustrates a 480 lb/day generator. This unit consists of 20 cells arranged so that there are two groups of 10 cells in series. However, each of these two groups is connected in parallel hydraulically.

The individual cells are arranged in pairs so that the first and last cells of any group are properly grounded to eliminate the possibility of stray current corrosion. Each cell is then electrically connected in series to the center pair. These anodes are connected to the positive power source. Since the cells are connected in series hydraulically, the strength of the hypochlorite solution produced increases from one cell to another.

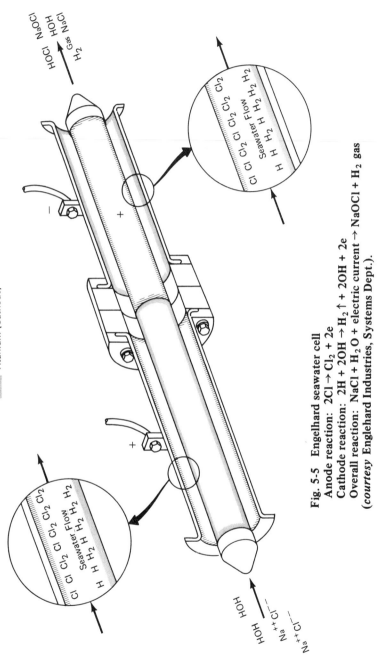

||||| Platinized Coating (Anode)

▓ Titanium (Cathode)

Fig. 5-5 Engelhard seawater cell
Anode reaction: $2Cl^- \rightarrow Cl_2 + 2e$
Cathode reaction: $2H + 2OH \rightarrow H_2 \uparrow + 2OH + 2e$
Overall reaction: $NaCl + H_2O + $ electric current $\rightarrow NaOCl + H_2$ gas
(*courtesy* Englehard Industries, Systems Dept.).

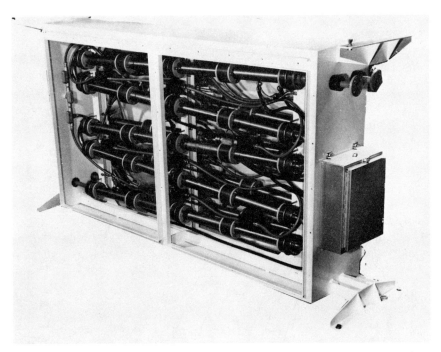

Fig. 5-6   Engelhard seawater cell system of 20 cells in series; capacity 480 lb/day available chlorine (*courtesy* Engelhard Industries, Systems Dept.).

Assuming a seawater with a total salt content of 32,000 mg/liter will contain about 19,000 mg/liter of chloride ions, each cell will generate about 100 mg/liter of hypochlorite. Therefore, each group of 10-12 cells produces a 1000-1200 mg/liter hypochlorite solution. At this concentration there is no need for product cooling or hydrogen gas venting (H_2 venting is required if a hypochlorite storage tank is used).

The electrolyte flow through any given cell is limited to 20 GPM. Therefore, assuming a seawater content of 19,000 mg/liter chloride ions, any saline water of this concentration will produce 1.0 lb Cl_2/hr/cell at the 20 GPM flow of electrolyte. Therefore, the Chloropac generator system is arranged so that the modules of hypochlorite production are hydraulically connected in parallel consisting of series connected units of 10 lb/hr capacity each. A 20 lb/hr system would consist of two groups of 10 cells in series arranged in parallel.

Chloropac Systems:

Once-through salt water system. This is the simplest system used for an unlimited supply of saline water containing from 1.5 to 4.0 percent salt and where a weak sodium hypochlorite solution is acceptable. Such a salt supply will result in hypochlorite solution strengths varying from 100 to 1000 mg/liter available chlorine. Typical users of this salt water source include: seaboard utilities; industrial condenser cooling water systems directly dependent upon a salt water source; offshore oil production facilities; desalting facilities; and sea-going vessels.

Once-through Brine System. This system is attractive if there is an abundant supply of either salt or brine, and the cost is lower than for the more efficient recycling system. Brine solution is prepared in a salt dissolver, then mixed with feedwater until it reaches the approximate salinity of sea water, 3–3.5 percent salt, which is the optimum strength for electrolytic decomposition. This solution is fed to the electrolytic generator and converted to a weak NaOCl solution as in the seawater system above. The typical user might be an inland industrial plant or waterflood facility located in an area of natural salt beds or strong brackish ground water.

Seawater Recycle System. In this system, the initial weak hypochlorite solution is discharged from the generator to a holding tank and then recycled to join the incoming brine flow. Recycling on a continuous basis gradually raises the strength and temperature of the solution. The maximum strength of solution is about 4000 mg/liter available chlorine. The rise in temperature of this process tends to speed up the decay of the hypochlorite solution, so a minimum storage time of the final product should be considered.

Seawater or Brine Recycle with Cooling of Product. Whenever recycling is involved, maximum extraction of chlorine is dependent upon the optimum temperature of the product. This requires a cooling system.

The system is designed to provide maximum economy of chlorine extraction from the brine. The brine is first diluted to nominal seawater salinity (about 19,000 mg/liter chlorides), and then it is recycled through the Chloropac generator until the hypochlorite generated reaches about 1.0 percent concentration. The amount shown in the Engelhard literature is 7520 mg/liter.[40] Figure 5-7 illustrates the operation of the brine recycle system. The entire process of brine dilution, product recycling, and product cooling is automatically controlled.

Automatic Dosage Control. The Engelhard systems are responsive to automatic control. The concentration of the hypochlorite solution can be controlled by a saturable reactor which responds to a 0–5 mV signal. This controller, which is a transformer within a transformer, can utilize a cascade system whereby a flow signal is combined through an electronic multiplier with a chlorine residual analyzer signal to produce a final control signal. This signal changes the power input which in turn changes the number of Faraday units applied to the constant

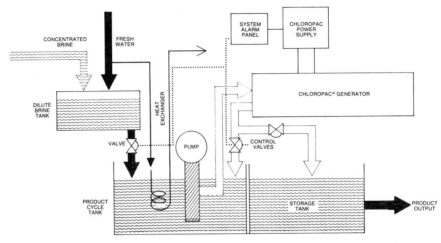

Fig. 5-7 Engelhard Brine System (*courtesy* Engelhard Industries, Systems Dept.).

flow of brine through the cells. This changes the amount of chlorine generated in accordance with the control signal.

Cost Comparisons. Engelhard claims 2.3 kWh/lb of chlorine at 70°F and 2.6 kWh/lb at 50°F, assuming a nominal seawater concentration of 19,000 mg/liter of chloride ion. The estimated salt requirement for the Engelhard brine system is 3 lb salt/lb of chlorine.

The Houston analysis by Matson and Coneway[42] compared on-site generation of a system producing an 0.8 percent hypochlorite solution and one producing an 8.0 percent solution. Both of these systems were evaluated on the basis of a concentrated brine system. Cost comparisons for individual systems are burdened with many unknown pitfalls; however, it is pertinent to note that the 8 percent hypochlorite system was by far the best buy for the dollars invested. This system demonstrated a 30 percent advantage over the 0.8 percent system for a 5-ton per day facility. This is not meant to be an indictment against the seawater systems but a warning to the designers of the problems involving system selection.

Sanilec[43] claims that using 80 percent strength seawater at 25°C will require 2 kWh to produce 1 lb of chlorine at full load production.

Much more operating experience is needed with both the seawater and brine systems to determine the chlorine production costs of these systems.

Sanilec Systems

Introduction. On site hypochlorite generating systems by this trade name have an interesting origin. The parent company of Sanilec is Diamond Shamrock, one of the leading producers of chlorine over the past 20 yr. Previously, it was known as the Diamond Alkali Co. The parent company must be credited with one of the most significant improvements in the chlor-alkali industry for many years: "the Dimensionally Stable Anode" (DSA). After acquiring the basic patents, trademarks, technologies, and laboratories of Electronor Corporation, Diamond Shamrock formed the Electrode Corp. of America and developed this new type of metallic electrode in cooperation with the DeNora interests in Milan, Italy.* Through Electrode Corp., Diamond Shamrocks has licensed over 55 percent of the North American chlorine capacity to use the energy saving DSA technology. The continuing development of chlorine production efficiency led to the development of cells for the production of chlorine from seawater. Diamond Shamrock is also responsible for the successful development of an ORP (Oxidation-Reduction Potential) control system used in the production of hypochlorite solutions.

With this background and demonstrated expertise, Diamond Shamrock launched the Sanilec systems for on-site generation of hypochlorite to capture whatever market was available on the basis of either expediency or safety.

Sanilec offers two types of systems for on-site generation. One is for the use of saline waters similar to seawater situations or various dilutions thereof, depending on local conditions. The other is the use of a concentrated brine as raw material. These systems have been described in detail by Bennett and Cinke.[41] These authors emphasize that many important facts distinguish electrolytic cells which use seawater from those which use prepared brine solutions. Therefore, Sanilec set about to design two distinctly different types of cells; one for seawater and one for pure brine.

Seawater System. These systems suffer from impurities inherent in seawater; therefore, the cell configuration must be designed accordingly. These impurities cause bulky deposits which interfere with the electrolyte flow. These deposits are a result of natural seawater hardness, which is about 1800 mg/liter, caused by calcium and magnesium in seawater containing 19,000 mg/liter chloride. The precipitates of these ions not only interfere with the hydraulics of the cell system but also act to insulate the cathode. This results in a reduction in the process efficiency. By using a turbulent flow regime through the cell Sanilec claims 3-6

*As of 1978 Oronzio de Nora is promoting their DE NORA SEACLOR system in the U.S.A. for on-site production of hypochlorite solution from seawater via bipolar flow through cells which can produce a maximum strength solution of 2500 mg/liter.

months continuous operation before cell cleaning is required. Cleaning consists of flooding the cells with a 10 percent muriatic acid solution for 1-2 hr. This therefore requires the availability of standby cells to accomplish this routine maintenance. The Sanilec seawater cell is shown in Fig. 5-8. The electrodes of the seawater cell are classed as dimensionally stable. The anode is the expanded metal type while the cathode is a thin solid sheet of metal which is a proprietary alloy.

The seawater system is shown in Fig. 5-9. This system requires filtration of the seawater before it is sent to the cells. Each cell module receives a constant flow of seawater. Hypochlorite production is controlled by current variation from the rectifier. Production control may be accomplished either manually or paced automatically by plant flow and trimmed by residual chlorine concentration analysis as shown in Fig. 5-9. The concentration of the hypochlorite solution produced from seawater is limited to a range of 200-1250 mg/liter hypochlorite as available chlorine (0.02-0.125 percent).

Hydrogen scavenging of the cathode is provided for each pair of cells. Otherwise the hydrogen formed at the cathodes would form an insulating layer which would result in an abnormal voltage rise across the electrodes.

The hypochlorite solution leaving the electrolytic cells passes to a gas release tank where the hydrogen produced in the electrolytic process is vented to the atmosphere by air dilution or is used as a by-product for its heat value.

These systems can provide hypochlorite solutions at up to 15 psi back pressure without the necessity of pumping the hypochlorite solution. The proper amount of seawater must be delivered to cells at a minimum of 40 psi pressure. Water flow varies with the size of the system. The Sanilec 1800 lb/day unit which is their Model S-1800 produces a 950 mg/liter available chlorine solution using six 300 lb/day cells and requires a seawater flow rate of 160 GPM. A 150 lb/day system requires two 75 lb/day cells and 20 GPM seawater flow.

Recycling is not used in the Sanilec seawater system. Although recycling can be used to lower cell voltage, this advantage cannot balance the accompanying inherent loss in current efficiency.

Sanilec points out one important factor in the production of hypochlorite from seawater: these solutions generated from seawater are inherently unstable. They deteriorate significantly within 48 hr because the seawater brine contains a vast array of ions which catalytically decompose the hypochlorite solution. Therefore, recycling seawater systems is not recommended by Sanilec.

The power consumption of a standard Sanilec system is a function of three variables; seawater salinity, water temperature, and percent of maximum production. A seawater salinity of 100 percent is generally defined as 18,900 mg/liter of chloride ion. In practice, seawater strength can be and is significantly diluted, especially in harbors or waterways with large inflows of surface water run-off. The Sanilec system is not designed to operate on seawater with less

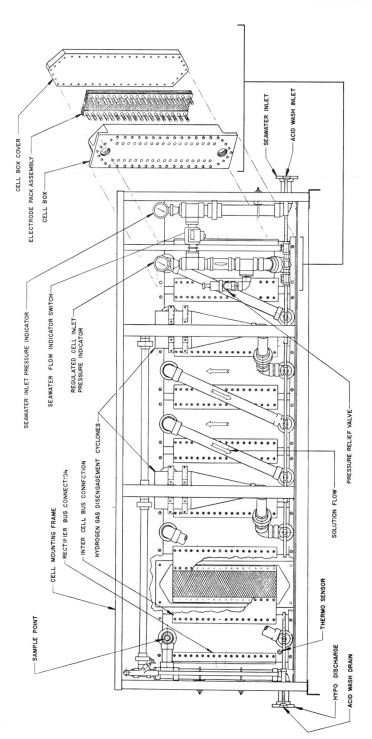

Fig. 5-8 Sanilec seawater cell (*courtesy* Diamond Shamrock, Electrode Corp.).

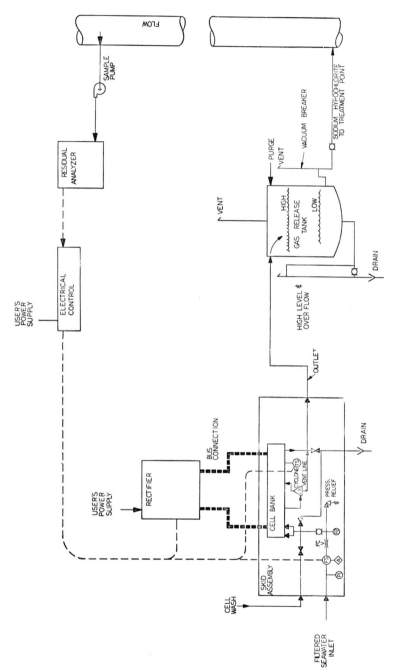

Fig. 5-9   Sanilec seawater system (*courtesy* Diamond Shamrock, Electrode Corp.).

than about 9500 mg/liter of chloride ion. The example given by Sanilec for a seawater containing 80 percent salinity (0.8 X 18,900) at a water temperature of 25°C will produce available chlorine at 2 kWh/lb chlorine at full load.

The use of seawater as cell feed contributes to a certain amount of chlorides and total dissolved solids to the plant effluent which may sometimes be a significant factor. Additionally, electrolysis of seawater will produce a certain amount of suspended solids in the form of magnesium and calcium hydroxides, and carbonates. For example, a 10 mg/liter dose of chlorine will add to the treated effluent; 189 mg/liter chlorides, 355 mg/liter TDS and 2 mg/liter SS.

Present operating experience indicates that one of the most crucial parts of any seawater system, whether it be Sanilec, Engelhard De Nora or others, is the seawater supply system. *First,* the seawater must be filtered; *second,* the pumping system and piping should be in duplicate because the intake piping is subject to marine life biofouling and has to be taken out of service for periodic cleaning; *third,* the piping system from the intake suction to the electrolyic cells must be designed to provide a flushing velocity of 4-5 ft/sec. Moreover, all of this piping should be PVC, Kynar, fiberglass, or saran or rubber lined steel pipe.

Sanilec Brine System. As in all electrolytic systems of chlorine and or hypochlorite production, prepared brine solutions rather than seawater or other brackish waters are much easier to deal with. This leads to a contrasting cell design. Electrolysis with selected grades of salts makes it practical to remove all hardness involved with cation resin exchangers. This allows lower electrolyte flow rates which results in higher concentration of hypochlorite solutions.

The Sanilec brine electrolysis system requires the use of purified salt (food grade) to be dissolved in softened water as do competitive systems. The resulting salt solution is optimum at 28 g/l NaCl. The hypochlorite solution produced by the electrolysis of this salt solution will contain about 8000 mg/liter (0.8 percent) available chlorine. The Sanilec cell used in the brine system is shown in Fig. 5-10. Owing to the design of the brine system cell and the low concentration of hypochlorite, no system cooling is required. This system produces hypochlorite directly in the cell. No molecular chlorine is evolved. Both of the electrodes are the expanded metal type. The anode is coated with a precious metal oxide. This cell is considerably different from the seawater cell.

The Sanilec purified brine system compares with other similar systems as follows: 3.5 lb of salt, 15 gal. of water, and 2.5 kWh of electrical power will produce 1 lb available chlorine.[41]

Choice of Systems

The primary consideration for an on-site generating system would seem to be availability of raw material. Availability of power is usually an accepted fact.

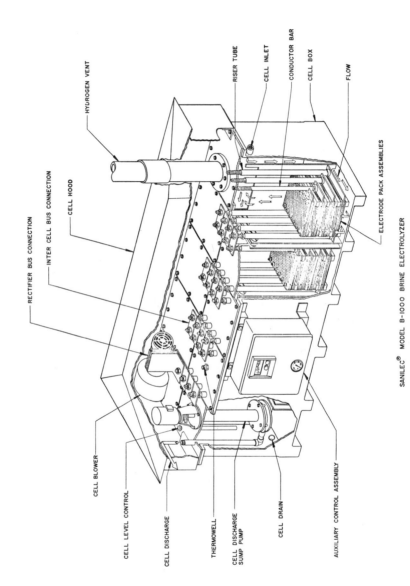

SANILEC® MODEL B-1000 BRINE ELECTROLYZER

Fig. 5-10 Sanilec brine cell (*courtesy* Diamond Shamrock, Electrode Corp.).

However, the decision between a brine system and a seawater system rests primarily with the resultant effect of TDS in the treated wastewater and the greater dilution factor when using seawater. For example, seawater should not be used in any water reuse situation. It is also questionable whether or not a seawater system should be considered for chlorine production when used for odor control in sewage collection systems. The large amount of seawater required, because of the low chlorine concentration in the hypochlorite, would undoubtedly cause an additional load on the sewage flow which in turn would promote the generation of hydrogen sulfide thereby causing more and more odors. Force mains can generate hydrogen sulfide in concentrations sufficient to require 20–30 mg/liter of chlorine for proper control. Seawater hypochlorite generators would be largely self-defeating in these cases.

ON-SITE MANUFACTURE OF HYPOCHLORITE

Introduction

This concept of a chlorination facility is entirely different in method and scope from that of on-site generation.

About 1957, the Sanitation Districts of Los Angeles county, California, pioneered a unique concept of chlorination for wastewater disinfection. This chlorination system was designed for the Joint Water Pollution Control Plant adjacent to South Figueroa St. in Carson, California. This treatment plant currently discharges about 400 mgd primary effluent through a sophisticated outfall system consisting of a special ocean outfall and dispersion piping system. The outfall tunnels under the San Pedro Hills and then discharges into the Pacific Ocean about two miles offshore from the Palos Verdes peninsula.

This hypochlorite generating and control system is unique in that it produces a stable hypochlorite solution (approximately 8000 mg/liter available chlorine) directly from liquid chlorine in tank cars without the use of intermediate vaporizing equipment. Conventional chlorination systems always require vaporization for the accurate control of chlorine feed rates.

This particular installation compelled the designers to consider a constant strength solution because the point of application of chlorine is at too great a distance from the point of generation to consider remote injectors.

This system's desirability is related in part to the incompressible characteristic of water. Therefore, regardless of chlorine solution flow variation, the impact of this change occurs instantaneously at the point of application several thousand feet away. This is accomplished by the production of a constant strength chlorine solution at the chlorine station.

System Description

The on-site manufacturing process is a simple one. It merely involves the simultaneous injection of liquid chlorine and lime slurry or caustic solution into a transmission line carrying the makeup water. The hypochlorite solution produced is about 8000 mg/liter in strength. It is normally poised at a pH of 7.5 to 8.5. Although this is an alkaline solution it is preferable to transport this solution in corrosion resistant piping such as PVC, KYNAR, or rubber lined, or saran-lined steel pipe. The secret of a successful installation of this type is the ability to achieve complete and immediate mixing of liquid chlorine to prevent evolution of any chlorine gas.

The complete system includes storage facilities for chlorine tank cars, a 175 psi air-pad system with air driers, liquid chlorine flow control system, liquid chlorine injectors, makeup water supply system, pH control system, hypochlorite solution lines, and lime or caustic facilities.

Chlorine Facilities

Chlorine Supply System. The layout of railroad siding and the design of chlorine headers, loading platforms, air padding, air drying, expansion tanks etc., is the same as for liquid-gas systems using conventional chlorination equipment described in Chapter 3 with one exception: the air pad on the tank car must be continuous at 175 psi for reasons described below.

Reserve Tank. The same applies here to the use of the reserve tank concept described in Chapter 3. This insures a continuous uninterrupted supply of liquid chlorine. Moreover, it eliminates the need for weighing devices.

It is also practical to use a resettable reverse totalizer on the readout signal provided by the liquid chlorine rotameter signal. With an accurate signal from the liquid chlorine rotameter to the flow meter totalizer, the reverse counter on the totalizer can be set to alarm at 500–700 lb left in the car. Then the operators should check with the totalizers on the flow meter to verify the amount left in the car. The alarm on the reserve tank serves as the final verification that the car is empty of liquid chlorine.

Chlorine Control System. It has been stated elsewhere in this text and in *Handbook of Chlorination* that it is impractical to measure liquid chlorine with conventional metering devices. The pressure drop, although very small, across the meter causes the liquid to flash to gas which destroys the accuracy of the meter. This is true at the pressures usually encountered in conventional practice. The flashing phenomenon can be eliminated if the liquid supply is pressurized to 175 psi. At this pressure, Los Angeles County operators have confirmed that the liquid chlorine rotameter reading is as accurate as a similar meter would be if measuring Cl_2 gas or water flow.

The chlorine control system can be designed to respond to either a plant-flow signal or a combination of flow and chlorine residual. This is the same as a conventional system. The major difference is that the hypochlorite makeup water flow rate, rather than the chlorine feed rate is the primary response to changes in plant flow and/or chlorine residual. The chlorine feed rate follows as a secondary response.

The flow rates of liquid chlorine and lime slurry (or caustic) are varied automatically to comply with changes in the makeup water flow, so as to maintain a constant concentration of hypochlorite solution going to the point of application.

Chlor-Alkali Mixing. This is the key to a successful operating system. In order to achieve a stable hypochlorite solution without the evolution of molecular chlorine, a constant back pressure of 35–40 psi must be maintained at the point of liquid chlorine injection. The alkali solution (CaO slurry or NaOH) is injected diametrically opposite to the chlorine injection. Immediately downstream from this injection point is a turbine mixer.

The high head loss from the artificially padded liquid chlorine line upon discharge from the special chlorine injection valve permits the chlorine to vaporize immediately inside the makeup waterpipe. If the pressure drop across the injection valve and the back pressure in the hypochlorite solution line are not maintained, a complete and stable solution will not be produced. The necessary back pressure can be sustained by the installation of an automatic pressure regulating valve in the solution line downstream from the mixer. If the hypochlorite solution is discharged through more than one diffuser, each diffuser header must be equipped with a pressure regulator. A single regulation in the mainline is not sufficient to keep the proper back pressure and provide uniform flow when more than one diffuser is involved.

Solution Line and Diffusers. These appurtenances should be constructed of the same materials and designed for the same hydraulic conditions as for a conventional chlorination system (See Chapter 3).

Although the alkaline hypochlorite solution is mildly corrosive, it will be necessary at prescribed intervals to purge the calcium scale deposit in the solution lines caused by the use of lime slurry. Removal of this scale deposit is easily accomplished by lowering the pH of the hypochlorite solution (temporarily) to about 4 or 5. Obviously these procedures require corrosion resistance piping, valves, and fittings.

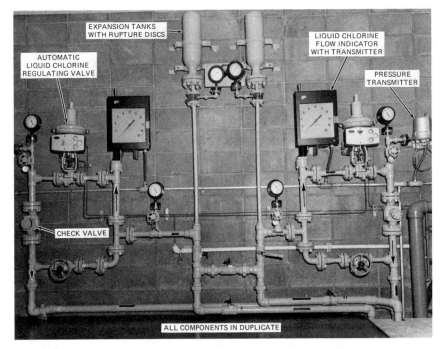

Fig. 5-11 On-site hypochlorite production systems showing metering and control system (*courtesy* County Sanitation Districts of Los Angeles County, California).

Makeup Water System. The water to make the hypochlorite solution should be chlorinated plant effluent, even if the ammonia nitrogen content is high. A small portion of chlorine injected into this water will be consumed immediately by the ammonia nitrogen in the breakpoint reaction. This reaction between Cl_2 and NH_3-N is complete in a few seconds at the pH and chlorine concentrations used in this method. For example, 350 GPM of makeup water containing 20 mg/liter NH_3-N will consume 834 lb/day chlorine in the breakpoint reaction. For an 8000 mg/liter hypochlorite solution this represents 3 percent chlorine loss with 20 mg/liter NH_3-N present in the makeup water.

A variable speed booster pump is essential for the makeup water system. Instantaneous response of chlorine dosage rates is made possible by holding a constant hypochlorite concentration in the solution line. Therefore, the rate of flow in the solution line is varied according to plant flow and residual changes. Regardless of solution line length, chlorine dosage changes at the chlorination station are reflected immediately at the point of application.

If the system is designed to produce a hypochlorite solution with a chlorine concentration of 7000–8500 mg/liter, the makeup water flow would have to be about 240–290 GPM/1000 lb/hr chlorine feed rate.

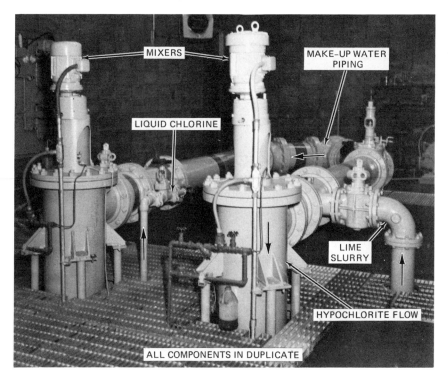

Fig. 5-12 On-site hypochlorite production system; makeup water system and chemical injection points (*courtesy* County Sanitation Districts of Los Angeles Co., California).

Essential instrumentation for this system requires that a flow meter be installed in the common discharge of the makeup water line. The flow signal from this meter is combined through a ratio station with the chlorine flow signal to provide a signal to the hypochlorite concentration flow recorder.

Alkali System. The chemical of choice is pebble lime; however, some systems designed for nitrogen removal where peaks cannot be handled by biological processes would be better suited to a caustic installation, particularly if the intermittent operation is 70–75 days or less per year. The choice is mainly one of economics, but from a chemical standpoint lime may be preferable because the calcium ion present provides a protective coating on the hypochlorite solution line.

Burned pebble lime costs about $40/ton as compared to caustic soda at about $170/ton. Even though the storage and handling system for lime is more expensive than for caustic, the cost-effectiveness is heavily in favor of lime. In the quantities required for this type of operation, quicklime is preferable to hydrated lime; however, source of supply and chemical cost would still govern the choice. Less mechanical equipment is required for the hydrated lime system, and the least mechanical equipment is required for the caustic soda system. However, the overall costs invariably favor the quicklime system.

Quicklime is readily available by truck or in railroad carload lots. It can be conveyed pneumatically to storage silos with ease and without dust or other air pollution problems. Proper quicklime (pebble) handling equipment includes the following:[36]

1. Air compressors for vacuum and pressure requirements.
2. Transfer piping.
3. Pressure-vacuum separation device (lime is removed from delivery vehicle by vacuum and delivered to storage by pressure).
4. Bulk storage facilities.
5. Filters on air intakes and discharges.
6. Monitoring, alarm, and control devices.

The pneumatic conveyor system for unloading quicklime should have interlocking safety devices to shut down the system if excessive vacuum or pressure develops. It should be further interlocked with level measuring devices in the lime storage tank to cause the conveyor to shut down when a full tank is reached. Low-level warning devices should also be installed in the storage tanks.

Feeding of bulk lime to the slaker is normally by gravity. The slaker package consists of a slaker, lime feeder, control panel, and accessory control valves. The lime feeder should be equipped with a variable speed drive motor so that it can be manually adjusted to run at any of the numerous lime feed rates. Automatic operation of the slaker-feeder combination can be achieved, by a level probe system in the lime slurry storage tank, to activate or deactivate the combination as low- or high-levels are attained. Manual adjustment of the lime feeder to the slaker is sufficient to supply slurry of the proper concentration. The slaker equipment normally includes grit removal components and the overall system must include belt or other conveying equipment to move this material to storage facilities for disposal. The bulk storage and slaking facilities can be eliminated by purchasing liquid slurry, but at greatly increased costs.

Lime slurry from the slakers flows by gravity to lime slurry storage tanks where the lime is kept in suspension by turbine mixers. Lime slurry from the storage tanks is pumped by a variable speed progressive cavity positive displacement pump into the hypochlorite pipeline for combination with chlorine. A piston or diaphragm positive displacement pump is not satisfactory for this application. The lime slurry pump, pumps lime as demanded by a pH control system, which samples the contents of the hypochlorite pipeline and signals the lime pump to pump as required to maintain a pre-set pH level required for satisfactory hypochlorite production. To maintain a slightly alkaline pH, approximately 0.80 ton of CaO are required per ton of Cl_2.

Reliability. The low initial cost of all the mechanical equipment and piping lends itself to complete duplication of the entire system. This provides a system reliability comparable to or better than most conventional systems. Power failures affect this system with equal results as a conventional system, whether it be on-site generation or imported hypochlorite.

Cost Comparison. The system described above was designed for the Sacramento County Central Plant (1974) as a nitrogen removal process of a secondary effluent to supplement the biological nitrification-denitrification system. This was considered necessary only during the cannery season when biological treatment could not handle the cannery load. The chlorination facility was designed for 110 mgd. The nitrogen removal process by breakpoint chlorination was estimated to require 70 days operation per year.

The comparison plant was designed for 60 mgd to produce a commonly nitrified secondary effluent using on-site generated chlorine for nitrogen removal.

Scaling the above two plants to a common size, the following is an economic evaluation of the on-site hypochlorite manufacturing (Los Angeles System) versus the *on-site generated* system.

This comparison is of considerable interest owing to the enormous dollar savings in capital cost of chlorination equipment, particularly where chlorine usage approaches tank car quantities.

The comparative costs are as follows:[37]

Amortized capital cost (25 yr. at 7 percent) plus operating* and maintenance costs based on 1974 prices—$/mg (annual average).

On-site hypochlorite manufacturing $37.00.
On-site hypochlorite generation $107.00.

A similar plant which would reduce the ammonia nitrogen (biologically) continuously by nitrification followed by denitrification, the following are the relative costs based on the Sacramento analyses.

	Annual	
	Cap. Cost $	Oper. Cost $
Nitrification-denitrification	5,640,000	1,250,000
Breakpoint Chlorination	32,000	1,295,000

Based on these figures, if nitrogen removal by chlorine is a requirement for a given effluent, or if the chlorine dosage requirements for disinfection approaches

---

*Costs are based on chlorine @ $144/ton and caustic soda @ $168/ton. A system using lime for caustic buffer would be even less costly.

a maximum of 10,000-12,000 lb/day, the *on-site hypochlorite manufacturing* system is of considerable interest. All of the cost figures for the Sacramento design (which was for only 70 days operation per year) were based upon the most expensive alkali system, i.e., caustic soda instead of lime. It should further be noted that this system of hypochlorite manufacture must be as described because it is not feasible without the addition of alkali.

Practical Considerations. Since only one operating system falls into this descriptive category, one may hesitate regarding its feasibility. The answer to this is straightforward and unequivocal. The first such unit was put into operation about 1958. This was a much smaller system than the prototype, however, all of the instrumental anomalies were sorted out in the first system so that the current system enjoys either continuous or intermittent operation, whichever is required, with the same flexibility that is enjoyed by conventional chlorination systems.

Chapter 5—Summary

Imported hypochlorite instead of liquid-gas chlorine is being used in large metropolitan areas to eliminate the potential hazard of shipping and storing liquid-gas chlorine.

Hypochlorite versus ton container liquid-gas chlorine is about 3 times more expensive.

High strength (10-15 percent) hypochlorite requires special attention to materials for storage tanks and piping systems.

All hypochlorite solutions are subject to deterioration with time, heat, exposure to sunlight, and the presence of iron or copper in solution.

Hypochlorite hydrolyzes in an aqueous solution to form hypochlorous acid the same as in the hydrolysis of liquid-gas chlorine.

High strength solutions tend to raise the pH of the treated water. Some investigators have contended that this enhances the germicidal efficiency of hypochlorite over the liquid-gas chlorine application. This has not been conclusively proved.

Of all the hypochlorite compounds available, the most practical is aqueous sodium hypochlorite. The granular or flake forms of hypochlorite should be shunned. They present too many metering, control, and maintenance problems.

The best choice for imported hypochlorite is 10 percent "trade" strength, unless there is a sufficient price incentive to buy 15 percent. The latter deteriorates more rapidly than the 10 percent so the user must practice strict monitoring procedures to be sure the product is in fact at 15 percent strength.

The control and application of hypochlorite as a process must follow the same principles as those for the chlorine liquid-gas systems. These are flow proportional control, plus residual control, monitoring, proper mixing, and sufficient contact time. The installations in the Chicago metropolitan area are probably the best in North America, of those using imported hypochlorite so they could serve as a model.

Delivering hypochlorite to the points of application can be by gravity, metering pumps, or an eductor system. The least desirable is the eductor method. The use of gravity eliminates the need for a pump but requires a clever choice for a wide range control valve. Metering pumps provide the most flexibility and "off the shelf" hardware is available for the most sophisticated of control systems; mainly, compound loop control.

On-site generation of hypochlorite has been practiced intermittently for a long time to accommodate areas where the chlorine or hypochlorite supply reliability is low. Increased interest in on-site generation of chlorine (hypochlorite) in the past decade has spawned a variety of systems for the user to choose from. One such system utilizes purified salt brine and an electrolytic cell comparable to those used for the commercial production of liquid-gas chlorine. As of 1978, this system is unique. It provides an 8 percent trade strength hypochlorite solution of high quality which translates to a long half-life.

Two other systems are available from highly reputable equipment manufacturers which address themselves only to the use of seawater or brackish water as a source of brine. These are commonly referred to as the open cell systems. They produce a hypochlorite of much lower concentration than the one which uses food-grade salt for the brine. These latter systems produce a hypochlorite solution which varies in strength from 200-2500 mg/liter (0.02-0.25 percent). The critical part of these brackish water systems is the quality of the brine makeup water. The poorer the quality the more severe and numerous are the operating problems.

The choice of seawater systems must be based on the effect of the high dilution ratio of seawater to the treated water flow. For example, a seawater or brackish water system would not be used in a water reuse situation.

It is almost a certainty that when seawater is used as the source of chloride to produce the on-site generated hypochlorite the bromide content (about 70 mg/liter in seawater) will be converted to bromine which will immediately hydrolyze to form hypobromous acid. This tends to make the hypochlorite solution generated from seawater a little more potent than electrolysis of pure salt. Just how significant this may be is unknown and needs verification.

On-site *manufacture* of hypochlorite solution is an attractive alternative when the chlorine consumption reaches 10,000 lb/day. If tank cars or tank trucks are available for chlorine delivery it can result in a system of much less first cost than conventional liquid-gas systems because evaporation and costly control equipment are not required. This system is particularly effective where long chlorine solution lines are involved. Lag time for controlling the rate of application is eliminated since the chlorine solution is always at a constant strength. This system involves the application of liquid chlorine and lime slurry or caustic simultaneously to the system makeup water without having to vaporize the liquid chlorine.

REFERENCES

1. White, G. C. *Handbook of Chlorination*, Van Nostrand Reinhold Co., N.Y., 1972.

2. Springs, J. D. "Hypochlorination For Slime Control." *Power* **102** (June 1957).

3. Bacon, V. W. "Chicago MSD Progress Report on Chlorination." *Water and Sewage Works* **114**: 350 (Sept. 1967).

4. Bacon, V. W. "How Chicago Saved $2.5 Million." American City 16 (Oct. 1967).

5. Dorolek, R. J. "Wastewater Plant Effluent Chlorination Made Easy and Inexpensive." *Water & Wastes Eng* **48** (Oct. 1968).

6. Lue-Hing, C., Lynam, B. T., and Zenz, D. R. "Wastewater Disinfection: A Case Against Chlorination." Paper Presented at Forum on Disinfection With Ozone, Chicago, Ill. (June 2–4, 1976).

7. Baker, R. J. "Characteristics of Chlorine Compounds." *J. WPCF*, **41**: 482 (1969).

8. "Chlorine Bleach Solutions." Solvay Div. Chem Co. Bull. 14, New York, 2nd Ed. (1960).

9. "Chlorination of Sewage With Hypochlorites." Dow Chem. Co. Form #125-1086-68 (1968).

10. Sawyer, C. N. "Hypochlorination of Sewage." *Sewage Ind. Waste* **29**, 978 (1957).

11. Fair, G. M., Morris, J. C., and Chang, S. L. "The Dynamics of Water Chlorination." *J. NEWWA* **61**: 285 (1947).

12. White, G. C. "Disinfection Practices in the San Francisco Bay Area." *J. WPCF* **46**: 89 (Jan. 1974).

13. Krusé, C. W., Olivieri, V. P., and Kawata, K. "The Enhancement of Viral Inactivation by Halogens." *Water & Sewage Works* **118**: 187 (June 1971).

14. White, G. C., Private communication. K. Fraschina and A. Bagot, City and County of San Francisco (1971).

15. Rietema, K. "Segregation in Liquid-Liquid Dispersions and its Effect on Chemical Reactions." *Chem. Eng. Sci.* 8: 103 (1958).

16. "Wallace and Tiernan 44 Series Solution Metering Pumps." Cat. File 440.100 Rev. 5–76.

16a. Straight, W. S. Personal Communication. (Oct. 12, 1976).

17. Resistoflex Corp. "Flexible and Rigid Piping Accessories." Bull. Sk-5 Roseland, N.J. (1975).

18. Barbolini, R. R. Private Communication. (Aug. 4, 1970).

19. Tech. Practice Committee, Water Pollution Control Federation MOP No. 4 "Chlorination of Wastewater." Wash. D.C. (1976).

20. Barbolini, R. R. Metropolitan Sanitary District of Greater Chicago, Personal Communication. (Nov. 4, 1976).

21. "Feedrator II Liquid Chemical Feed Dispenser." Fischer and Porter Co. Warminster, Pa., Cat. 70 FR 1000 (1977).

22. Davoust, N.J. Deputy Commissioner of Water Operations, City of Chicago, Private Communication. (Nov. 29, 1976).

23. Olin Corporation, Oak Brook, Ill., Private Communication. (Feb. 1977).

24. Knox, D. Wallace and Tiernan Div. Pennwalt Corp. Private Communication. Houston, Texas (Jan. 24, 1977).

25. Georgia Pacific Co. Private Communication. Richmond, Calif. (1977).

26. Michalek, S. A., and Leitz, F. B. "On-Site Generation of Hypochlorite." *J. WPCF* 44: 1697 (Sept. 1972).

27. Leitz, F. B. "On-Site Hypochlorite Generator for Treatment of Combined Sewer Overflows: Report No. 11023 DAA 03/72 EPA Wash., D.C. (1972).

28. Sweeney, O. R. and Baker, J. E. "An Electrolytic Apparatus for the Production of Antiseptic Sodium Hypochlorite Solution." Iowa State College Bulletin 111, Ames, Iowa (Jan. 1933).

29. Van Peursem, R. M., Pospishu, B. K., and Harris, W. D. "Antiseptic Hypochlorite by Electrolysis." Iowa State College, *J. Sci.* 4: 37 (1929).

30. Griffith, I. "The Dakin or Carrel-Dakin Solution." *Am. J. Pharm.* 89: 497 (1917).

31. "W & T Electrolytic Chlorinator Type EVC-M," Wallace and Tiernan Company, Inc., Tech. Pub. No. 201 Newark, N.J. (1941).

32. Dahl, S. A. "Chlor-Alkali Cell Features New Ion-Exchange Membrane." *Chem. Eng.* p. 60 (Aug. 18, 1975).

33. Iammartino, N. R. "New Ion-Exchange Membrane Stars in Chlor-Alkali Plant." *Chem. Eng.* p. 86 (June 21, 1976).

34. Peterson, R, BIF, Private communication. Walnut Creek, Calif. (1977).

35. Pulsafeeder Catalog No. 7120 "Pulsa 7120." Rochester, N.Y. (1976).

36. Nagel, Carl, Private communication. Sanitation Districts of Los Angeles County (1972).

37. In-house analysis by Brown & Caldwell Engrs. and G. C. White acting for Sacramento Area Consultants, County of Sacramento Central Plant (1975).

38. D'Elia, R. A. Mfg. Rep. Ionics, Inc. Private Communication. San Mateo, Calif. (1977).

39. Baur, F. "Contract Granted for Sodium Hypochlorite Generator." *Water and Sewage Works* **119**, 76 (June 1972).

40. "Engelhard Brine Chloropac®." Catalog 75.001 Union, N.J. (1977).

41. Bennett, J. E. and Cinke, J. E. "On-Site Hypochlorite Generation for Water and Wastewater Disinfection." Electrode Corp. Chardon, Ohio (1975).

42. Matson, J. V. and Coneway, C. R. "Economics of Disinfection." A Paper Presented at the IOI Forum on Ozone Disinfection. Chicago, Ill (June 2–4, 1976).

43. Seawater Data Systems Tech. Information Bulletin; E-SC-21 Diamond Shamrock-Sanilec Systems, Electrode Corp. Chardon, Ohio (1976).

6

Chlorine dioxide

INTRODUCTION

Until recently, chlorine dioxide has been considered a rather expensive chemical treatment process to be used only in special situations for the removal of tastes, odors, color, iron, and manganese in the production of potable water. However, the revelation in 1974 by the United States Envrionmental Protection Agency that potable waters disinfected with chlorine, particularly those of poor quality, may contain carcinogenic chloro-organic compounds, has stimulated interest in other disinfectants. Of the various chloro-organic compounds formed during chlorination, the one of greatest concern is chloroform. Research work sponsored by the EPA in 1975 has revealed that chlorine dioxide does not produce chloroform as does chlorine when applied at similar dosage levels.[1,51] Therefore, chlorine dioxide becomes an interesting alternative to chlorine, particularly as a replacement for prechlorination of potable water and for the disinfection of reuse water for industrial and agricultural purposes. Chlorine dioxide also possesses another desirable characteristic as a wastewater disinfectant: it does not react with ammonia nitrogen.

PHYSICAL AND CHEMICAL PROPERTIES

Chlorine dioxide is an unstable gas which is explosive at temperatures higher than $-40°C$;[2] therefore, it must be generated at the point of use. The major use of chlorine dioxide is for bleaching pulp used in the manufacture of paper. It is used rather sparingly in special situations for potable water and wastewater. Chlorine dioxide is almost never used commercially as a gas because of its explosiveness. Concentrations of the vapor over about 11 percent in air may give mild explosions or "puffs," whereas concentrations over about 4 percent in air will sustain a decomposition wave set off by an electric spark.

Chlorine dioxide explodes when its temperature is being raised, when it is being exposed to light, or when allowed to come into contact with organic substances. This tendency to explode at the slightest change in environment is even greater when it is in the compressed liquid state. It has been observed that the mere transfer from one container to another can cause an explosion. Furthermore, these explosions are of the same magnitude as those of hydrogen-oxygen mixtures.[3] Dilution of ClO_2 gas with inert gases reduces the explosion hazard. Industrially it is handled by mixing with air so that the chlorine dioxide content is between 8 and 12 percent.[4]

Exposure of the gas to light results in photochemical decomposition.[5] At first exposure to light, a large quantity of red liquid forms on the walls of the containing vessel. Continued exposure renders the liquid colorless. The products of this type of decomposition are chlorine heptoxide (Cl_2O_7), chlorine monoxide (ClO), chlorine (Cl_2), and possibly oxygen.

In potable water and wastewater treatment processes, chlorine dioxide is handled only as an aqueous solution. In the major industrial uses for bleaching pulp and textiles, a mixture of chlorine dioxide (produced on site) and air is put into an aqueous solution in an absorption tower that will yield a chlorine dioxide solution strength of 6-10 mg/liter.

Chlorine dioxide vapor has a deeper shade of green than chlorine. The odor, while somewhat similar to that of chlorine and chlorine monoxide is detectably different to experienced personnel. It is more irritating and toxic than chlorine. The odor is evident at 14-17 ppm in air, and at 45 ppm it is irritating. A gas at ordinary temperatures, it can be compressed to a liquid. It has a density of 2.4 (air = 1), a boiling point of $11°C$, and a melting point of $-59°C$.[4] Exposure to chlorine dioxide gas has, in addition to the usual toxic effects of chlorine gas, an effect of producing violent headaches and general fatigue lasting for several days.

The water solubility of ClO_2 depends on temperature and pressure. At room temperature ($20°C$) the gas is soluble in water to the extent of 4.0 gm/liter at 40 mmHg partial pressure.[2] This compares to chlorine at 7.0 gm/liter for the same conditions. In chilled water the solubility of ClO_2 increases to more than 7 gm/liter and at 80 mmHg partial pressure, the solubility increases to 6.2 and 15 gm/liter respectively.[44]

Although chlorine dioxide is soluble in water it does not react chemically with water as does chlorine. For example, it is easily expelled from solution in water by blowing a small amount of air through the solution. This phenomenon together with the difference in solubility of ClO_2 and Cl_2 at certain temperatures is the basis upon which these two gases can be separated in some manufacturing processes. For this reason ClO_2 is not stable while in solution in an open vessel. Moreover, the strength of solution deteriorates rapidly under these circumstances. This inherent instability of chlorine dioxide solutions requires that solution lines be designed so *that there is no possibility of ClO_2 gas coming out of solution.*

It is significant to note that as the chlorine dioxide leaves the solution, the yellowish green color of the solution fades and finally becomes colorless. The remaining solution will contain negligible amounts of chloric acid ($HClO_3$), indicating a very weak acid reaction with water. Aqueous solutions of ClO_2 are also subject to some photodecomposition. This reaction is a function of both time and intensity of the ultraviolet light source, and the products are chloric and hydrochloric acid.[6] The rate of photodecomposition from an ultraviolet light source is low as compared to decomposition from its volatility, described above.

Aqueous solutions of ClO_2 will retain their strength for several months if properly stored in the dark.

The vapor pressure of ClO_2 is sufficient to be troublesome. Even the weakest of solutions gives off an objectionable odor. For example, a 1 mg/liter solution will provide a vapor pressure of 10 mg/liter in the air above the solution at equilibrium. A concentration this low gives an objectionable odor.[2]

Chlorine in dilute aqueous solutions has virtually no vapor pressure nor odor since it reacts with the water to form chloride ion, hypochlorous acid, and hypochlorite ion. In strong acid solutions, however, chlorine dioxide is much more soluble than is chlorine.[34]

The high volatility of chlorine dioxide in an aqueous solution gives it a characteristic which is most desirable in the field of odor control of foul air. This is often associated with wastewater treatment and certain odoriferous industrial processes which contribute to air pollution.

METHODS OF GENERATING CHLORINE DIOXIDE

Introduction

The methods used for generating on-site ClO_2 are sharply divided into the classifications large and small. For large production units such as the bleaching of paper pulp and textiles, ClO_2 is generated from sodium chlorate ($NaClO_3$). For small production units such as potable water, wastewater or industrial processes, ClO_2 is generated from sodium chlorite ($NaClO_2$). Since sodium chlorite is commercially produced by the reaction of ClO_2 with H_2O_2 in an NaOH solution, it becomes apparent that the sodium chlorate route is the most economical for large-scale operations.

For large-scale production, four different processes are currently used in North America. They are generally named for the company which developed them or the reducing agent used. They are as follows:

1. The Mathieson or SO_2 process.
2. The Solvay or methanol process.
3. The Hooker R-2 process.
4. The Hooker SVP® process.

Most of the chlorine dioxide produced in North America is by the first two processes.[2] Increasing tonnage requirements call for changes in the process which will reduce capital and operating costs. To this end the chloride reduction processes (Olin chloride reduction, Hooker R-2 and Electric Reduction Co. ER-2) are becoming a major factor.[2]

The Mathieson or Sulfur Dioxide Reduction Process

This process consists of blending a 45 percent solution of sodium chlorate with 66°Bé sulfuric acid in the top of a reaction vessel.[8] Air containing 10 percent SO_2 is blown into a diffuser at the bottom of this vessel and chlorine dioxide plus air is extracted at the top of the vessel.

The basic reaction is:

$$2\,NaClO_3 + H_2SO_4 + SO_2 \longrightarrow 2\,ClO_2 + H_2SO_4 + Na_2SO_4 \qquad (6\text{-}1)$$

Side reactions also take place and include:

$$2\,NaClO_3 + 5\,SO_2 + 4\,H_2O \longrightarrow Cl_2 + 3H_2SO_4 + Na_2SO_4 \qquad (6\text{-}2)$$

Figure 6-1 illustrates the Mathieson process. This process produces chlorine dioxide in an air or nitrogen mixture and is not contaminated by the presence of chlorine. The exit gases flow through a scrubber (against a portion of the chlorate feed) to remove any sulfur dioxide leaving the generator.

The Solvay or Methanol Process

Methanol is used as the reducing agent in this process. The overall reaction is written as follows:

$$2NaClO_3 + CH_3OH + H_2SO_4 \longrightarrow 2ClO_2 + HCHO + Na_2SO_4 + 2H_2O$$

$$(6\text{-}3)$$

THE MATHIESON PROCESS

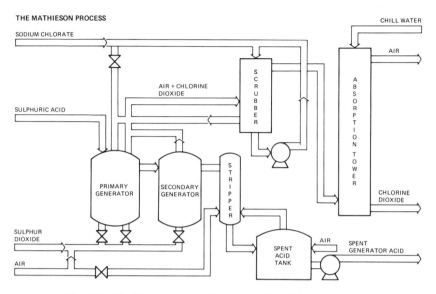

Fig. 6-1   Mathieson process for generating chlorine dioxide.

an intermediate reaction is:

$$2NaClO_3 + CH_3OH + 2H_2SO_4 \longrightarrow ClO_2 + \tfrac{1}{2}Cl_2 + CO_2 + 2NaHSO_4 + 3H_2O$$

(6-4)

The reaction efficiency may be determined by analyzing the exit gas or the chlorine dioxide solution for relative quantities of ClO_2 and Cl_2. Since Cl_2 should not be present as a reaction product, its level is an indirect measure of the reaction efficiency. This procedure may not be applied to the Mathieson system, however, since any SO_2 escaping will destroy the Cl_2 and give erroneous results.

The operation of the Solvay process is quite similar to the Mathieson process, except that the methanol is added to the incoming stream of the 45 percent sodium chlorate solution immediately ahead of the generator. In this process methanol and 75 percent H_2SO_4 are fed to the secondary generator. This drives the reaction to completion. As a result the acidity is increased. The ClO_2 is recovered by the same technique used in the Mathieson process. The reaction rates in the Solvay process are considerably slower than in the Mathieson system, and therefore high operating temperatures are required. The primary generator is run at about 60°C and the secondary at 63°C.[9]

Chloride Reduction

This process consists of blending 66°Bé sulfuric acid with a 20 percent sodium chlorate solution and a 10 percent sodium chloride solution in a reaction vessel.[8]

Air is blown into the bottom of the vessel driving off the gases formed which are diluted by the incoming air. The gas is composed of 1 part chlorine dioxide, $\frac{1}{2}$ part chlorine, and air. This goes to an absorption tower with water at 42°F which absorbs the ClO_2 and produces a solution of about 8 gm/liter. The chlorine gas is extracted from the top of the absorption tower and is sent to another absorber. The basic reaction of this process is:

$$2NaClO_3 + 2NaCl + 2H_2SO_4 \longrightarrow 2ClO_2 + Cl_2 + 2Na_2SO_4 + 2H_2O$$

$$(6\text{-}5)$$

Jaszka-CIP Process

There are additional routes to chlorine dioxide generation which are tailored to substantially lower production rates lower than those described above. The Canadian International Paper Process has fallen into disuse because of its somewhat lower efficiency.[44] However, for substantially lower production rates this process is an attractive alternative. In this process the sulfuric acid required is generated on site by the overreduction of sodium chlorate.

The reaction is:

$$2NaClO_3 + SO_2 \longrightarrow 2ClO_2 + Na_2SO_4 \qquad (6\text{-}6)$$

In this process, SO_2 is introduced into the base of a packed column and flows countercurrent to a stream of concentrated sodium chlorate solution. The flow of chlorate solution is relatively low in contrast to the volume of gas; therefore, the process may be started and stopped quickly.

Jaszka of Hooker (U.S. Patent No. 3,950,500; 1976) made a significant improvement in this process by introducing chlorine with the sulfur dioxide in the packed tower. This reduces the chlorate requirement to oxidize the sulfur dioxide to sulfuric acid. Therefore, increased efficiencies are obtained.

It is of interest to note that the Societe Universelle de Prodicts Chimiques et d'Appareiliages of Paris has described a chlorine dioxide generator based on the sodium chloride reduction of sodium chlorate in sulfuric acid solution.[9] This unit has a capacity of up to 1000 lb/day ClO_2 which makes it very interesting for wastewater or potable water applications.

Laboratory Preparation

Chlorine dioxide is best prepared in the laboratory by passing a mixture of 4-6 percent chlorine in air through tubes packed with sodium chlorite (U.S. Patent 2,309,457; Jan. 1943).[2] The exit gas may be tested to see if chlorine is coming through by the use of ammonia fumes similar to testing for chlorine apparatus leaks. As one tube is exhausted, the flow is switched to another tube, then through a fresh tube replacing the spent tube. The reaction proceeds as follows:

$$Cl_2 + 2NaClO_2 \longrightarrow 2NaCl + 2ClO_2 \qquad (6\text{-}7)$$

The ClO_2 gas may be dried and then condensed in tubes set in fluorotrichloromethane which is cooled with dry ice (solid carbon dioxide). Cooling solutions of methanol, acetone, or other flammable liquids cooled with dry ice should not be used. The preparation of ClO_2 should be performed in a subdued light, preferably a faint red light. If solutions in fluorocarbons or in carbon tetrachloride are used they should be kept in the dark.

For additional information on laboratory preparation of chlorine dioxide, consult the work of Granstrom and Lee,[10] Bernarde et al.,[11] and Hood.[54]

Methods of Generating Chlorine Dioxide for Wastewater and Water Reuse Systems

Olin Chlorine–Chlorite System.   All of the commercial methods previously described are for productions of large quantities of ClO_2; on the order of 100–500 tons/day. It would not be available for use in water-treatment processes if it were not for the availability of solid sodium chlorite ($NaClO_2$). The Mathieson (now Olin) Chemical Corporation first made this compound available in 1940; it was first used in water treatment in 1944.[14] The chlorine-chlorite process reaction is as follows:

$$2NaClO_2 + Cl_2 \longrightarrow 2\,ClO_2 + 2NaCl \qquad (6\text{-}8)$$

Thus, 1.34 lb pure sodium chlorite ($NaClO_2$) will react with 0.5 lb chlorine to produce 1.0 lb chlorine dioxide. The technical sodium chlorite used in this process is only about 80 percent pure, so that in practice 1.68 lb chlorite will be required for the above reaction, which is usually performed at the point of use. This is done by merging the chlorine solution discharge of a solution feed-type gas chlorinator with a solution of sodium chlorite from a diaphragm pump. Both chemical solutions are fed to the base of a glass reaction tower filled with porcelain (Raschig) rings. As this mixture flows upward, with a contact time of approximately 1 min., chlorine dioxide forms. When it is forming properly, the mixture in the reaction tower will turn a greenish yellow. Absence of color indicates that no chlorine dioxide is forming. The solution strength of the injector discharge from the chlorinator should be not less than 500 mg/liter. The maximum concentration possible for a conventional chlorinator, 500 lb/day capacity, is about 3000 mg/liter. Under ideal conditions this would provide a metering range of 6:1. Wallace and Tiernan (Division of Pennwalt Corp.) who pioneered the on-site generation of ClO_2 with conventional chlorination equipment have always recommended that there be an excess of chlorine to increase the rate of reaction and to insure the complete activation of the chlorite to give the highest possible yield of ClO_2.[10,13] At molar ratios of less than 1:2 of chlorine to chlorite, the activation will not be complete, and some chlorite will show as an end product. Since chlorite is expensive (75-80¢/lb), this is an undesirable situation. Therefore, two conditions must be met for the maximum ClO_2 yield in the relatively short reaction time available. 1) Proper chlorine-chlorite ratio (preferably 1:1 by weight), 2) proper pH (preferably 4). This means an HOCl concentration of not less than 500 mg/liter in most waters. The surface area provided by the porcelain rings and good mixing in the reaction tower also play important parts in the overall efficiency of the reaction. This method of generation will produce an aqueous solution which consists of about 70 percent HOCl and 30 percent ClO_2. Figure 6-2 illustrates how this method can be used to provide automatic flow-proportional generation of chlorine dioxide using conventional equipment. This system can also be arranged using all electronic instrumentation. Unless conditions are within the limits described above, all installations operating on a flow-proportional basis should be provided with acid injection directly upstream from the chlorinator injector in order to maintain optimum pH ($<$4) for the reaction.

Olin Water Services of Kansas City provides a special service for promoting the sale of their sodium chlorite product. They provide either a complete generating unit as is shown in Fig. 6-3 or just the reacting tower and sodium chlorite pump on the proviso that the customer purchase the sodium chlorite from Olin. In addition to the complete unit shown in Fig. 6-3, which utilizes liquid-gas chlorine, their Model 350 Dioxolin is complete for generating ClO_2 from hypochlorite. This system is equipped with 3 metering pumps, one for hypochlorite, one for sodium chlorite, and one for sulfuric acid to acidify the two reacting solutions to the optimum pH. It is rated at 192 lb/day capacity in terms of ClO_2 production. This method is described in detail on p. 599 *Handbook of Chlorination.*[14]

The Olin system provides the customer with a 25 percent solution of sodium chlorite which has been cut back from the hot (85°F) 50 percent solution. This is delivered in 55 gal. drums and eliminates the handling problems of the solid sodium chlorite and the hot solution. The Olin system is also provided with an optional acid injection point to cover situations where the existing or proposed chlorine gas injection system is liable not to be able to meet the conditions of optimum pH. Owing to the short reaction time and limited HOCl concentration in the chlorinator injector discharge, the best ClO_2 yield of this method is on the order of one part ClO_2 to 2.5 parts of chlorine. Since this system does not put out a "pure" solution of chlorine dioxide, proof of the results of the process are greatly complicated by the difficulty of isolating the true chlorine dioxide residual from the various species of chlorine residuals forming as a result of the predominant species (HOCl) in the solution discharging from the ClO_2 generating tower.

There are limitations to the Olin once-through generating system. It is not par-

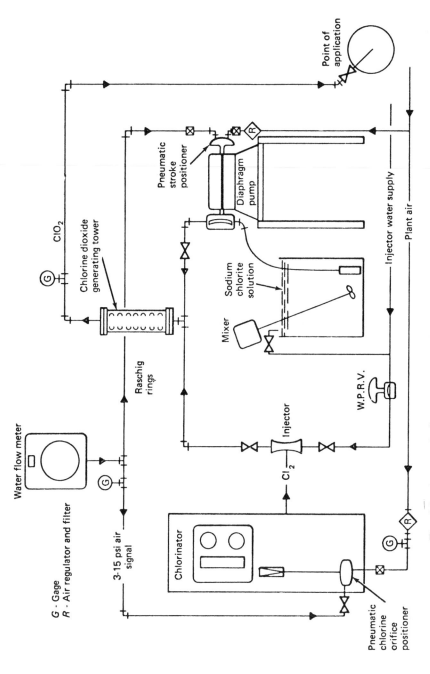

Fig. 6-2 Automatic flow-proportional chlorine dioxide once-through system.

Fig. 6-3 Olin water services model 150 Dioxolin chlorine dioxide generator. (*Courtesy* Olin Water Services Co.)

ticularly adaptable to potable water or wastewater-treatment processes where the chemical demand varies over a wide diurnal range (on the order of 6–8 to 1, based on both chemical demand and flow change). It is most suited to situations of constant demand such as cooling tower circuits and food-plant process waters.

CIFEC Method (Paris, France). This is also a chlorine-chlorite system similar to the Olin method, except that it uses an enrichment loop in the chlorine system and a longer reaction time in the ClO_2 reaction tower. This system closely approaches a 100 percent stoichiometric reaction.[15] Based on many analyses, this system can reliably produce 95-98 percent pure chlorine dioxide solution.[16] Recent analytical investigations have shown that the purity is more on the order of 98-99 percent.[17] Figure 6-4 illustrates the completely automatic CIFEC generator. This system was designed primarily for varying flow and chemical demand situations which characterize potable water and wastewater treatment. The unit illustrated is capable of maintaining a pure chlorine dioxide solution under the severest conditions of both chlorine dioxide demand and flow variation. Referring to Fig. 6-4 the recirculating pump powers the chlorinator injector and delivers the solution to the point of application. The continuous recirculation of the injector discharge solution maintains an HOCl concentration between 4000 and 5000 mg/liter at a pH usually less than 2. This insures the purity of the final chlorine dioxide solution. The size of the reacting vessel varies with the capacity of the system. Currently, these systems are available in capacities of 1-10 lb/day up to 50-1000 lb/day ClO_2.[16] These units are equipped with alarms to warn the operator of any malfunction and to automatically shut the system down in case of loss of chlorine, chlorite, or makeup water.

This equipment is adaptable to flow pacing plus residual control. Any 4-20 ma analog signal from both a flow meter with ratio station and a chlorine residual analyzer fed into a multiplier can be properly characterized through controllers to regulate the three variables chlorine, sodium chlorite, and makeup water.

The manufacturers of this system claim to have installed in excess of 50 of these units in 1976[11] in Western Europe.

Owing to the rigorous requirements of the recirculating pump and the sensitivity of the chlorine dioxide solution to vapor pressure conditions, this system is limited to applications where the pressure at the point of injection is not more than 6-8 psi.[16] Moreover, the solution line between the generating vessel and the point of application must be hydraulically designed so that there is no appreciable release of gaseous ClO_2. Also to utilize the advantage of the purity of solution, the solution lines should include short pieces of glass pipe as inspection windows for the operator to visually confirm the formation of ClO_2 solution by its characteristic color.

Acid-Chlorite Method

Combining sodium chlorite with an acid has long been a popular method for generating pure chlorine dioxide gas in the laboratory for bench scale experiments. Recently, however, this method has been used in Italy and Switzerland for the on-site generation of a chlorine dioxide solution. The system installed at the Lengg water treatment plant, Zurich, Switzerland (put into operation 1974) is described by Valenta and Gahler.[52] At two other water treatment plants in Zurich the chlorine-chlorite method is being used.

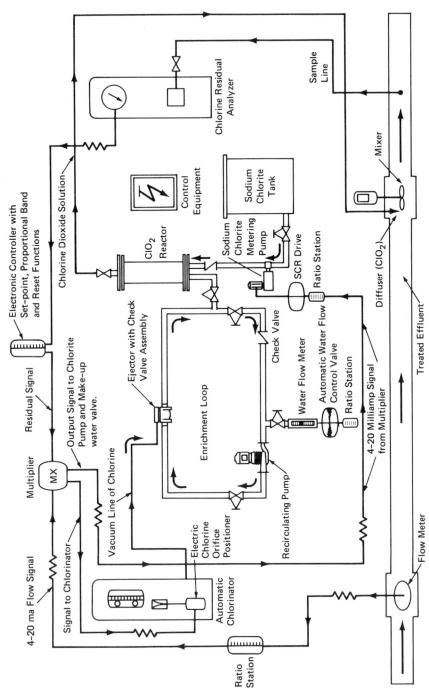

Fig. 6-4 CIFEC automatic residual control chlorine dioxide system. (*Courtesy* CIFEC, Paris France.)

This method is based upon the stoichiometric reaction described by the following equation:

$$5NaClO_2 + 4HCl \longrightarrow 4ClO_2 + 5NaCl + 2H_2O \qquad (6\text{-}8a)$$

Comparing this reaction to the chlorine-chlorite reaction (Eqs. 6-7 and 6-8) the acid process requires about 25 percent more sodium chlorite than the chlorine-chlorite method to produce the same amount of chlorine dioxide.

The advantages claimed for this method over the chlorine-chlorite methods are: 1) no danger of the simultaneous formation of free chlorine in significant amounts; and 2) it eliminates the hazard of transporting and storing liquid-gas chlorine. At this early date and with meager field operating experience the advantages and disadvantages of this system cannot be assessed; however, it does have significant possibilities as a reliable method for on-site generation of chlorine dioxide.

Fischer and Porter Company is currently (1978) marketing in Europe an on-site generator based upon this method. It is designated as Series T70G1000 and is limited as of now to about 300 lb/day chlorine dioxide. An adaptation of this unit for on-site generation is illustrated in Fig. 6-4a. The Fischer and Porter method[53] differs significantly from the Zurich method[52] in two respects: 1) it is a continuous process as compared to the Swiss process which makes the chlorine dioxide solution on a batch basis; and 2) it uses an injector to provide the time-tested procedure of vacuum-operated solution feed devices for metering highly volatile chlorine solutions, whereas the Swiss method pumps the concentrated chlorine dioxide solution. Since chlorine dioxide, while highly soluble in water, is easily stripped from solution by aeration or negative hydraulic gradients, any system which operates on the vacuum system for transporting the chlorine dioxide solution through the metering system is to be preferred.

Fischer and Porter obtained reaction yields of 94 percent in a recent shakedown of their 70G1000 reactor.[53] However, they claim there is no free chlorine in the generated solution.

Conversely Valenta and Gahler[52] in their shakedown of the system placed into operation at Lengg make claims similar to Fischer and Porter, i.e., chlorine dioxide yield varies from 91-96 percent. However, they list free chlorine production of 4.0-7.5 percent. This is the basis of the controversy about methods which has inspired the Fischer and Porter research and development arm of the company to investigate the most reliable and practical methods for quantifying as well as qualifying chlorine dioxide measurements.[53]

Figure 6-4a is a schematic of the Fischer and Porter acid-chlorite method described by their Series 70 G1000 ClO_2 generator. It is not, however, an exact replica of their system because of certain proprietary considerations.

It is neither possible nor prudent to conclude which of the systems currently available is the most practical for the on-site generation of chlorine dioxide.

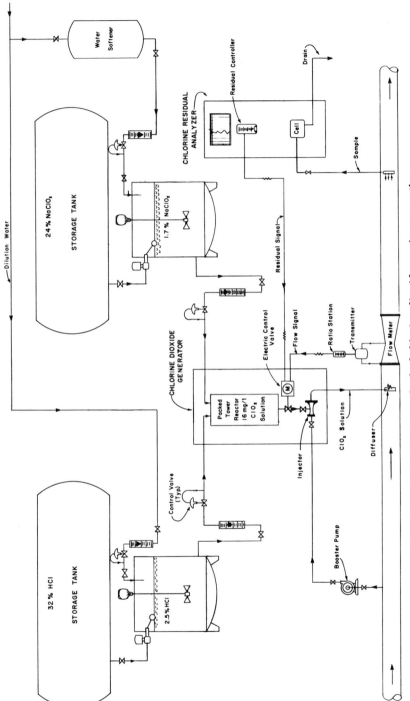

Fig. 6-4a Acid-chlorite method of chlorine dioxide on-site generation.

Comparisons of cost and availability of chlorine versus hydrochloric acid combined with considerations of safety in chemical handling require examinations for each prospective installation.

The system having the most field experience and widest selection of chlorine dioxide capacity is the CIFEC system.* It has also been adapted to compound loop control. This is not to say that other systems will not be able to provide the same performance.

Other methods and/or modifications of those described above are bound to develop so long as the interest in on-site generation of chlorine dioxide continues at its current pace. Whatever these methods will be, that with the brightest future will be the one capable of producing a near-perfect solution of chlorine dioxide with little or no free chlorine.

CHEMISTRY OF CHLORINE DIOXIDE

Introduction

The chemistry of chlorine dioxide in water treatment is not completely understood. In many instances it has been thought of as a more powerful oxidant than chlorine, but this is not so as it applies to the narrow pH ranges encountered in water or wastewater-treatment practices. Theoretically, chlorine dioxide has about 2.5 times the oxidizing power of chlorine. However, this oxidizing capacity is not all used in water or wastewater practice because most of its reactions with substances in water only cause the chlorine dioxide reduction to chlorite:

$$ClO_2 + e^- = ClO_2^- \qquad (6-9)$$

The only time the full oxidation potential is realized by chlorine dioxide is when it reacts with phenols. Then the chlorine atom undergoes five valence changes in reduction to the chloride ion:

$$ClO_2 + 5e^- = Cl^- + 20 = \qquad (6-10)$$

There may be compounds in wastewater other than phenols that utilize this same oxidation potential.

End Products

The chlorine dioxide treatment of natural waters or wastewaters will probably react to form end products which would include chlorite (ClO_2^-), chlorate (ClO_3^-) and chloride (Cl^-). The possibility of a chlorite ion residual of significant proportion is of interest because its toxicity has been questioned on the

---

*CIFEC systems range in capacity from 10 to 1000 lb/day chlorine dioxide.

basis of its reaction with hemoglobin to form methemoglobin.[18,19] This is not a desirable circumstance. It is also capable of red blood cell rupture. The toxicity of chlorate formation has also been questioned. Laboratory studies have demonstrated its formation of methemoglobin at concentrations sufficient to cause death.[20] It, too, is capable of red blood cell rupture.

A recent study by Miltner suggests that when generating chlorine dioxide at pH 2.55 using a molar ratio of 1:2 chlorine to chlorite, four chloro-species were determined (Palin's DPD procedure) in the following amounts:[18]

Initial chlorine-146 mg/liter.
Initial chlorite-556 mg/liter.
Final ClO_2 formed-397 mg/liter (71 percent yield).
Final chlorate formed (ClO_3^-) = 226 mg/liter.

The results are generally in agreement with Granstrom and Lee.[8] Both chlorine and chlorite are likely to be present in a solution generating ClO_2 under these conditions.

Montiel of Paris has reported the following analysis as a typical situation found in current potable water treatment practice.[21] Chlorine dioxide dosages from 1.3-4.0 mg/liter after 3 days at 25°C show the following species:

50 percent ClO_2 (Chlorine dioxide)
25 percent ClO_2^- (Chlorite)
9 percent ClO_3 (Chlorate)
14 percent Cl^- (Chloride)

The remaining 2 percent is probably due to analytical errors or an imprecision of the procedure. Chlorine dioxide dosage in the Paris water supply of 2 mg/liter will generally show chlorine dioxide residuals of 0.25 mg/liter and chlorite of 0.125 mg/liter at the end of 48 hr.[17] This has been standard practice for many years. It also shows the superior persisting residual power of chlorine dioxide. These findings are based upon their use of the CIFEC method of chlorine dioxide generation. It could be postulated that if the chlorine dioxide solution generated is pure, the formation of chlorate is liable to be at a minimum and chloride at a maximum.

Important Reactions with Water Systems

Chlorine dioxide as it relates to potable water and wastewater treatment has the following significant properties.

1. Although extremely soluble in water it does not react chemically with water, and it does not dissociate nor disproportionate as do chlorine solutions.[22,23]

2. ClO_2 reacts rapidly with oxidizable material without reacting with the ammonia nitrogen present.

3. Between pH 5 and 9 an average of 5.2 parts by weight chlorine dioxide oxidizes one part by weight of the sulfide ion instantaneously, and in one step to the sulfate ion, thereby avoiding the formation of colloidal sulfur.[24]

4. Under ordinary wastewater conditions it does not react with the following classes of compounds: alkanes, alkenes, alkynes, alcohols, glycols, diols, aldehydes, ketones, ethers, acids, primary aliphatic amines, and unsubstituted aromatics.[24]

5. When used as a pretreatment chemical in potable water supplies it does not contribute to the formation of chloroform (a known carcinogen).[1]

6. It displays a "chlorine dioxide demand" as does chlorine, but because of its potentially greater oxidizing power it displays a proportionally higher demand than chlorine in the same environment.

7. The predominate reaction in water or wastewater treatment is the reduction of chlorine dioxide to the chlorite ion. Since the chlorite ion may be toxic to man or aquatic life, it is of considerable interest that chlorine dioxide followed by ozone application may prevent the formation of the chlorite ion.[25] However, the compatibility of ozone and chlorine dioxide is not known.

8. It is specific for the destruction of phenols.

9. It will display some bleaching effect upon the organic color in water or wastewater.

For additional details on the chemistry of chlorine dioxide see pages 600–602 Ref. 14.

Dechlorination

Chlorine dioxide reacts stoichiometrically with both sulfur dioxide and sulfite solutions similarly as does chlorine.

The overall reactions can be expressed as follows:

Sulfur dioxide solution:

$$SO_2 + H_2O \longrightarrow H_2SO_3 \tag{6-11}$$

$$5H_2SO_3 + 2ClO_2 + H_2O \longrightarrow 5H_2SO_4 + 2HCl \tag{6-12}$$

Similarly with sulfite compounds:

$$5Na_2SO_3 + 2ClO_2 + H_2O \longrightarrow 5Na_2SO_4 + 2HCl \tag{6-13}$$

From the above equations it can be seen that it will require *2.5 mg/liter SO$_2$ for each mg/liter of chlorine dioxide* residual (expressed as ClO_2) to dechlorinate a sample to zero residual. This is the stoichiometric relationship. In practice it would be well to design on the basis of 2.7 to 1. It is expected, however, that chlorine dioxide residuals would be much lower than those currently used in chlorination practice.

Some proponents for chlorine dioxide as an alternative to chlorine claim that dechlorination will probably not be necessary owing to the rapid die away of chlorine dioxide residuals in wastewater. There is no evidence to substantiate this claim as of June 1978.

Chloro-Organic Compounds

Very little is known about the chloro-organic residues of chlorine dioxide water treatment processes, whether they be potable water or wastewater. Stevens et al. reported in 1976, that chlorine dioxide plus aldehydes might enter into a reaction which would produce a chlorite, which in turn might form compounds which would regenerate chlorine dioxide.[26] While trihalomethanes are not formed by the reaction of ClO_2 with organics in water, the reaction products might be predominantly aldehydes, carboxylic acids, ketones, and quinones. Chlorine dioxide could possibly form epoxides under certain conditions. These are known carcinogens. Further studies on this subject are underway (1977) at the Water Supply Research Division, EPA, Cincinnati, Ohio.

GERMICIDAL EFFICIENCY

In the years since the introduction of chlorine dioxide (1944) for the treatment of water supplies, several investigations have been made to compare the disinfecting power of chlorine dioxide with that of chlorine. These studies have resolved very little since the investigators did not have adequate analytical techniques to differentiate between ClO_2 and the various chlorine residuals available. Furthermore, the methods used for the laboratory preparation of chlorine dioxide most probably introduced interfering substances which would contribute to decomposition of chlorine dioxide, but would also yield enormously high values on iodometric analysis. If this is true, the chlorine dioxide concentrations used in these studies were probably lower than the reported values, and in these cases the efficiency of chlorine dioxide as a germicide would suffer accordingly when compared with chlorine.

It was not until 1965 that a critical bactericidal comparison of ClO_2 and chlorine was made by Bernarde, Israel, Olivieri, and Granstrom.[11] This work was made possible by the availability of proper analytical techniques and the detailed physiochemical findings of Granstrom and Lee[10] in 1958, and Granstrom et al. (unpublished data), also in 1958.

Bernarde et al. used an improved method of generating ClO_2 of much greater purity than that produced by the methods of earlier investigators. For quantitative analysis of stock solutions, dosages, and residuals, they used the most sophisticated device known, a spectrophotometer. Their data are not only the most reliable available but the most specific so far on this subject. These studies attempted to critically compare the bactericidal efficiency of chlorine and that of chlorine dioxide when applied to a sterile, unchlorinated sewage effluent to which a known cell density of *E. coli* had been added. The effluent was poised at pH 8.5 and had a BOD of 160 mg/liter. It is significant to note that a 5 mg/liter dose of chlorine gave a 90 percent kill after 5 min., compared to 0.9 mg/liter of chlorine dioxide. This rate of kill is of major significance when considering chlorine dioxide for wastewater disinfection.

It was found that at pH 4.0, 6.45, and 8.42 the chlorine dioxide molecule remained unaltered and intact, and that it must therefore be the bactericidal compound. This confirms the hypothesis that chlorine dioxide does not react with water to hydrolyze, as does chlorine.

Another important achievement of this investigation was establishing the disinfecting ability of ClO_2 and its relative efficiency as a function of pH. Figure 6-5 illustrates the relative efficiency of both ClO_2 and chlorine with respect to pH. At pH 6.5, chlorine appears to be more efficient at the lower dosages. Both compounds are equally efficient at an initial dose of 0.75 mg/liter. Increasing the pH to 8.5 shows a dramatic change in efficiency. As would be expected, the chlorine efficiency drops, because at this pH level only 8.72 percent of the residual is HOCl while at pH 6.5 the residual is 89.2 percent HOCl. As the chlorine

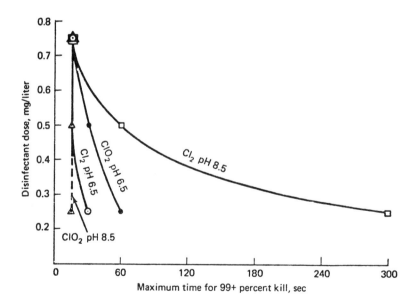

Fig. 6-5   Relative germicidal efficiency of chlorine and chlorine dioxide.[9] (Dosage in mg/liter vs. time required for a 99+ percent destruction of *E. coli* at the pH indicated.)

efficiency drops with the increase in pH, the chlorine dioxide efficiency increases. A 99+ percent destruction of *E. coli* in 15 sec with a 0.25 mg/liter dose of ClO_2 is noted at pH 8.5, as compared with a 0.75 mg/liter dose required for the same destruction by chlorine.

Additional work about the same time by Bernarde et al.[27] investigated the mechanism of kill by chlorine dioxide. This work so far indicates that ClO_2 does not react sufficiently with amino acids to alter their characteristic structures, thus eliminating the possibility of a reaction within the cell. They found, however, that, by some unknown fashion, ClO_2 abruptly inhibits protein synthesis, which is probably the mechanism by which it destroys these vegetative organisms.

In 1947, Ridenour and Ingols[28] concluded that chlorine dioxide was at least as effective as chlorine, and in contrast to chlorine the bactericidal efficiency of chlorine dioxide was relatively unaffected by pH values between 6 and 10. Ridenour and Armbruster[29] found in 1949 that less than 0.1 mg/liter chlorine dioxide destroyed the common water pathogens *Eberthella typhosa*, *Shigella dysenteriae*, and *Salmonella paratyphi B* at temperatures between 5°C and 20°C at pH values above 7 with a 5-min. contact period. An increase in pH brought about an increase in germicidal efficiency.

Ridenour and Ingols[30] reported in 1949 that *chlorine dioxide was clearly superior to chlorine in the destruction of spores on an equal OTA residual basis.* Using a 5-min. contact period, a 99.9 percent reduction of *B. subtilus* in a chlorine demand-free chlorine suspension required a 1.0 mg/liter ClO_2 residual, as compared with a 3.5 mg/liter free chlorine residual. They believed that in the case of the spores, the cell wall was penetrated by the use of the five valence changes in the oxidation of chlorine dioxide, but that vegetative cells do not utilize this phenomenon.

Bedulvich et al.[31] concluded in 1953 that the bactericidal efficiency of chlorine dioxide towards *E. coli*, *Salmonella typhosa*, and *Salmonella paratyphi* was as great or greater than that of chlorine.

In 1953 Hettche and Ehlbeck[32] investigated the action of chlorine dioxide on poliomyelitis virus and reported that it is more effective than either ozone or chlorine. This superior effectiveness may be explained by a chemical characteristic of chlorine dioxide reported by Ingols and Ridenour in 1948.[33] They found that chlorine dioxide (but not chlorine) reacted with peptone in amounts that followed the laws of adsorption. Since viruses have a protein coat, it may be assumed that chlorine dioxide will be adsorbed by the virus coat. This would cause higher local concentrations on the surface of the virus than would be expected from the measured residual and may account for the claimed effectiveness of the virucidal effect of chlorine dioxide.[34] This theory appears to be compatible with the findings of Benarde et al.[27] described above.

In 1976, Lambert and Bernard reported on a field investigation of the disinfection of sewage effluent from two different treatment plants in France.[35] One effluent was from a physical-chemical treatment plant at Le Lavandou, Var, and the other from a secondary treatment plant at Tergnier, Aisne. Their results indicate that chlorine dioxide is superior to chlorine for the reduction of total coliforms in these two effluents. The total coliform reduction was from 10^6/100 ml MPN to 10^2/100 ml MPN in 10 min. with 2.0 mg/liter ClO_2 dosage, compared to an 8.0 mg/liter dosage of chlorine. The ClO_2 residual at the end of 10 min. contact was on the order of 0.05 mg/liter while the total chlorine residual was 7.5 mg/liter as measured by DPD.

In 1977, Vilagines et al reported on the successful use of chlorine dioxide as the disinfectant for the Paris, France water supply.[36]

The germicidal efficiency of chlorine dioxide is probably superior to chloramines and may even be superior to free chlorine. The use of chlorine dioxide is being thoroughly investigated as a disinfectant for wastewater and water reuse situations by a Stanford University research group. Results will be available late in 1978.[55]

CURRENT PRACTICES

Potable Water Treatment

In the United States and Canada it can be postulated that the use of chlorine dioxide has been to assist the chlorination process in the control of tastes resulting from the raw water being contaminated with phenolic substances and algae. Chlorine dioxide has also been used successfully in reducing consumer complaints where it was necessary to carry a chlorine residual in the distribution system.[14] It has been used whenever phenols are the cause of taste complaints *because ClO_2 is specific for the destruction of phenols.* It is also used effectively in the removal of iron and manganese in potable water supplies, owing to its high redox potential.

Gall[9] and Sussman et al[37] reported in 1976 that there were some sixty water-treatment plants using chlorine dioxide and until the Hamilton, Ohio installation, chlorine dioxide had never been used as a potable water disinfectant.[38]

Western Europe, particularly those regions which attempt to produce potable water from grossly polluted portions of rivers (e.g., the Seine et al in France and the Rhine in Germany) has been gradually switching over to chlorine dioxide as a substitute for prechlorination. This is primarily because it has been found that the use of chlorine dioxide will not produce chloroform as a by-product as does chlorine. This was reported upon recently by Vilagines et al.[36]

In the same regions chlorine dioxide has been used regularly along with chlorine, ozone, and activated carbon for the control of tastes and odors. It has proved to be very effective especially when used with ozone.[25]

Chlorine dioxide is also used extensively in Switzerland, but to a lesser degree in the British Isles and in the Middle East. Obviously there will now be a renewed interest in its application for potable water production, particularly in the western hemisphere since it has been announced that ClO_2 will not contribute to the formation of chloroform in potable waters.

Chlorine dioxide is being used effectively in the disinfection of high pH waters. Its germicidal efficiency is not affected in the pH range of 6-10. Lime-softened waters or others of natural high pH can be far more effectively disinfected with chlorine dioxide than with chlorine.[14]

Chlorine dioxide residuals have been known to persist for long periods of time* if not subjected to sunlight.[39] For this reason the use of chlorine dioxide in a distribution system has distinct advantages in both potable water and water reuse systems.

Wastewater and Water Reuse

There is very little field experience with the application of ClO_2 as a disinfectant for either secondary or tertiary effluents. Because of its unique chemical characteristics, it should have some advantages over chlorine in special situations. Its germicidal efficiency is about on a par with free chlorine which means it is probably more effective than chloramines. Since it will not react with ammonia nitrogen and is germicidally effective between pH levels of 6-10, chlorine dioxide should have a great potential as a superior disinfectant for tertiary effluents, particularly where the final coliform requirement is 2.2/100 ml. At an operating installation in France, with chlorine dioxide applied to a secondary effluent, a dosage of 2 mg/liter resulted in an MPN coliform reduction from 10^6/100 ml to 200/100 ml with only 10 min. contact time.[39] This is roughly comparable to the 1965 findings of Bernarde.[11]

There is considerable interest in California as to the use of chlorine dioxide where treatment plant effluents must meet the most severe MPN coliform requirement of 2.2/100 ml. These situations usually involve high quality effluents having MPN coliform concentrations of 50,000-100,000/100 ml and containing about 2-10 mg/liter ammonia nitrogen. Chlorine dioxide should be able to achieve this disinfecting requirement for less money than chlorine because it has a much higher germicidal efficiency than chloramines.

Moreover some proponents claim that the residual die-away is much faster than for chlorine so dechlorination of chlorine dioxide treated effluents will probably not be necessary. There is no evidence to substantiate or to refute this claim.

---

*This is the case where the water treated is of good quality thereby reflecting a low chlorine dioxide demand.

Industrial Processes

Documentation by Olin Water Services has shown that the selective chemical characteristics of chlorine dioxide are superior to other oxidants and disinfectants when dealing with difficult situations of biofouling and chemical oxidation.[9,24,37,40,41,43,47] These situations include: ammonia synthesis plants; petroleum refining cooling systems; reclaimed wastewater for cooling water purposes; and reclaimed water in food-processing plants.

It is important to note the case of an ammonia synthesis plant (recirculated cooling water system) which was attempting to control biofouling with an intermittent chlorine dose of 6 mg/liter but could not achieve proper microbiological control. However, with a dosage of 2 mg/liter of ClO_2, a residual (ClO_2 by DPD method) of 0.9 mg/liter was observed and biofouling was controlled. The total organism plate count dropped from 15,000/ml with chlorine to 5/ml with chlorine dioxide.[42] This indeed recommends the use of chlorine dioxide for use in biofouling treatment processes.

Odor Control

Again because of its selectivity and because it is easily aerated from solution, chlorine dioxide is extremely effective in deodorizing foul air containing mercaptans and other odoriferous organic compounds. In the case of hydrogen sulfide destruction, the end product is the sulfate ion provided enough ClO_2 is applied. (Also see p. 608 "AIR POLLUTION" Ref. 14.)

RESIDUAL MEASUREMENTS

One of the serious drawbacks to the acceptance and use of chlorine dioxide has been the difficulty in accurately measuring the chlorine dioxide fraction in a mixture of other species such as free chlorine, combined chlorine, chlorite, and chlorate. The 14th Ed. Standard Methods[29] beginning on p. 349 describes three methods for determining chlorine dioxide: iodometric, orthotololidine-oxalic acid, and amperometric.

Orthotolidine-Oxalic Acid

This is a colorimetric flash test requiring great speed for proper execution of the procedure. The oxalic acid is used to tie up the free chlorine. This method is not recommended because it suffers from the same problems as does orthotolidine for chlorine residual measurement in wastewater.[14]

Iodometric

This is used primarily for standardizing solutions. Chlorine dioxide releases free iodine from a potassium iodide solution when the sample is acidified. This method does not differentiate the various species of residuals.

Amperometric

The amperometric method can differentiate free chlorine, chloramines, chlorine dioxide, and chlorite present in any combination. If all the above fractions are present four titrations are required (See p. 613 Ref. 14 and pp. 357 and 358 Ref. 44).

H-Acid

This is a colorimetric procedure wherein the H-Acid produces a bluish pink color with chlorine dioxide residuals between 0.05 and 1.5 mg/liter. It is not specific for chlorine dioxide; free chlorine and chloramines interfere with this test. It must be carried out at a carefully controlled pH (4.1–4.3) and requires the use of a spectrophotometer. (See page. 614 Ref. 14.) This was the method adopted by Hamilton, Ohio.[43]

DPD

Introduction. Although not described as an analytical procedure for the determination of chlorine dioxide residuals in Standard Methods, this method is in wide-spread use for measuring all the various chlorine residual species and has been well documented in the literature.[18,44,45,46,47,48] Moreover, it is listed as the only method specified in the new EPA drinking water regulations.[50] This method using the FAS titrimetric procedure is a favorite with the researcher because of the rapidity, repeatability, and accuracy of the measurements. The reader is cautioned, however, that the tablet method for wastewater effluents must be carried out with care, and most of all, the tablets must be *absolutely fresh.* Current field experience dictates this judgment.

The following is taken from the latest procedure for all chlorine species as developed by Dr. Palin[48] which he revised in 1977.[50]

The differentiation of this test procedure is the superior results which have been attained by using glycine instead of malonic acid. The glycine converts free chlorine instantaneously into chloraminoacetic acid which has no effect on chlorine dioxide. This new reagent is presented in the following method.

Principle. This procedure is an extension of the standard DPD method for determining free and combined chlorine in water or wastewater. Chlorine dioxide appears in the first step of this procedure but only to the extent of one-fifth of its total available chlorine content. This corresponds to the reduction of chlorine dioxide to chlorite ion (see Eq. 6-9). If the sample is then acidified in the presence of iodide, however, the chlorite also is caused to react. When neutralized by the subsequent addition of bicarbonate, the color thus produced corresponds to the available chlorine content of the chlorine dioxide. If chlorite (ClO_2^-) is present in the sample, it will be included in the step involving acidification and neutraliziation. In evaluating mixtures of these various chloro-compounds it is necessary to suppress any free chlorine before reacting the residual in the sample with the DPD reagent. This is why glycine is used.

Reagents. It has been observed that the iron present in the FAS titrant (ferrous ammonium sulfate) may interfere with the first end point titration. So for complete suppression of this effect a pinch (about 0.2 g) of disodium EDTA should be added at the start with the DPD powder. (Additional EDTA is not required in colorimetric procedures.)

Other reagents required are as follows:

1. Standard ferrous ammonium sulfate (FAS) solution.

$$1 \text{ ml} = 0.100 \text{ mg available chlorine.}$$

2. DPD No. 1 powder.
3. Potassium iodide crystals.
4. Sulfuric Acid Solution 5 percent V/V.
5. Sodium bicarbonate solution 5.5 percent W/V.
6. Glycine solution 10 percent W/V.

The DPD No. 1 powder is a specially formulated reagent which combines the DPD indicator reagent with a phosphate buffer compound. If liquid reagents are used the buffer solution and DPD solution must be kept separate.

Details for the preparation of the DPD solution, the phosphate buffer solution, and the FAS solution are given on pages 330 and 331 Standard Methods 14th Ed. 1975.[44]

Procedure for Chlorine Dioxide (ClO_2). To 100 ml of sample add 2 ml glycine solution and mix. Add about 0.5 g DPD No. 1 powder, mix and titrate immediately with standard FAS solution. This will be reading G.

Procedure for Free Chlorine (HOCl). To a *second* 100 ml sample add about 0.5 g DPD No. 1 powder, mix and titrate immediately. This is reading A.

Procedure for Combined Chlorine: All of the Chloramines—Organic and Inorganic. Add to this *second* 100 ml sample about 0.5 g (one small crystal) of KI. Mix and continue the titration with FAS immediately. This is reading B.

Then add 0.5-1.0 of KI (several crystals), mix to dissolve and allow to stand for 2 min. Then titrate with FAS as before. This is reading C.

Procedure for Total Chlorine Residual—Including Chlorite. To this very same second 100 ml sample add 1 ml sulfuric acid solution. Mix and allow to stand for 2 min. Then add 5 ml of the bicarbonate solution, mix and titrate as before. This is reading D.

Calculations. For a 100 ml sample, 1 ml of standard FAS solution = 1 mg/liter available chlorine.

In the absence of chlorite (ClO_2^-):

Chlorine dioxide = 5G

Free chlorine = A - G

Monochloramine = B - A

Dichloramine = C - B

Total chlorine = C + 4G

If the step leading to reading B is omitted, mono- and dichloramine are obtained together, so the total *combined* chlorine = C - A.

To check whether or not chlorite is present in the first sample it is necessary to obtain reading D. The presence of chlorite is indicated if reading D is greater than C + 4G.

In the presence of chlorite (ClO_2^-):

Chlorine dioxide = 5G

Chlorite = D - (C + 4G)

Free Chlorine = A - G

Monochloramine = B - A

Dichloramine = C - B

Total Chlorine = D

If B is omitted combined chlorine = C - A.

DPD for Wastewater

The measurement of chlorine dioxide residuals becomes greatly simplified if: 1) the generating equipment produces a pure or almost pure ClO_2 solution; and 2) if there is sufficient ammonia nitrogen to capture any stray chlorine residual as combined chlorine. In either case the operator would only be interested in the chlorine dioxide measurement (reading G), without the necessity to use glycine. This procedure would be used to calibrate the chlorine residual analyzers which respond equally as well to a chlorine dioxide residual as they do to free chlorine. This, of course, changes the operation of these units to the "free chlorine" analyzer concept which uses far less chemical than the analyzers recording total chlorine residuals, particularly in wastewater effluents. Occasional analyses by the operating staff may be required to measure chlorite ion (ClO_2^-); this is, D - (C + 4G). This requires making an additional determination for D and C (see above).

If the CIFEC generating principle is used it would only be necessary to perform one step to determine the chlorine dioxide residual, namely, step 1 but *without* the use of glycine.

If the generating equipment discharges a solution containing a significant concentration of free chlorine (HOCl), then in wastewater applications this would be immediately converted to chloramines, so that the additional step for combined chlorine would have to be performed. If such a situation were encountered in a nitrified effluent, where the free chlorine portion of the chlorine dioxide solution proceeds to react as part of the residual, then the entire procedure described above would have to be performed. This brings up the question of the Olin Water Services system. This system does indeed perform quite well in food plants and cooling water systems because the water involved is usually of potable water quality. However, for the usual wastewater secondary effluent and most tertiary effluents its efficacy as a method for disinfection by chlorine dioxide leaves much to be desired. The French "enrichment loop" system is to be preferred owing to the assured purity of the chlorine dioxide solution.

CHLORINE DIOXIDE FACILITY DESIGN

General Considerations

In the disinfection of wastewater or water for reuse, a major consideration is the additional cost of the sodium chlorite. Chlorine costs about 15 ¢/lb versus about $1.00/lb for chlorine dioxide (about a 7:1 factor). The extra cost for the necessary additional equipment for on-site generation is not so significant. Therefore, chemical consumption is of prime importance, and prior knowledge of the "halogen" demand is the only practical way of choosing the method of sequential addition of chlorine and chlorine dioxide. A good rule of thumb would be to choose sequential application, if the 15 min. chlorine demand were in excess of 3 mg/liter.

Sequential Addition of Chlorine and Chlorine Dioxide

This system utilizes a chlorine control system identical to the chlorination system described in Chapter 3 except that the chlorine dosage will be much lower. Using the same control systems (compound loop), the residual set point on the 2-3 min. chlorinated sample would be about 0.5 mg/liter, or one which would decay to 0.2–0.3 in 15 min. contact time. This would have to be predetermined by laboratory demand studies.

The chlorine dioxide would then be applied immediately downstream from the chlorine control sampling point. Therefore, a second mixing chamber for the ClO_2 addition would be required.

The need to make a small-scale laboratory study of chlorine demand and coliform kill, with respect to dosage and contact time, with chlorine dioxide *cannot be overemphasized*. A well polished effluent might be able to meet the severest of coliform requirements with considerably lower dosages and shorter contact time than are now currently used with chlorine. Moreover, the chlorine demand information will confirm the need for the sequential addition of chlorine and chlorine dioxide or whether to let chlorine dioxide do the entire job of disinfection plus the halogen demand of the wastewater.

Once-Through System

This has been described elsewhere as the Olin system. This concept can be applied to most existing chlorination facilities. It consists of installing a sodium chlorite pump and a reaction tower (see Fig. 6-2). The chlorine solution piping must be changed to merge with the sodium chlorite solution immediately upstream from the reaction tower. The injector characteristics must be arranged so that the chlorine solution strength never goes below 500 mg/liter. In highly alkaline waters it may be necessary to inject acid into the injector water supply just upstream from the injector so that the pH of the chlorine solution is always less than 4.0.

The solution discharging from the reaction tower is not a pure chlorine dioxide solution. It is about 60–70 percent chlorine. The remainder is chlorine dioxide.

Facilities for handling and storing the additional chemicals must also be provided. Otherwise the details for such a facility are the same as for a conventional chlorination system.

This system works best for situations where the required dosage is constant (i.e., most industrial processes). Its effectiveness and reliability for proportional flow systems with changing chemical demand, as in most wastewater plants, is unknown at this time.

CIFEC Recirculation System

This system provides a practically pure chlorine dioxide solution at a reliably controlled pH less than 2.0; therefore, it is readily adaptable to automatic control. Moreover, it is readily adaptable to existing automatic chlorination systems. Figure 6-4 illustrates how the CIFEC system is arranged for automatic control by both flow and residual.

Owing to the purity of the chlorine dioxide solution produced by this system, conventional chlorine residual analyzers may be used for residual control.

The solution discharge system must be carefully designed because of the high concentration (4000–5000 mg/liter) and the purity of the solution. While the system can tolerate back pressures up to about 7 or 8 psi, the solution lines must be designed so as to prevent the occurrence of a negative hydraulic gradient. Such a situation could cause the formation of pockets of chlorine dioxide gas which in turn could lead to a mild explosion and rupture in the solution line. Long solution lines should be avoided. The chlorine dioxide reaction tower, sodium chlorite pump, solution recirculating pump, and chlorine injector should be located as close to the point of application as possible. It is also desirable, but not mandatory, to have the chlorine control apparatus adjacent to the above components.

Otherwise the features of a complete system are the same as for a conventional chlorination system.

Dechlorination Facility

The reaction of chlorine dioxide residuals (measured as chlorine) have been described above by Eqs. 6-11 to 6-13 which demonstrate that it will require about 2.5 lb SO_2 for each lb chlorine dioxide residual measured as a chlorine residual.*

The facility for chlorine dioxide dechlorination (if required) is identical in every respect to the chlorination-dechlorination facilities described in Chapter 3.

CHEMICAL HANDLING AND STORAGE

Granular Sodium Chlorite

This is an orange colored powder with a loose bulk density of about 53 lb/ft^3. The packed bulk density is 69 lb/ft^3. It is generally available in 25 lb pails and 100 lb stainless steel drums. Technical-grade sodium chlorite is only about 80 percent pure. This means that it requires 1.68 lb technical grade $NaClO_2$ to 0.50 lb chlorine in order to produce 1.0 lb ClO_2 assuming a 100 percent stoichiometric reaction.

Special considerations should be given to storage facilities for granular sodium chlorite. Insurance experts should be consulted. Storage should be in an outside building, preferably detached from any main structure, which should be con-

*It is presumed that the controlling analyzer for the dechlorination process would be calibrated in terms of chlorine residual. However, since residuals using chlorine dioxide can be expected to be about one-half those residuals presently used in chlorination practice, the designer is advised accordingly.

structed wherever possible with noncombustible materials, such as corrugated steel, cement block, or brick. If it is located where a fire may occur, water should be available to keep the sodium chlorite area cool enough to prevent deterioration from heat.

Sodium chlorite should be handled with a minimum of spilling. To clean up spillage, never use a vacuum cleaner; always flush down a drain with plenty of water. Never use too little water, as this may be worse than no water at all. If drains are not available in the area of spillage to flush the spillage away, the spilled flakes should be carefully swept into a metal container and flushed down a drain with a large amount of water. The floor area should then be neutralized as follows:

1. Sprinkle the area with powdered anhydrous sodium sulfite, one-half lb for every ten ft^2.
2. Mop or brush over the area with the minimum water required to distribute evenly and dissolve the sodium sulfite.
3. Wait fifteen min., and repeat steps 1 and 2.

If sodium chlorite solution is spilled on nonwatertight floors, the area should be neutralized as follows:

Sprinkle the area wetted by the chlorite solution with powdered anhydrous sodium sulfite, ten lb for each gal. of solution spilled, but not less than one-half lb for every ten ft^2. Mop or brush as before, and again after fifteen min. Repeat as before.

Gloves should not be worn when handling sodium chlorite. If any of the material should come into contact with clothing or other combustible material, such clothing or combustible material should immediately be soaked in water to remove all trace of the sodium chlorite or it should be taken outside and burned without delay.

Liquid Sodium Chlorite

Solutions shipped in tank cars and trucks are hot. These solutions are at a concentration of about 50 percent. As soon as this solution cools to 85°F it crystallizes very rapidly. Therefore, liquid sodium chlorite is usually diluted to 25 percent before shipment to prevent crystallization. The same precautions should be observed as described above for solutions.

RELIABILITY

The need for continuous and dependable disinfection has been explained. The chlorine dioxide system can fail for the same reasons a chlorination system can fail. These causes have been delincatcd in Chapters 3 and 4.

Some additional factors must be considered in the case of a chlorine dioxide facility.

Both the Olin and CIFEC systems pump a sodium chlorite solution. This equipment should be in duplicate, including the solution tanks. The CIFEC system utilizes a solution recirculating pump which also acts as an injector pump. This pump should be in duplicate.

The Olin system usually requires an injector pump. This pump should also be provided with a standby unit. Whenever this sytem requires an acid injection pump, this should also be in duplicate.

Auxiliary power for operating the injector pumps and solution pumps should be available in case of power failures.

COSTS

The recent report on costs for various disinfectants by the EPA[51] shows the comparison between chlorine and chlorine dioxide.

For a 2 mg/liter dose, and a 15 min. contact time; chlorine at 15¢/lb and sodium chlorite at 70¢/lb the cost of disinfectant in cents per 1000 gal. is as follows:

Plant Size (mgd)	1	10	100	150
Chlorine	4.0	1.0	0.7	0.6
Chlorine dioxide	4.0	2.0	1.0	1.0

Chapter 6—Summary

The role of chlorine dioxide as a wastewater disinfectant is uncertain because of lack of field experience. It has three important chemical characteristics: it does not react with water to form a compound as does chlorine (HOCl); it does not react with ammonia nitrogen; and it does not react with any natural or manmade precursors to form chloroform—a known carcinogen—as does chlorine.

With what little evidence is available, all indications are that pure chlorine dioxide residuals may be more bactericidal and more virucidal than equivalent free or combined chlorine residuals. This might dictate chlorine dioxide as the disinfectant of choice for tertiary effluents and/or water reuse systems. Furthermore, the germicidal efficiency of ClO_2 does not appear to be affected by pH in the range of 7–10.

Beginning about 1975, a great many potable-water producers in Western Europe began to change prechlorination facilities over to chlorine dioxide to eliminate the formation of trihalomethanes caused by prechlorination. Therefore, in a very short time there should be available more information on the practicality of chlorine dioxide. About 60–70 of these installations were made in 1975 and 1976.

The toxicity of chlorine dioxide residuals to aquatic life is not known, but it is reasonable to assume that it is equally as lethal as the other chlorine species.

In closed systems the residuals can persist for a long time but decompose in sunlight. This suggests the use of a closed conduit (outfall) for wastewater contact chambers, and a holding lagoon for the dissipation of residuals instead of dechlorination.

Most existing chlorination facilities can be easily converted to generate chlorine dioxide. If the solution generated is a pure chlorine dioxide solution, then existing analyzers can be calibrated for control in terms of either chlorine or chlorine dioxide residuals by either the amperometric, starch iodide, or DPD methods.

There is still doubt regarding the best method of on-site generation. Granstrom and Lee[10] found the highest yield at pH 3-4 but did not succeed in producing a solution strength higher than about 80 percent ClO_2.

CIFEC[16] claims 98-100 percent ClO_2 solution by maintaining a 5000 mg/liter chlorine solution at a pH of 1. However, Granstrom and Lee claim that at pH levels below 3, the chlorite ion ClO_3^- formation may predominate. This would mitigate against achieving a stoichiometric yield as claimed by CIFEC at pH 1. Obviously more study is needed to establish the optimum conditions for generating a pure chlorine dioxide solution.

The relative instability of chlorine dioxide in aqueous solution can create a potentially undesirable and hazardous situation in the solution lines downstream from the ClO_2 reaction chamber. Chlorine dioxide will come out of solution as a gas if the pressure in the solution lines approaches atmospheric pressure. Negative pressures in these lines must be avoided because pockets of ClO_2 gas are liable to undergo spontaneous explosion which can rupture the solution lines.

Design of installations using conventional chlorination equipment should be arranged for automatic pH control of the chlorine solution downstream from the injector. This requires establishing a pH profile with respect to chlorine solution strength for each individual injector water supply.

In all cases the injector selection should be the one which uses the least amount of water for the maximum production of chlorine dioxide.

It is still questionable whether or not chlorine dioxide will ever be a popular substitute for the chlor-dechlor method for the disinfection of wastewater and/or water for reuse. To displace chlorine it would have to demonstrate a higher germicidal efficiency which could be translated into lower dosages and less contact times or the residual die-away would have to be such that dechlorination would not be required. A great deal more evidence is needed to substantiate either of these factors.

REFERENCES

1. Symons, J. M. "Interim Treatment Guide For The Control of Chloroform and Other Trihalomethanes." A report prepared by the Municipal Environmental Research Lab., EPA, Cincinnati, Ohio (June 1976).

2. *Kirk-Othmer Encyclopedia of Chemical Technology,* 2nd Ed. Vol. 5, Interscience, N.Y. (1964).

3. Kesting, E. E. "The Manufacture and Properties of Chlorine Dioxide." *TAPPI* **36**: 166 (1953).

4. Sconce, J. S. "Chlorine: Its Manufacture, Properties and Use." Reinhold, N.Y. (1962).

5. Booth, H. and Bowen, E. J. "Action of Light on ClO_2 Gas." *J. Chem. Soc. London* **127**: 510 (1925).

6. Bowen, E. J. and Cheung, W. M. "Photodecomposition of Chlorine Dioxide Solution." *J. Chem. Soc. London*, p. 1200, Part I (1932).

7. Rosenblatt, D. H. "Chlorine Dioxide: Chemical and Physical Properties." A paper presented at the IOI Workshop, Cincinnati, Ohio (Nov. 17–19, 1976).

8. Shera, B. L. Consulting Engineer, Private Communication. Tacoma, Wash. (1974).

9. Gall, R. J. "Chlorine Dioxide: An Overview of its Preparation, Properties and Uses." A paper presented at the IOI meeting, Cincinnati, Ohio (Nov. 17–19, 1976).

10. Granstrom, M. L. and Lee, G. F. "Generation and Use of Chlorine Dioxide in Water Treatment." *J. AWWA* **50**: 1453 (1958).

11. Bernarde, M. A., Israel, B. M., Olivier, V. P., and Granstrom, M. L. "Efficiency of Chlorine Dioxide as a Bactericide." *Appl. Microbiol.* **13**: 776 (Sept. 1965).

12. Synan, J. F., MacMahon, J. D., and Vincent, G. P. "Chlorine Dioxide, A Development in the Treatment of Potable Water." *Water Works and Sewage* **91**: 423 (1944).

13. "Chlorine Dioxide," Cat. File 60.300 Rev. 7-75. Wallace and Tiernan Div. Pennwalt Corp. Belleville, N.J. (1975).

14. White, G. C. *Handbook of Chlorination,* Van Nostrand Reinhold Co., N.Y. (1972).

15. Lambert, M., for Compagnie Industrielle de Filtration et d'Equipment Chimique, U.S. Patent No. 3,975,284; "Process for the Manufacture of Solutions of Halogens." (Aug. 1976).

16. "Chlorine Dioxide Generator: The French Method with Enrichment Loop." Cat. 167 CIFEC. Paris, France. (Nov. 1976).

17. Montiel, A., Chef de Laboratoire de Physico-Chimie Service de Controle des Eaux City de Paris. Private communication. (1977).

18. Miltner, R. J. "Measurement of Chlorine Dioxide and Related Products." A paper presented at the AWWA Water Quality Technology Conf. (Dec. 6, 1976).

19. Musil, J., Knoteck, Z., Chalupa, J., and Schmidt, P. "Toxicological Aspects of Chlorine Dioxide Application for the Treatment of Water Containing Phenol." Scientific Papers from Inst. Chem. Tech., Prague, Checkoslovakia 8: 327 (1964).

20. Bodansky, O. "Methemoglobenemia–Producing Compounds." *Pharm. Rev.* 3: 144 (1951).

21. Montiel, A. "Application of Chlorine in the Treatment of Water." A report prepared for the Prefecture De Paris, Service de Controle Des Eaux, Paris IV, France (1976).

22. Feuss, J. F. "Problems in Determination of Chlorine Dioxide Residuals," *J. AWWA* 56: 607 (1964).

23. Dodgen, H. and Taube, H. "The Exchange of Chlorine Dioxide with Chlorite Ion and with Chlorine in Other Oxidation States." *J. Am. Chem. Soc.* 71: 2501 (1949).

24. Ward, W. J. "Chlorine Dioxide, a New Development in Effective Microbio Control." A paper presented at the Ann Mtg., Cooling Tower Inst. Houston, Texas. (Jan. 19–21, 1976).

25. Coudert, M. Private communication. Compagnie General des Eaux, Choisy-Le-Roi, France (1974).

26. Stevens, A. A., Seeger, D. R., and Slocum, C. J. "Products of Chlorine Dioxide Treatment of Organic Materials in Water." A paper presented at the IOI Conference, Cincinnati, Ohio (Nov. 17–19, 1976).

27. Bernarde, Melvin A., Snow, W. B., Olivieri, Vincent P., and Davidson, B. "Kinetics and Mechanism of Bacterial Disinfection by Chlorine Dioxide." *Appl. Microbiol.* 15: 2, 257 (1967).

28. Ridenour, G. M. and Ingols, R. S. "Bactericidal Properties of Chlorine Dioxide." *J. AWWA* 39: 561 (1947).

29. Ridenour, G. M. and Armbruster, E. H. "Bactericidal Effects of Chlorine Dioxide." *J. AWWA* 41: 537 (1949).

30. Ridenour, G. M., Ingols, R. S., and Armbruster, E. H. "Sporicidal Properties of Chlorine Dioxide." *Water Works Sewage* 96: 276 (1949).

31. Malpas, J. F. "Disinfection of Water Using Chlorine Dioxide." *Water Treat. Exam.*, 22: 209 (1973).

32. Hettche, O. and Ehlbeck, H. W. S. "Epidemiology and Prophylaxes of Poliomyelitis in Respect of The Role of Water in Transfer." *Arch. Hyg. Berlin*, 137: 440 (1953).

33. Ingols, R. S. and Ridenour, G. M. "Chemical Properties of Chlorine Dioxide." *J. AWWA* 40: 1207 (1948).

34. Dowling, L. T. "Chlorine Dioxide in Potable Water Treatment." 23: 190 (1973).

35. Lambert, M. and Bernard, C. "Disinfection des Eaux Residuaires Urbaines par le Chlore et le Bioxyde de Chlore." A report on Plant Scale studies of wastewater discharges at Bormes le Lavandou, Var and Chauny la Fere à Tergnier, Aisne, France, the report; compliments of senior author (1976).

36. Vilagines, R., Montiel, A., Derreumaux, A., and Lambert, M. "A Comparative Study of Halomethane Formation During Drinking Water Treatment by Chlorine or its Derivatives (ClO_2) in a Slow Sand Filtration Treatment Plant and in Wastewater Treatment Plants." A Paper presented for the Disinfection Seminar at the Annual AWWA Conf. Anaheim Calif. (May 8, 1977).

37. Sussman, S. and Rauh, J. S. "Use of Chlorine Dioxide in Water and Wastewater Treatment." Paper presented at the IOI Meeting, Cincinnati, Ohio, (Nov. 17–19 1976).

38. Augenstein, H. W. "Use of Chlorine Dioxide to Disinfect Water Supplies." J. AWWA 66: 716 (1974).

39. Derreumaux, A. Private communication. CIFEC, Paris, France (1976).

40. Wheeler, G. L. and Yau, F. "Chlorine Dioxide: A Selective Oxidant for Industrial Wastewater Treatment." A paper presented at the Industrial Pollution Conf. Houston, Texas (Mar. 30–April 1, 1976).

41. Rauh, J. S. "Chlorine Dioxide a Cooling Water Microbiocide." A paper presented at the Ann. Mtg., Water and Wastewater Equip. Mfg. Assoc., Houston, Texas. April (1976).

42. Ward, W. J. "Chlorine Dioxide: Multi-Purpose Oxidant for Water and Wastewater Treatment." A paper presented at the 36th Ann. Mtg. of the Int. Water Conf., Pitt., Pa. (Nov. 4–6, 1975).

43. Ward, W. J. "Chlorine Dioxide: A New Selective Oxidant/Disinfectant for Wastewaters." A paper presented at the Int. Ozone Inst. Forum on Ozone Disinfection, Chicago, Ill. (June 2–4, 1976).

44. "Standard Methods for the Examination of Water and Wastewater." 14th Ed., APHA, AWWA, WPCF, Wash. D.C. 1975.

45. Adams, C. B., Carter, J. M., Jackson, D. H., and Ogleby, J. W. "Determination of Trace Quantities of Chlorine, Chlorine Dioxide, Chlorite, and Chloramines in Water." Proc. Water Treat. Exam., England, 15: 117 (1966).

46. Palin, A. T. "Determining Chlorine Dioxide and Chlorite," J. AWWA 62: 483 (Aug. 1970).

47. Palin, A. T. "Methods for the Determination, in Water, of Free and Combined Available Chlorine, Chlorine Dioxide and Chlorite, Bromine, Iodine, and Ozone, Using DPD." Inst. Water Eng. 21: 537 (Aug. 1967).

48. Palin, A. T. "Analytical Control of Water Disinfection with Special Reference to Differential DPD Methods for Chlorine, Chlorine Dioxide, Bromine, Iodine & Ozone." *J. Inst. Water Eng.* 28: 139 (1974).

49. Drinking Water Regulations and Proposals: Status as of May 1, 1976. U.S. Environmental Prot. Agency *Water Sewage Works* 123: 54 (June 1976).

50. Palin, A. T. Private communication. Ref. 48 (Rev. May 1977).

51. Love, O. T., Jr., Carswell, J. K., and Symons, J. M. "Comparison of Practical Alternative Treatment Schemes for Reduction of Trihalomethanes in Drinking Water." A paper presented at the IOI Conf., Cincinnati, Ohio (Nov. 17–19, 1976).

52. Valenta, J. and Gahler, W. "Chlordioxanlage." *Gaswasser, Abwasser* 9: 566 (1975).

53. Stevenson, R. G., Jr., Daily, L. L., and Ratigan, B. J. "The Continuous Analysis of Chlorine Dioxide in Process Solution." paper presented at ACS Conference, Anaheim, California (March 13, 1978).

54. Hood, N. J. "A Laboratory Evaluation of the DPD and Levco Crystal Violet Methods for the Analysis of Residual Chlorine Dioxide in Water." Master of Science thesis Virginia Polytechnic Institute, Blacksburg Virginia, Oct., 1977.

55. Roberts, P. V. Private communication (May 1978).

7

Bromine, bromine chloride, iodine, and UV radiation

BROMINE (Br_2)

Occurrence

Bromine was discovered in seawater, in 1826, by Antoine J. Balard. It derives its name from its offensive odor: the Greek word bromos is stench. It does not occur in nature as a free element. It exists primarily in the bromide form, and is found widely distributed in relatively small proportions. Bromides available for extracting bromine occur in the ocean, salt lakes, brines or salt deposits left after these waters evaporated during earlier geological periods, and from the mineral bromyrite. The bromide content of seawater is about 70 mg/liter. The total bromine content in the earth's crust is estimated at 10^{15} to 10^{16} tons.[1]

The Dead Sea in Israel is one of the richest sources of bromine in the world, containing nearly 0.4 percent at the surface and up to 0.6 percent at deeper levels. In Western Europe the most significant source is in the salt deposits at Strassfurt, Germany. Principal sources in the United States are the brine wells in Arkansas (Arkansas produces about three fourths of the national bromine output), Ohio, Michigan, and

West Virginia with a bromide content ranging from 0.2 to 0.4 percent. Those in Michigan underlie a large area of the Great Lakes region and occur in various sandstone strata at depths of 700–8000 ft. Here the bromide contents vary from 0.05 to 0.3 percent and are generally higher in the deeper levels.

Bromine Production

The recovery of bromine from seawater was first achieved on a commercial scale in 1924 by the Ethyl Corporation.[2] This process involved treatment of the seawater with chlorine and analine. The first successful bromine plant was put into operation at Kure Beach, North Carolina, in 1933, and was capable of extracting 3000 tons of bromine per year. In this plant, the process consisted of adjusting the pH of seawater to 3.5 with sulfuric acid, followed by the application of chlorine. The bromine, liberated by the chlorine, was removed as a dilute bromine gas with a current of air and absorbed in a sodium carbonate solution, from which it was recovered by acidification and stripping with steam. The critical part of this type of bromine extraction is the control of the pH at 3.5.

Oxidation of bromide to bromine can be accomplished either chemically or electrochemically. The electrochemical methods are no longer significant for commercial production. Chemical oxidation can be effected by either chlorine compounds, or oxygen containing compounds such as manganese dioxide, bromate, or chlorate.

The extraction of bromine from bromide compounds requires four steps: 1) oxidation of bromide to elemental bromine (Br_2); 2) separation of the bromine from solution; 3) condensation and isolation of the bromine vapor; 4) purification. Current bromine production methods are based on the modified Kubierschky steaming-out process and the H.H. Dow blowing-out process.

Kubierschky Process. In this process the raw brine is preheated to about 90°C, treated with chlorine in a packed tower, then placed in the steaming-out tower into which steam and additional chlorine are injected. The outgoing brine is neutralized with caustic and used to preheat the raw brine. From the top of the steaming-out tower, the halogen and steam vapor passes into a condenser and then into a gravity separator. Vent gases from the separator return to the chlorination system, the upper water layer containing Br_2 and Cl_2 is returned to the steaming-out tower, and the lower layer containing crude bromine passes on to a stripping tower. From the stripping column, bromine is purified in a fractionating column which produces a 99.8 percent pure liquid bromine as the final product.

The H.H. Dow Process. This process utilizes air instead of steam for the "blowing-out" step in the extraction of bromine. It is a more economical extracting agent than steam, especially when the bromine source is as dilute as in seawater. In the process, the halogens are absorbed from the air in a sodium carbonate solution, or by sulfur dioxide reduction.[1]

$$Br_2\ (Cl_2) + SO_2 + 2H_2O \longrightarrow 2\ HBr\ (2HCl) + H_2SO_4 \qquad (7\text{-}1)$$

Bromine can then be separated by chlorinating the mixed acids in a blowing-out tower. The theoretical yield is 2.2 tons of bromine per ton of chlorine.[3]

From 1973 to 1975, the estimated total annual bromine production in the United States was about 220,000 tons.[4]

In the years following World War I, the demand for bromine was for pharmaceutical bromides, the organic chemical industry, and photography. However, the biggest boon to the bromine industry was the discovery of tetraethyl lead as an antiknock ingredient in gasoline to accommodate the more powerful high-compression automobile engines. But this ingredient posed a serious problem: deposits of lead in the engine. It was found that a mixture of ethylene dibromide and ethylene dichloride added to the tetraethyl lead was an excellent scavenger which prevented lead deposition in the engine. These lead halides were sufficiently volatile to be expelled in the engine exhausts. It is estimated that about 70 percent of the 1973-1975 production of bromine was used to make ethylene dibromide for gasoline. However, future air pollution controls are almost certain to ban the use of tetraethyl lead in gasoline, thereby wiping out the major market for bromine production. This might be a boon to the water pollution control industry. This will be discussed later in this chapter. Bromine at a lower price becomes a most interesting disinfectant, particularly for water reuse situations.

Physical and Chemical Properties

Bromine is a dark brownish red, heavy mobile liquid. It gives off, even at ordinary temperatures, a heavy, brownish red vapor with a sharp, penetrating, suffocating odor. The vapor is extremely irritating to the mucous membranes of the eyes, nasal passages, and throat, and is extremely corrosive to most metals. Liquid bromine is likewise corrosive and destructive to organic tissues. In contact with the skin, it produces painful burns which are slow to heal.

Bromine (Br_2; atomic number, 35; molecular weight, 159.832; specific gravity, 3.12) weighs 26.0 lb/gal, and has a boiling point of 58.78°C. Of the metals used to handle bromine, lead is the most versatile.[3] Bromine reacts with lead to form a dense superficial coating of lead bromide, which, if not disturbed, prevents further attack. This is similar to the reaction of chlorine and silver. Tantalum is completely resistant to bromine; wet or dry, at temperatures up to 300°F.

Nickel and its alloy, monel, resist dry bromine and are especially useful as a material for shipping containers. Other nickel alloys, including the hastelloys, are less suitable. Iron, steel, cast iron, stainless steel, and copper are attacked by bromines, either wet or dry. Silver withstands dry bromine.

Bromine handled in lead, nickel, or monel containers should be dry (less than 0.003 percent moisture)[3] and should be protected from ordinary air, from which it can readily absorb enough moisture to make it severely corrosive to these materials.

Bromine is 3 times as soluble as chlorine (i.e., 3.13 g/100 ml water at 30°C). This is an important characteristic when considering the physical aspects of applying bromine to a process stream. Dispersion and diffusion is made "easier" and diffuser design to prevent off-gassing is less of a problem than with chlorine.

Chemistry of Bromine in Water and Wastewater

Bromine is unique in being the only nonmetallic element which is liquid at ordinary temperatures. It reacts with ammonia compounds in solution to form bromamines and displays the breakpoint phenomenon similar to chlorine.

Bromine in water hydrolyzes:

$$Br_2 + H_2O \rightleftharpoons HOBr + H^+ + Br^-$$ (7-2)

for which the equilibrium constant is 5.8×10^{-9}.

Depending upon the pH, the proportion of dissociation of hypobromous acid (HOBr) and hypobromite is:

$$\frac{[OBr^-] \, [H^+]}{[HOBr]} = K = 2 \times 10^{-9}$$ (7-3)

Like chlorine, bromine reacts with ammonia forming bromamine. Both Galal and Morris[5] and Johnson and Overby[6] have identified and studied the rate reactions of the compounds of NH_2Br, $NHBr_2$, and NBr_3 using the ultraviolet absorption spectrophotometry technique. They reported rapid formation of all bromamine species, however, once the bromamines have formed, a series of decomposition reactions take place. The major chemical difference between the bromamine species and the chloramine species is that the formation of the bromamine species is reversible from monobromamine through dibromamine to tribromamine and back again by fast reactions simply by changing the pH of the solution.

Bromine displays a breakpoint similar to chlorine, and it is the decomposition of the dibromamine which is the basis for this reaction. Tribromamine is the major species of combined residual bromine present beyond the breakpoint. In the pH range of 7-8 it decomposes in accordance with the following equation:

$$2NBr_3 + 3H_2O \longrightarrow N_2 + 3 \, HOBr + 3Br^- + 3H^+$$ (7-4)

La Pointe, Inman and Johnson[7] have shown that the breakpoint occurs when the bromine to ammonia nitrogen molar ratio is 1.5. This is precisely the stoichiometric amount of bromine required to oxidize all of the ammonia to nitrogen gas.

For wastewater disinfection, it is of considerable practical significance that the predominate species of bromine compounds is dibromamine over a pH range of 7–8.5. This is because dibromamine has a germicidal efficiency almost equal to free chlorine. Moreover, its rapid decomposition results in treated effluents that are low in toxic residuals. There must be an excess of ammonia nitrogen present for this to occur which is usually the case in all but highly nitrified effluents.

Free bromine residuals* (HOBr), which would occur in highly nitrified effluents, do not decompose nearly as rapidly as the bromamines. Their persistence increases with decrease in halogen demand of the environment.

Current Practices

The use of bromine for disinfection has been and still is quite limited. As described elsewhere in this text, it is not a popular disinfectant owing to its considerable cost and its difficulty in handling. The most widespread use of bromine as a disinfectant began in Illinois during World War II.

The Illinois State Department of Health, under the direction of C. W. Klassen, began an investigation of the use of liquid bromine in the elemental form as a substitute for chlorine when the latter became scarce during the war years. After a trial run in a few swimming pools for a couple of years, it was concluded by bacteriological results that it was doing an efficient job of disinfection. Permission to use elemental bromine was granted to a number of additional pools in 1947, and before the end of that year there were about twenty-five outdoor and thirty indoor pools where bromine was applied instead of the difficult-to-obtain chlorine. However, in order to avoid the hazards of handling liquid bromine, bromine in stick form was introduced about 1958. The stick is a compound of both bromine and chlorine known as bromo-chloro-dimethyl hydantoin (Dihalo). The use of this chemical has been documented by Brown et al.[8] It is thought that this compound, which hydrolyzes to form hypobromous acid, has an additional capability in that it also releases some HOCl by hydrolysis, which reacts with the reduced bromides to form more hypobromous acid. This form of bromine can produce satisfactory results for indoor and very small outdoor swimming pools. The current cost of the bromine sticks is $1.60/lb

---

*Current laboratory techniques, described in this chapter, cannot distinguish between free bromine and bromamine residuals.

(1975); therefore, its use on large outdoor pools would be decidedly uneconomical.

The use of bromine in wastewater or water reuse situations is unknown in the United States or Canada. The only known use in municipal potable water treatment was at Irvington, California about 1938. It was discontinued after a reasonable trial period because it did not solve the distribution system problem of water quality degradation. The bromine applied reacted so quickly and completely with the zoological slimes on the walls of pipes that it was impossible to obtain a residual downstream from the point of application. It also imparted a high intensity medicinal taste to the water.[2]

As early as 1955, significant efforts were being made to produce solid or dry granular disinfectants using the best attributes of bromine. U.S. Patent 781,730 was issued to the Diversey Corporation of Chicago, Illinois for the invention of a stable dry product composed of hypochlorite and alkali metal bromide. This product was claimed to have extraordinary disinfectant properties when placed in aqueous solution due to the formation of the hypochlorite-hyprobromite mixture.

In 1967 and 1969 patents were issued to Jack F. Mills et al. of the Dow Chemical Co.[9,10] The invention described in these patents represents a process for treating water with elemental bromine obtained from the polybromide form of an anion exchange resin. An effective method for the preparation of this resin is to pass an essentially saturated solution of bromine in aqueous sodium bromide slowly up through a bed of quarternary ammonium anion exchange resin. The resulting polybromide resin in wet form contains about 48 percent bromine. Development of the polybromide resin system as a practical means for disinfection has been concentrated on units capable of treating small quantities of water:—potable water for household use and swimming pools.[11] The Everpure Co. of Chicago has developed disposable cartridges containing bromine impregnated resin to feed predetermined amounts of bromine into water for disinfection.[12] The polybromide resin is sealed permanently into the cartridge to prevent its excape into the water system. Polybromide resins with bromine loadings of 25 percent have a very low acute oral toxicity. Direct contact with undiluted 25 percent resin is only moderately irritating to the skin but is capable of producing uncomfortable irritation upon direct contact with the eyes.

The disposable cartridge-type brominator has been installed and operated aboard offshore oil well drilling rigs, some remote land stations, and on ocean going vessels which use seawater distilling systems as a source for their potable water supply.

Bromine Facility Design

Current practice of the use of bromine in either stick form or resin impregnation systems is not practical for wastewater treatment. Molecular bromine (either liquid or vapor) could be a practical disinfectant if the hazards of handling and the difficulties in metering were not so enormous. These difficulties stem from the unique characteristic of bromine which allows it to be in the liquid form at room temperature and pressure. This is one of the major reasons why BrCl is being pursued as a practical means for the bromination of wastewater. BrCl has a vapor pressure of 30 psi at room temperature (20°C). This simplifies the packaging and handling problems.

Owing to the undesirable handling characteristics of molecular bromine there can be no valid recommendations for a facility to handle molecular bromine.

BROMIDES: ON-SITE GENERATION OF Br_2

System Description

The on-site oxidation of bromide salts in an aqueous solution to form free bromine (Br_2) is an old concept. In 1948, a patent was issued to Marks and Strandkov of Wallace and Tiernan Co. which involved on-site production of either iodine or bromine by the oxidation of their respective salts by either free or combined chlorine at pH 7–8.[13] Chemically the iodine release was predictable and quantitative, but the bromine release was not quantitative or predictable at this pH. The iodine system failed because control was difficult. However, in recent years there has been a renewed interest in the on-site generation of bromine from bromide for disinfection of reclaimed water. A patent was issued in 1973 to A. Derreumaux (France) for such a system.[14] This system is different from the one proposed by Marks in that the bromide salt solution is oxidized by a chlorine solution at a carefully controlled pH of 1–2. It conforms to all the established kinetics of bromine chemistry. Free bromine is produced in the Derreumaux process by injecting a bromide salt solution of known concentration into the injector discharge of a conventional chlorinator. The reaction between the chlorine and bromide ion to convert the bromide to hypobromous acid takes place in a reactor where the pH for the reaction is kept below 2. The reaction proceeds as follows:

$$HOCl + NaBr \longrightarrow HOBr + NaCl \qquad (7\text{-}5)$$

Thus 1.45 lb of pure sodium bromide (NaBr) will react with 1.0 lb chlorine to produce 2.25 lb bromine (Br_2).

The patented Derreumaux system is illustrated in Fig. 7-1. It consists of conventional chlorination equipment, a reaction vessel, and a metering pump for the addition of the bromide salt. The success of this system is its ability to consistently produce an almost pure solution of hypobromous acid at a controlled pH of less than 2. This insures maximum quantitative conversion of the bromide ion in accordance with Eq. 7-5. In cases where the chlorinator injector water is plant effluent (i.e., high NH_3 concentration), there will not be any

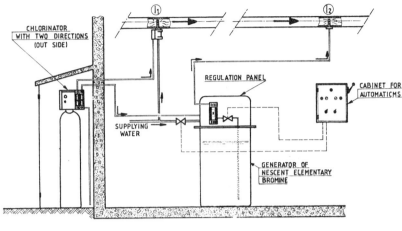

(I_1) : FIRST POINT OF INJECTION: CHLORINE WATER

(I_2) : SECOND POINT OF INJECTION: CHLORINE AND NESCENT BROMINE

Fig. 7-1   The Derreumaux bromine generating system (*courtesy* CIFEC Co.).

chlorine lost in side reactions with ammonia nitrogen at this low pH and with the reaction time involved.[2] After the bromide is oxidized to free bromine it will remain as Br_2. Therefore, at this controlled low pH in the reactor none of the free bromine will be wasted on side reactions with the ammonia nitrogen present (see Fig. 9-6 Ref 15).

The most important feature of this process is the sequential addition of chlorine followed by bromine. In practice the chlorine is added first to satisfy the immediate halogen demand (3–5 min. contact time) with a conventional compound loop residual control system (see Fig. 3-13). Then bromine is added on a flow proportional basis relying on manual control of the bromine dosage.

This procedure serves three purposes: 1) it reduces significantly the amount of bromine required; 2) it improves the disinfection efficiency[16]; 3) it solves the problem of trying to cope with bromine demand changes normally found in wastewater effluents. Field experience has shown that owing to the rapid die-away of bromamine residuals it is not practical to control the dosage by residual control.[17] In a comparative situation it was shown that chlorine is far more able to meet the diurnal variations in halogen demand, particularly those caused by the usual unpredictable and uncontrollable plant upsets. This is to be expected because of the much greater reactivity of bromine with the various constituents in wastewater.

The system shown in Fig. 7-1 would be modified considerably for use in the United States. For example, the application of chlorine would be on a flow paced—residual control compound loop system (see Fig. 3-13). However, the set point control would be as low as practical, probably 0.3-0.5 at 3-5 min. contact. This would be a completely separate piece of equipment. The second chlorinator, which discharges to the bromine reactor and the bromide salt metering pump, would operate from a flow proportional signal with manual preset dosage discharging at the second point of application. This would provide the necessary amount of bromine to achieve the desired level of disinfection.

The presumed advantages of bromine in wastewater treatment are considered to be: 1) the presence of ammonia nitrogen in wastewater converts the bromine applied to the dibromamine which is nearly as germicidal as free chlorine; 2) dibromamine is so highly reactive with wastewater constituents that the residual dies away very rapidly; 3) owing to these two characteristics, contact times may be reduced from current chlorination practices of 30-45 min to perhaps 15-20 min and possibly the necessity to debrominate might be eliminated.

However, the process must be cost effective. Most of the bromide salts being marketed today in the United States and Canada are very high grade and are used primarily in photographic emulsions so their cost is relatively high; in excess of 70¢/lb for available bromine.

Dow Chemical produces a low grade calcium bromide used primarily as a well pack fluid. It is marketed in two forms: 1) a 53-54 percent calcium bromide solution (specific gravity of 1.7-1.72 g/cc); 2) an 80 percent calcium bromide flake. The bromide ion content of the solution is about 42-43 percent and in the solid form is about 64 percent. Assuming the following reaction:

$$2 \, HOCl + Ca \, Br_2 \longrightarrow 2 \, HOBr + CaCl_2 \qquad (7\text{-}5a)$$

it would take approximately 1.4 lb of pure calcium bromide plus 1 lb of chlorine to make 2.25 lb of bromine.

Tank-truck lots of the solution currently cost 28.5¢/lb FOB Midland, Michigan and the solid form is 63.75¢/lb in 3500 lb lots FOB Ludington, Michigan (1977 prices).

Comparison with Other Methods

Advantages. This system of on-site generation of free bromine has the following advantages over the application of molecular bromine and bromine chloride.

1. It can be adapted to conventional chlorination equipment and controls.
2. Conventional metering equipment (diaphragm pumps) can be used for feeding the bromide salt solution.
3. Readily adaptable to conventional compound loop control utilizing existing chlorine residual analyzers.
4. The equipment specific for this process has been proved satisfactory by sufficient field experience.

5. Eliminates hazards of handling liquid molecular bromine or bromine chloride.
6. Eliminates equipment problems of corrosion and the chemical problem of dissociation encountered in the use of bromine chloride and molecular bromine.

Disadvantages

1. The major disadvantage of this process is chemical cost. The least expensive high-grade bromide salt costs about $0.70/lb in 500-lb lots (1976). The low-grade $CaBr_2$ may be the answer.
2. The availability of bromine compounds are limited. However, since there has been a mandatory elimination of leaded gasoline for automobile engines in the United States on account of air pollution, this eliminates the production requirements of bromine compounds which are used as lead scavengers in gasoline. In 1973, 70 percent of the total United States bromine production was for this purpose.
3. The reaction of bromamines in wastewater effluents with organics and the subsequent formation of undesirable bromoorganics is unknown at this time.
4. There is insufficient field experience to properly evaluate bromine as a disinfection process.

BROMINE CHLORIDE, BrCl

Physical and Chemical Properties

Bromine chloride is classified as an interhalogen compound because it is formed from two different halogens. These compounds resemble the halogens themselves in their physical and chemical properties except where differences in electronegativity are noted. Bromine chloride at equilibrium is a fuming dark-red liquid below 5°C. It can be withdrawn as a liquid from storage vessels equipped with dip tubes under its own pressure (30 psig 25°C). Liquid BrCl can be vaporized and metered as a vapor in equipment similar to that used for chlorine.

Bromine chloride is an extremely corrosive compound in the presence of low concentrations of moisture. Although it is less corrosive than bromine[18] great care must be exercised in the selection of materials for metering equipment. Like chlorine and bromine, it may be stored in steel containers. This is based on the assumption that BrCl or Br_2 is packaged in an environment where the air is dry (i.e., a dew point not higher than -40°F).

When in contact with skin and other tissues liquid BrCl, like Br_2, causes severe burns. Low concentrations of the vapor are extremely irritating to the eyes and respiratory tract.[1]

Bromine chloride exists in equilibrium with bromine and chlorine in both gas and liquid phases as follows.[1,6,15]

$$2\,BrCl \rightleftharpoons Br_2 + Cl_2 \tag{7-6}$$

There is little information on the equilibrium in the liquid state. A recent study by Mills[1,19] and a previous investigation by Cole and Elverum[20] indicates less dissociation in the liquid (20 percent) than in the vapor state. The equilibrium constant for the vapor phase dissociation of BrCl is close to 0.34 which corresponds to a degree of dissociation of 40.3 percent at 25°C.[1,20]

The density of BrCl is 2.34 g/cc at 20°C.

Figure 7-2 compares the vapor pressure temperature curves for bromine chloride, bromine, and chlorine.

The solubility of BrCl in water is 8.5 g/100 cc at 20°C. This is 2.5 times the solubility of bromine and 11 times that of chlorine.[1]

Bromine chloride forms a yellow crystalline hydrate, $BrCl \cdot 7.34\,H_2O$ at 18°C and 1 atm. This compares to the formation of chlorine hydrate $(Cl_2 \cdot 8\,H_2O)$ which forms at 9.6°C.[2] This is a significant characteristic because the formation of these hydrates causes operational problems in metering equipment.

Preparation of Bromine Chloride

Bromine chloride is prepared by adding an equivalent amount of chlorine (as a gas or liquid) to bromine until the mixture has increased in weight by 44.3 percent.

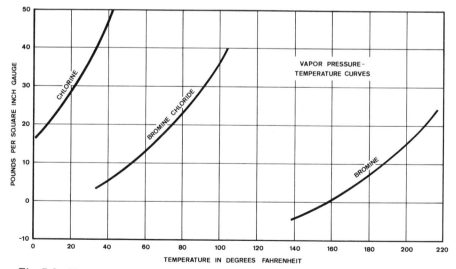

Fig. 7-2 Vapor pressure vs. temperature curves for bromine, bromine chloride, and chlorine (*courtesy* Capital Controls Co.).

$$Br_2 + Cl_2 \rightleftharpoons 2\ Br\ Cl \qquad (7\text{-}7)$$

Bromine chloride may also be prepared by the reaction of bromine in an aqueous hydrochloric acid solution. In the laboratory it can be prepared by oxidizing a bromide salt in a solution containing hydrochloric acid.[1] This produces the following reaction:

$$KBrO_3 + 2KBr + 6\ HCl \longrightarrow 3BrCl + 3KCl + 3H_2O \qquad (7\text{-}8)$$

Chemistry of Bromine Chloride in Wastewater—

Bromine chloride vapor appears to hydrolyze exclusively to hypobromous acid and hydrochloric acid.

$$BrCl + H_2O \rightleftharpoons HOBr + HCl \qquad (7\text{-}9)$$

Whereas bromine vapor (or liquid) hydrolyzes to hypobromous acid and hydrogen bromide.

$$Br_2 + H_2O \rightleftharpoons HOBr + HBr \qquad (7\text{-}10)$$

The formation of HBr represents a significant loss in the disinfecting potential of the expensive bromine molecules. In the hydrolysis reaction of BrCl, any HBr formed by the dissociation of elemental bromine is presumed to be oxidized quickly to HOBr by the HOCl remaining in solution:

$$HBr + HOCl \longrightarrow HOBr + HCl \qquad (7\text{-}11)$$

However, in the case of wastewaters, the ammonia nitrogen present would immediately convert any HOCl in solution to chloramines in near neutral pH environments. Chloramines cannot oxidize any HBr formed to hypobromous acid (HOBr) at these pH levels.[21] Therefore, the problem of dissociation of bromine chloride is significant in the presence of ammonium ions.

It is of practical significance that the hydrolysis constant for BrCl in water is 2.94×10^{-5} at $0°C$, compared with the same constant for chlorine which is 1.45×10^{-4} at $0°C$. It is paradoxical that BrCl is several times more soluble in water than chlorine, yet it hydrolyzes by a factor of 10 times slower. The hydrolysis constant of molecular bromine is 0.7×10^{-9} at $0°C$ which is significantly different from bromine chloride.

Bromine chloride combines with ammonia the same as does molecular bromine to form bromamines. At the usual pH levels encountered in wastewaters (7–8.5), the dominant species will be dibromamine.[15] Typical reactions are as follows:

$$NH_3 + HOBr \rightleftharpoons NH_2Br + H_2O \qquad (7\text{-}12)$$

$$NH_2Br + HOBr \rightleftharpoons NHBr_2 + H_2O \qquad (7\text{-}13)$$

$$NHBr_2 + HOBr \rightleftharpoons NBr_3 + H_2O \qquad (7\text{-}14)$$

Bromine chloride is presumed to have a higher speed of reactivity than bromine. In wastewater, the formation of bromamines is probably much faster than the formation of chloramines. Of greater practical significance is the rapid dieaway of the bromamine residuals. It has been estimated that the half-life of bromamine residuals in secondary wastewater effluents is less than 10 min.

This would not be true in a highly nitrified effluent where the predominant species of residual might be HOBr. This free bromine compound persists for a significant length of time in a low chlorine demand environment.*

Bromine Chloride Facility Design

Current Practice. The first attempt at plant scale use of BrCl was at Grandville, Michigan[22] in 1974 and 1975. This was an EPA sponsored project intended to evaluate several wastewater disinfectants. It was not a strict exercise in disinfection because the coliform reduction ratio was too low. The MPN of total coliforms before disinfection ranged between $8 \times 10^4/100$ ml and $3 \times 10^6/100$ ml. The disinfection requirement was to achieve $10^3/100$ ml MPN.† According to the model developed by Collins et al.[23] the required chlorine residual-contact time for the destruction of the lower range of coliform (8×10^4) is 7 and for the higher range (3×10^6) is 23.5. Therefore, if mixing at the point of application is adequate and if 30 min. contact time is available and assuming 80-90 percent plug flow conditions in the contact chamber, the predicted chlorine residual requirement would range from 0.23-0.8 mg/liter to meet the 1000/100 ml total coliform disinfection requirement. For comparable quality effluents this level of chlorine residual is considered very low.

The chlorine dosage at the beginning of this project was 2.9 mg/liter resulting in a 2.0 mg/liter residual at the end of 30 min. This demonstrates a chlorine demand of 0.9 mg/liter in 30 min which appears to be unreasonably low for such an effluent.

At the middle of the project the chlorine dosage was lowered to 2.3 mg/liter which produced a 1.0 mg/liter residual at the end of 30 min. This residual compares to the mathematical model prediction of 0.8 mg/liter for a total coliform concentration before chlorination (y_o) of $3 \times 10^6/100$ ml. This is the high range of y_o for the above project.

The BrCl dosage was lowered from 3.6 mg/liter to 3.0 mg/liter early in the project and about the middle of the project it was lowered to 2.0 mg/liter where

---

*An example of this is the practice of power plant condenser cooling water chlorination, where seawater is used. If the ammonia nitrogen content is less than 0.2 mg/liter and the chlorine dose is 1 to 2 mg/liter the chlorine applied immediately converts (stoichiometrically) an equivalent amount of the bromide ions present into HOBr. The resulting bromine residual persists in the condenser discharge plume nearly as long as would comparable chlorine residuals.

†Good disinfection is usually considered as a 4- to 5-log reduction of initial total coliform content.

it remained for the rest of the project. Because of the rapid die-away of the BrCl residual, this process could only be controlled by dosage, while the chlorination process was controlled by residual.

The performance of BrCl was disappointing. Chlorine succeeded in meeting the requirements on the Grandville project 90 percent of the time while BrCl met the requirements only 80 percent of the time. This was largely the result of equipment failures. The reasons are described later in this chapter.

Storage and Handling. As in all liquid-vapor exchange chemical systems, the problem of reliquefaction is of importance. This problem increases with decreasing vapor pressure equilibrium of the liquified gas. Sulfur dioxide reliquefies much more easily than chlorine because of its low vapor pressure (i.e., 35 psi at 68°F). This compares with a vapor pressure of 30 psi at 68°F for BrCl. In the case of BrCl it is more significant because dissociation occurs during reliquefaction and all of the chemical advantages of bromine chloride are lost.

Reliquefaction can be prevented by either raising the temperature of the liquid-vapor container (bromine cylinder) or by external pressure padding with compressed air or nitrogen.

Padding the container to about 50 psi and withdrawing the liquid to an evaporator similar to the practice of handling sulfur dioxide is a preferred procedure. This requires that the bromine chloride containers be fitted with dip tubes and outlet valves similar to chlorine or sulfur dioxide ton containers. One connection would be for pressurization of the vapor and the other for liquid withdrawal. If this were the case then the container storage and handling system would be the same as for chlorine or SO_2.

Evaporators. It is generally conceded that the most practical way to handle bromine chloride is to withdraw the liquid and vaporize it under controlled conditions.

Evaporators used for both chlorine and sulfur dioxide can be used without modification for bromine chloride. The piping arrangement and accessories would be the same as for chlorine. The downstream reducing valve would be set at 25 psi and the containers pressurized to 50 psi. The evaporator should be large enough to provide 30°F superheat.

Materials of Construction. Although bromine chloride is much less corrosive than bromine,[18] great care must be taken in the selection of materials throughout the design of this facility. The material for handling the liquid and/or the vapor under pressure is the same as for chlorine. Therefore, Sch. 80 seamless steel pipe can be used between the containers and evaporator, and between the evaporators and metering system.[24]

Bromine chloride vapor or liquid cannot be handled in PVC or ABS plastics, even at very low pressures, as can chlorine and sulfur dioxide. The preferred piping material for moist BrCl vapor is Kynar, a vinylidenefluoride resin. Similarly, Kynar valves should be used when required with this piping material.

Metering and Control Equipment. A limited field experience quickly revealed that a new species of equipment would have to be designed to solve the complex problems of construction materials which could stand up to the corrosivity of bromine chloride.

At the Grandville, Michigan project[22] the equipment chosen was a modified Wallace and Tiernan pressure gas-feed chlorinator fitted with an injector. This model (No. 20-055) was chosen because it is constructed of materials thought to be able to withstand bromine chloride. Many problems concerning corrosion and dissociation of the vapor were encountered requiring various modifications to the equipment.

At about the same time the Public Service Electric and Gas Co. of New Jersey entered into a project at one of their power stations* to evaluate the difference between chlorine and bromine chloride using conventional chlorination equipment for both gases on a side by side installation. It took only a few hours operation to observe that conventional chlorination equipment is not suitable for bromine chloride.[24] Most of the plastic and rubber materials were found to be entirely unsuitable because of permeation and/or direct chemical attack leading to nearly complete failure of this material. A successful experimental model was constructed on the direct gas-feed approach using slightly modified current production components such as a stainless steel indicating rotameter, chlorine line valves, and chlorine pressure reducing valves. Kynar pipe and valves were used adjacent to the injector where moist BrCl vapor might be encountered. The "suck-back" phenomenon was dealt with by using a modified chlorine check valve adjacent to but not part of the injector assembly.

Figure 7-3 illustrates a system subsequently developed by Capitol Controls Colmar, Pennsylvania. This system is essentially the same concept as the Public Service Co. experimental model described by Cole.[24]

Injectors. The injectors for use with BrCl may require modifications. The one used on the project described by Cole[24] was a 3-in. Wallace and Tiernan model A452. This unit was about 10 yr. old and did not require any modification. The one used on the Grandville project did require some modification.

Solution Lines and Diffusers. The material and the design for this part of the system can be of the same material used for chlorine.

*This was a once through cooling water system.

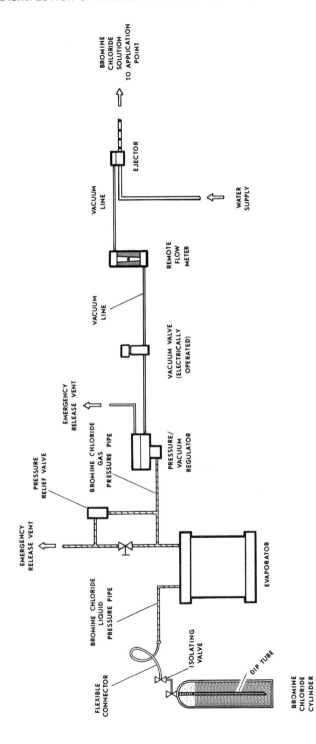

Fig. 7-3 Bromine chloride metering and control equipment (*courtesy* Capital Controls Co.).

Automatic Controls. The system shown in Fig. 7-3 is readily adaptable to automatic control by simply automating the control valve.

Conventional amperometric analyzers may be calibrated to record bromine residuals, but because of the evanescent characteristics of bromine residuals, automatic residual control might not be practical. Field experience is too limited to make a judgment on this facet.

Safety Equipment. Chlorine leak detectors can be readily adapted to detect BrCl in the atmosphere.

Chlorine container kits can be used on BrCl containers if the BrCl is packaged in containers of similar design.

Comparison With Other Methods

Advantages
1. Use of BrCl eliminates the hazards of handling Br_2.
2. It is presumed to have greater germicidal efficiency than Br_2 and chloramines.
3. At the 1976 price of 20¢/lb it is cheaper than on-site generated bromine.
4. Contact time should be less than for chlorine so contact chambers would be smaller.

Disadvantages
1. Equipment for metering and injection of BrCl is still in the experimental design stage.
2. Limited field experience exists in wastewater applications.
3. If used alone, residual control will be difficult due to rapid die-away of bromine residuals.
4. It is subject to reliquefaction which causes dissociation of the vapor. This defeats the purpose of bromine chloride.

Costs. The following is the estimated cost for wastewater disinfection using bromine chloride as determined by the 1976 EPA Task Force Report.[56]

Plant Size (mgd)	1	10	100
Capital Cost ($)	47,000	129,000	414,000
Disinfection Cost (¢/1000 gal.)	4.52	3.04	2.65

GERMICIDAL EFFICIENCY (Br_2 AND BrCl)

A review of the literature from 1945–1976 reveals a considerable lack of agreement on the germicidal efficiency of bromine compounds; i.e., free bromine (Br_2), hypobromous acid (HOBr), monobromamine (NH_2Br_2), and dibromamine

(HHBr$_2$). The major difficulty in sorting out the literature was to determine which species was in fact being investigated. This has to do with bromine chemistry. The publication edited by Johnson[33] is the most informative on the germicidal chemistry of bromine. The most informative investigations of the germicidal efficiency of bromine are those by Wyss and Stockton,[26] Marks and Strandkov,[27] Johannesson,[28] McKee et al.,[29] Koski et al.,[30] Schaffer and Mills,[31] Sollo et al.,[32] and Johnson and Sun.[15]

Although this work was done in the laboratory under ideal conditions, the following pertinent conclusions can be made with respect to wastewater disinfection.

In a wastewater effluent containing an ammonia nitrogen content of 10–30 mg/liter and a pH from 7–8 the dominant species of bromine will be dibromamine.[15]

Dibromamine appears to be almost as germicidal as hypobromous acid (free bromine) but not as germicidal as molecular bromine (Br$_2$) which only exists below pH 7.

Therefore, the germicidal efficiency of dibromamine is almost comparable to free chlorine as it has been demonstrated that hypobromous acid and hypochlorous acid have about the same germicidal efficiency.[6,15,33] This is significant in wastewater treatment because the application of bromine to wastewater will result in the formation of dibromamine which is *theoretically* many times more germicidal than chloramines.[33] The difficulty in resolving these comparisons is the total lack of field experience. As described above the Grandville, Michigan work[22] revealed disappointment in the germicidal efficiency of bromine. This is probably due to the fact that bromine is such a powerful oxidant, much of it is lost in side reactions with organic matter in the sewage.

Koski and Stuart[30] using the AOAC (Association of Official Agricultural Chemists) method for evaluating disinfectants, reported that liquid bromine at 1.0 ppm is as effective as the 0.6 ppm of available chlorine control, with *E. coli* as the test organism, but that 2.0 ppm liquid bromine is necessary to provide activity equivalent to the 0.6 ppm available chlorine control when *S. faecalis* is the test organism. McKee et al.[29] found that bromine in settled sewage at fifteen-min. disinfectant contact times was less effective than either chlorine or iodine. It is to be presumed that this phenomenon is a result of bromine becoming irretrievably bound up with the organic matter in sewage, or forming bromamines that decompose rapidly into components that do not have any disinfecting power. The interesting point here is that the residual bromine, which is a combination of HOBr and bromamine, appears to be as colicidal as chlorine, but the difficulty is that not much of the bromine becomes a stable residual. This is based on milliequivalents per liter of the halogen. On a dosage basis, however, chlorine is far superior to either bromine or iodine because much of the bromine or iodine is utilized in satisfying the halogen demand. Consequently, they are the least effective of the halogens where sewage is to be disinfected.[29] Eight mg/liter of chlorine will achieve disinfection comparable to 45 mg/liter of bromine or iodine. This is why the literature on this subject stresses the necessity of first satisfying the halogen demand of the sewage with the old reliable less-expensive chlorine. Kott in 1969[34] and subsequent investigations has demonstrated the superior germicidal efficiency of a combination of chlorine and bromine. First, chlorine is applied to satisfy the immediate halogen demand, followed by the application of bromine.

This corresponds to the work of Sollo et al.[16,32] and to the field experiences of CIFEC.[35] The latter reports that a 3 mg/liter dose of chlorine in a wastewater effluent at Bormes, France, containing 90 mg/liter ammonia nitrogen followed by an 8 mg/liter dose of bromine gave a 100 percent reduction of 10^6 coliforms with only 10 min. contact.

According to Mills,[1] bromine chloride is superior in germicidal efficiency to molecular bromine. This notion is derived from the fact that the chemical reactivity of BrCl is much faster than molecular bromine. So far there is too little information to substantiate this supposition.

TOXICITY OF BROMINE RESIDUALS

The acute toxicity tests with chlorobrominated effluent reported by Ward et al.[22] on the Grandville, Michigan project indicates that bromine residuals are not as lethal to fish as those of chlorine compounds. This supports the finding of Mills[19] who concluded that chlorobromination produced a less toxic effluent because bromamines are less stable than chloramines and thus do not persist as long. This would be true regardless of how the active bromine compounds were applied. The active ingredient would be a mixture of mono- and dibromamine at the pH normally encountered in wastewater treatment.

BROMO-ORGANIC COMPOUNDS

There is continuing concern over the use of halogens as disinfectants of potable water, wastewater, and water for reuse. This is a direct result of the recent (1975) detection of undesirable halogenated organic compounds in the nation's drinking water.

In addition to the chloro-organic compounds, a variety of bromo-organic compounds were also detected. These included bromodichloromethane, dibromochloromethane, and bromoform. The source of these compounds, generally referred to as precursors, is uncertain, but appears to be distributed by nature as a natural phenomenon. The precursors of these compounds have been found in soil and in the chlorophyll of blue-green algae. It was recently suggested that the occurrence of halogenated bromine compounds was the result of fallout from automobile exhausts. It was postulated that since bromine is used as a lead scavenger in leaded gasoline, these bromine compounds were deposited as a residue on paved roads and therefore reached our water supplies by natural runoff. This possibility is probably overemphasized because bromides are found almost everywhere in nature. The phasing out of leaded gasoline will certainly curtail this source as a precursor, however.

It is important to recognize that all of the bromine species are better oxidizing agents than their analogous chlorine species. For example, the oxidation of cellulose by hypobromous acid is much faster than by hypochlorous acid, and the oxidation of glucose to gluconic acid by hypobromite is 1300 times faster than by hypochlorite. Therefore, it is not surprising that bromine is a lesser halogenating agent than chlorine simply because it is a more potent oxidizing agent.

There are many organic reducing agents in wastewaters. These include organic alcohols, aldehydes, amines, and mercaptans. They are completely oxidized by bromine chloride resulting in the formation of inorganic chloride and bromide salts as the major by-products.

The carbon-bromine bond is less stable than the carbon-chlorine bond and there is a possibility that in addition to metabolism by hydroxylation, well established for chlorobiphenyls, a reductive debromination may be a degradation pathway of brominated aromatics. In general, compounds which are more readily brominated are also more susceptible to either hydrolytic or photochemical degradation. This reduces the incidence of bromo-organic compound formation. Moreover, the chemical bond strengths show that bromine bonds are weaker than chlorine bonds. Thus bromine compounds are generally more unstable than those of chlorine and are therefore easier to chemically degrade to innocuous inorganic compounds. This is substantiated by the fact that in potable water or wastewater, bromine residuals die away in a few minutes as compared to chlorine residuals which may last for hours. Therefore, it is reasonable to conclude that the formation of bromo-organic compounds may be more easily degraded and less obnoxious than their chlorinated analogs.[57]

MEASUREMENT OF BROMINE RESIDUALS

Regardless of the method of bromine application, whether it be generated on-site, injected as molecular bromine, as an aqueous solution of either bromine or bromine chloride, the resulting residuals respond the same to all analytical procedures.

At the Grandville, Michigan project[22] the bromine residuals were determined by a spectrophotometric procedure developed by Mills of Dow Chemical Co.[36] Other researchers used the DPD FAS titrimetric method[32] and Johnson[6,15,33] used a thiosulfate-iodide amperometric titration which measured the oxidation of iodide to iodine at pH 7 by the bromine residuals. Larsen and Sollo[37] used the bromcresol purple colorimetric procedure.

There are two basic considerations for deciding on a method of analytical procedure to control the bromine disinfection process. For wastewater where there is usually an excess of ammonia nitrogen, the most difficult situation would result from the addition of chlorine followed by bromine. For control purposes the operator might be required to calibrate the final analyzer for bromine residuals only. This would mean differentiating between total chlorine and total bromine. This can be done by Palin's latest DPD method.[38]

In most cases the measurement of total residual (chlorine + bromine) will be all that is required. This can be accomplished by the iodometric method with either the starch-iodide, or the amperometric end point or the DPD colorimetric method. If the residuals are to be expressed as bromine, multiply the results obtained from procedures for "available chlorine" by 2.25. For measuring total residual (i.e., chlorine plus bromine, the operator has the option of using either the forward or the back titration procedures). (See pp. 316–325 in 14th Ed. Standard Methods (1975) Ref. 39).

DPD Differentiation Method[38]

The following procedure assumes free chlorine, combined chlorine, and free and combined bromine residuals. Hereafter the term "residual bromine" is taken to mean free bromine plus bromamines.

The reagents required are: 1) standard ferrous ammonium sulphate (FAS) solution (1 ml = 0.100 mg available chlorine); 2) DPD No. 1 powder (a combined buffer-indicator reagent); 3) potassium iodide crystals.

Free Chlorine. To a 100 ml sample add 0.5 g DPD No. 1 powder, mix rapidly to dissolve and titrate immediately with the FAS solution. This is reading A. (If free chlorine is not present this step may be omitted.)

Combined Available Chlorine. Add to the previous sample several crystals (approximately 1.0 g) of potassium iodide, mix to dissolve and after standing for two min. continue titration with FAS solution. This is reading C.

Residual Bromine. To a *second* 100 ml sample add 2 ml of 10 percent wt/v glycine solution and mix.* Then add approximately 0.5 g DPD No. 1 powder, mix and titrate with the standard FAS solution. This is reading Br.

*If it has already been determined that free chlorine is absent this step may be omitted.

Calculations. For a 100 ml sample, 1 ml FAS solution = 1 mg/liter available chlorine.

Residual Bromine = Br
Free Available Chlorine = A – Br
Combined chlorine = C – A

If it is desired to report the residual bromine in terms of bromine, multiply the Br reading result by 2.25. (This is the ratio of the molecular wt of bromine to chlorine.)

IODINE

Occurrence and Production

Iodine is always found combined, as in the iodides. It is prepared from kelp, well brines, and from crude Chile salt peter. Historically, the United States has depended upon imports for almost all of its iodine needs. Originally from Chile, it is produced as a coproduct from natural nitrate production. In recent times, however, Japan has become the dominant factor in iodine world trade. The 1974–1976 United States consumption of iodine is on the order of 6–7 million lb. During 1975, United States imports totaled 5.3 million lb, of which 93 percent was from Japan and the remainder from Chile.[40] Chile's production appears to be stabilized at a maximum of 4.4 million lb; this is from natural sources.

The small United States output in 1975 came entirely from the Dow Chemical Co. in Midland, Michigan which recovered iodine and several other chemical products from a mixed-salts-type of natural well brine. There are some reportedly high-grade brines in northern Oklahoma which may be exploited by several companies. In the fall of 1975, Houston Chemical Co. broke ground for its 2 million lb/yr iodine plant at Woodward, Oklahoma.

Uses

Iodine and its compounds are used as catalysts in the chemical industry (production of synthetic rubber) food products, pharmaceutical preparations, stabilizers (as nylon precursors), antiseptics, medicine (treatment of cretinism and goiter), inks and colorants, and industrial and household disinfectants.

Iodine was officially recognized by the Pharmacopoeia of the United States as tincture of iodine.[41] The first official United States tincture was a 5 percent solution of iodine in alcohol.

Iodine has been used as an alternative disinfectant for small or individual water supplies. During World War II, a series of studies at Harvard University by Chang, Morris, et al.[42,43] led to the development of globaline tablets for disinfecting small or individual supplies for the U.S. Army. The globaline tablets superseded the Halozone tablets for troop use, because only one globaline tablet per quart of water is required to destroy the cysts of *Endamoeba histolytica* in ten min. as compared to six Halazone tablets.

Water Supplies. Iodination of water supplies has been limited largely to emergency treatment by the military. As a disinfectant for potable water, it has been recognized for a long time but has never generated enough interest to displace the popular use of chlorine.

Swimming Pools. Iodine has been tried as an alternative to chlorine compounds in the treatment of swimming pools without much success. Iodine application, if not carefully controlled, imparts a pink tinge to the water and it has proved to be unsatisfactory as an algicide.

Wastewater. Iodine has never been tried on a plant-scale basis as a wastewater disinfectant, but only in the laboratory for purposes of investigating its germicidal efficiency. The cost and nonavailability of iodine mitigate heavily against its use as a disinfectant for wastewater.

Cost

The 1976 price for crude iodine crystals in 100-lb drums was $2.60/lb and USP granular iodine was $4.00–$5.00/lb in 100-lb drums. This compares to chlorine at 10–15¢/lb and bromine chloride at 20¢/lb.

Germicidal Efficiency

It is the consensus of many investigators[2] that elemental iodine (I_2) and hypoiodous acid (HIO) are the two most powerful disinfecting agents among the titrable iodine species. In order to relate these differences in the germicidal efficiency of these two species Chang[44] has prepared Fig. 7-4 which shows the relation of titrable iodine and contact time for the 99.9 percent destruction of cysts, virus, and bacteria by I_2 and HIO at 18°C. In practice these parameters allow a good margin of safety. Satisfactory bactericidal results can be obtained with 0.1–0.2 mg/liter in less than four min. Virus destruction by I_2 lies beyond the practical limit of 1–2 mg/liter, but for HIO it is well within these limits. The I_2 cysticidal curve lies close to the upper limit of practical range: 1.0 mg/liter in 30 min. to 2 mg/liter in thirteen min.

Compared with free chlorine residuals, iodine is definitely inferior. Free chlorine is five times more cysticidal than HIO, two hundred times more virucidal, and two times more cysticidal than I_2.

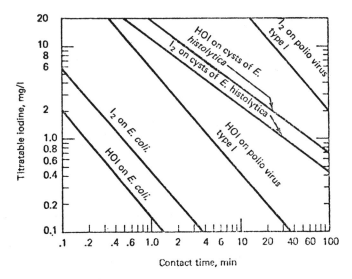

Fig. 7-4 Germicidal efficiency of iodine. From S. L. Chang.[44]

The 1960 report by McKee et al.[29] demonstrates that iodine is an effective disinfectant in wastewater; however, wasted side reactions due to the halogen demand of wastewater and its heavier molecular weight *requires 45 mg/liter iodine* to achieve the same level of disinfection as *8 mg/liter of chlorine*.

Conclusions

Iodination is not a practical means of disinfection of wastewater or water for reuse primarily because of its cost. The supply of iodine is limited, and there is insufficient field experience to even suggest iodine as an emergency measure. However, if the reader should desire further information on iodination chemistry and the design of an iodination facility see pp. 687–697 Ref. [2].

ULTRAVIOLET RADIATION

Introduction

The radiation energy of UV rays can be used to destroy microorganisms. In order to kill, the electromagnetic waves of ultraviolet irradiation must actually strike the organism. In this process some of the radiation energy is absorbed by the organism and other constituents in the medium surrounding the organisms. There is sufficient evidence to conclude that if sufficient dosages of UV energy reach the organisms, water or wastewater can be disinfected to any degree required. The germicidal effect of ultraviolet energy is thought to be associated with its absorption by various organic molecular components essential to the functioning of cells. Energy dissipation by excitation, causing disruptions of unsaturated bonds, particularly of the purine and pyridimine components of nucleoproteins, appears to produce a progressive lethal biochemical change.[45] The UV treatment does not alter water chemically; nothing is being added except energy, which produces heat, resulting in a temperature rise in the treated water.

The UV rays penetrate the cell walls of microorganisms. In accordance with the laws of photochemical action, the energy which kills a bacterium is that absorbed by the organism. The only radiant energy effective in killing bacteria is that which reaches the bacteria; therefore, the water must be free of any particles that would act as a shield. This is one of the main disadvantages of the UV process; other disadvantages are the lack of a field test which readily establishes the efficiency of the process, and its inability to provide any residual disinfecting power.

Current Practices

Potable Water. This process is being used to disinfect some small water supplies where the water is highly polished, such as filtered and demineralized water, or distilled waters produced on ocean-going vessels. There are also industrial and product water applications which use UV; e.g., breweries, pharmaceutical manufacturers, fish hatcheries, and aquariums. Most of these systems using UV radiation are in the 25–50 GPM range even though commercial units are available in sizes up to 1000 GPM. The cost of power for operating the UV lamps as well as the capital cost for the equipment make the largest units practically prohibitive.

In 1975 the Vermont State Dept. of Health embarked on a demonstration grant from the EPA to investigate and compare chlorine, ozone, and UV as practical methods for disinfecting small water supplies.[46] This report was definitely unfavorable to UV as a practical alternative to chlorine. The specific complaints were that both UV and ozone units are many times more costly than chlorination units, require more operation and maintenance, and are less reliable. Most important, though, was that neither ozone nor UV could provide a residual disinfectant to protect the water in the distribution system. For this project the capital cost for the UV system was $1995 versus $550 for a hypochlorinator and the annual operation and maintenance cost for a 20,000 GPD system and 2 mg/liter chlorine dosage was $141/yr for chlorine and $458/yr for UV.

Wastewater. The UV process for disinfecting wastewater has not been studied extensively. The Department of Environment of Canada has made some laboratory and scaled-up laboratory studies because of the fear that the discharge of chlorinated sewage effluents might result in acute toxicity to the aquatic ecosystem.[47,48] Unfortunately these experiments only ventured to a 99 percent reduction of coliforms which is 2 logs removal. This can hardly be considered disinfection. Assuming 6×10^6 coliforms/100 ml MPN as y_o the final coliform concentration would be 6×10^4 for a 99 percent removal. This low-level removal distorted the conclusions reached as to the cost of UV for disinfection.

However, the possibility of UV was demonstrated at a 40,000 GPD activated sludge plant at St. Michaels, Maryland.[49] The effluent from this plant discharges into a shellfish growing area which means that the coliform concentration must not exceed 70 MPN/100 ml. During the investigation of the UV process the effluent was discharged into a tertiary settling tank and chemicals were added to improve the suspended solids removal. The UV disinfection process proved to be highly dependent upon the quality of the activated sludge plant effluent. It did not respond successfully to plant upsets during which time the effluent had to be recycled. It was determined that the BOD of the activated sludge effluent, before tertiary treatment, should be not more than about 20 mg/liter for consistent disinfection results. This signifies that UV would probably never succeed as the sole method of disinfection but would need a backup process such as chlorine to provide reliability.

The EPA continues to support research for investigating the use of UV as a disinfectant for wastewater. Such a study has been reported by Petrasek et al.[50] in 1977. This work was conducted as a feasibility study to evaluate UV disinfection of secondary effluents as a practical method to meet the standard of 200 fecal coliforms/100 ml, based on a geometric mean. This work was carried out at the Henry J. Graeser Environmental Engineering Research and Training Facility in Dallas, Texas. This work compared the operation of equipment from two different manufacturers mainly to test the reliability of the process and to get some operating experience with sewage effluent.

The EPA is continuing its support of research into the UV process as much more operating experience is required to properly evaluate this complicated subject of irradiation.

Germicidal Efficiency

It has been determined that the germicidal effect of ultraviolet rays is maximal at wavelengths of 2500-2600 Å.[51,52] There is an abrupt decrease at 2900-3000 Å and a continual decrease to visible light which is 5000 Å. Various investigations have shown a wide range of sensitivity of different microorganisms to ultraviolet energy. In 1959, Kawabata and Harada[53] reported the following contact times required to achieve a 99.9 percent kill (3-log reduction) at a fixed UV intensity for the following organisms:

E. coli	60 sec
Shigella	47 sec
S. Typhosa	49 sec
Streptococcus faecalis	165 sec
B. subtilus	240 sec
B. subtilus spores	369 sec

Ultraviolet radiation has also been shown to be effective in the inactivation of viruses. In 1965, Huff et al.[45] reported satisfactory results which included studies of several strains of polio virus, ECHO 7, and Coxsackie 9 viruses. The intensities varied from 7000 to 11,000 μW-sec/cm^2.

There is no reference to the ability of ultraviolet radiation to destroy cysts. It is unlikely that UV would have any effect on this type of organism, *as it is the consensus that UV radiation will not kill any organism which can be seen with the naked eye.*

The St. Michaels investigation used a 130-sec exposure time to achieve a final coliform count of 70 MPN/100 ml. The initial coliform concentration in the tertiary effluent ranged from 1×10^6 to 8×10^6 per 100 ml. The absolute dose of UV to achieve a 5-log reduction was on the order of 210 μW/cm^2. This is equal to 27,300 μW-sec/cm^2 — considerably greater than previous laboratory investigations. The work by Petrasek et al.[50] at Dallas tends to confirm this UV intensity dose. They found that for a 3.5–4.0-log reduction in coliforms, which was necessary to meet the previously defined disinfection standard, a dose of 30,000 μW-sec/cm^2 would be adequate, provided a unit similar to the UPS model E-50 is used.

There are several factors which can affect the dose required. These are described below.

Interfering Factors

Ultraviolet Absorption. This may also be described as "UV demand." There are various constituents in wastewater which affect the rate of absorption of the ultraviolet energy as it passes through the wastewater in the disinfection chamber. The depth of liquid in this chamber should be such that no more than 90 percent of the UV intensity of radiation can be absorbed by the time the ultraviolet rays reach the bottom of the chamber. At St. Michaels the UV absorption phenomenon was studied in relation to the amount of turbidity and organic matter in the sewage. For each observation an absorption coefficient was calculated from the equation:

$$\frac{I}{I_o} = e^{-Kd} \tag{7-15}$$

Where

I = radiation intensity at the bottom of disinfection chamber.
I_o = initial radiation intensity.
K = absorption coefficient.
d = distance in cm between points of measurement of I and I_o.
e = natural log base.

During the St. Michaels study the absorption coefficient varied from 0.052 to 0.722, a factor of 14 depending on the quality of the effluent.

Turbidity. There is no doubt that turbidity has some deleterious effect on the absorption of ultraviolet radiation by sewage, but the St. Michaels experience indicates the amount of COD might be a much better indicator. The turbidity ranged from 1.5 to 17 JTU and the COD from 15 to 65 mg/liter. When the turbidity was 10 JTU and the COD 65 mg/liter, the absorption coefficient was at a high of 0.722; but at a turbidity of 17 JTU and a COD of 35 mg/liter the absorption coefficient was only 0.356. This confuses the issue of turbidity as an effluent quality parameter.

Transmittance. Petrasek et al.[50] reasoned that since the wavelength of ultraviolet light is 2357 Å at a maximum germicidal efficiency, turbidity measurements based on visible light are of little significance because the wavelength of visible light is about 5000 Å. For this reason percent transmittance data were obtained (instead of absorption coefficient) in order to correlate effluent water quality parameters. The data developed confirmed the St. Michaels work thusly; turbidity and suspended solids do not have the pronounced effect on the light of transmission characteristics as might be expected at the UV wavelength used. Their data concluded that the UV dose required was a function of the transmittance factor.

Furthermore, it was observed by Petrasek et al.[50] that both organic nitrogen and ammonia nitrogen interfered with the transmittance of UV light. These data indicate that the more polished the effluent the higher the transmittance factor; and hence, the more efficient disinfection will be by the UV light. A polished effluent is one low in ammonia nitrogen (1.0-2.0 mg/liter) and organic nitrogen (<2.0 mg/liter).

Photoreactivation. The St. Michaels project compared the coliform regrowth after radiation by exposing one set of replicate samples to visible light while a duplicate set was stored in the dark. From the results it is indicated that bacterial multiplication in the aliquot exposed to sunlight immediately after UV exposure, continued at an exponential rate. The density of the bacteria in the aliquot kept in the dark declined, but then multiplied after exposure to sunlight. The calculated relative dose of UV radiation in the photoreactivation study described above was 14,514 μW-sec/cm^2 in the first and 7811 μW-sec/cm^2 in the second. In each case it was obvious that some of the bacteria had not received a lethal dose of UV light, and that exposure to visible light favored bacterial regrowth.

Two experiments were carried out to determine the dose at which subsequent exposure to visible light would not favor photoreactivation. The results of these indicated that a dose of at least 33,000 μW-sec/cm^2 would be necessary to prevent photoreactivation. Similar tests are yet to be developed at Dallas.

Toxicity of Radiation. Radiation technology has been used to destroy substances known to be responsible for foul tastes in potable water. Beta radiation is effective in the destruction of geosmine and phenols at a penetration distance of 10 cm; however, gamma radiation has been discovered to produce mutagenesis in the surviving viruses during potable-water and water reuse-treatment.[54] This could also be true in the case of UV radiation. Obviously there is much more work to be done to establish UV radiation as a practical disinfectant.

Facility Design

In 1967, the U.S. Public Health Service made a policy statement on the criteria for acceptability of a UV-disinfecting unit[55] (also see p. 701 Ref (2)). All of the requirements set forth are still valid for potable water, but for well polished wastewaters the minimum dosage requirement should be doubled.

The germicidal or UV lamp is the most essential part of the system, and the UV equipment supplier is dependent entirely upon the lamp manufacturer over whom he has little control.

At present the deficiency the design engineer has to contend with is the lack of performance standards which are nationally recognized for UV-radiation units to be used for either potable water or wastewater. Another consistent problem is with the UV-intensity measuring device.[46] This component is essential for inferential proof of disinfection. If there is poor reliability of measuring UV-radiation intensity in the contact chamber there is little chance that disinfection can be reliably accomplished.

Therefore, the designer will be wholly dependent upon the integrity of the equipment supplier as long as there is an absence of performance standards for both the germicidal lamps and the monitoring intensity meters.

Since UV radiation does not provide a persisting disinfecting residual, there must be effluent by-pass or recycling requirements provided as an inherent part of the installation. This can be a large holding structure. Since the usual contact time in UV practice is on the order of 2–3 min, a holding chamber equivalent to one hr at PWWF would be considered adequate in the event there was a UV equipment failure. During any equipment failure the effluent discharge should be arranged so it can be recycled through the plant or be supplemented by standby chlorination equipment.

Reliability

From the most recent and intensive field investigations, which are too few to make a complete judgment, it must be concluded that the reliability of the UV process is grossly dependent upon the quality of the water being treated.

Costs

The following is the estimated cost for wastewater disinfection by UV radiation as determined by the 1976 EPA Task Force report.[56]

Wastewater Disinfections Costs by UV Radiation

Plant Size (mgd)	1	10	100
Capital Cost ($)	71,000	360,000	1,780,000
Disinfection Cost (¢/1000 gal.)	4.19	2.70	2.27

Conclusions

From the current information available on the UV-radiation process as a wastewater disinfectant it is clear that many more reasons than cost, or possible elimination of toxic compounds, will have to be presented to put the UV process in any sort of contention with conventional methods of disinfection.

Chapter 7—Summary

Bromine is considered the most reactive oxidant of all the halogens, except fluorine which is not considered here. For this reason it has some desirable chemical characteristics. The hydrolysis of bromine in an aqueous solution is almost identical to chlorine. Bromine added to water reacts to form hypobromous acid and in the presence of ammonia nitrogen in the quantities usually found in wastewater bromamines are formed. The germicidal efficiency of these bromamines in a pH environment similar *to that of wastewater is practically equal to that of free chlorine.*

Bromine is highly reactive; therefore, its halogen demand is greatly distorted beyond its usefulness. So any system in wastewater treatment attempting to exploit the features of bromine should be arranged to use chlorine to minimize the "bromine demand."

Bromine as Br_2 is a difficult and hazardous chemical to handle because it is in liquid form at room temperature. This led to the development of bromine chloride which has the same vapor pressure characteristics as sulfur dioxide. This makes BrCl much more desirable than elemental bromine from the standpoint of chemical handling facilities.

Bromine Chloride has been investigated thoroughly as a practical substitute for elemental bromine and liquid-gas chlorine installations. Those investigations have shown conclusively that a new species of equipment will be required for the metering and control of BrCl. The factors of increased corrosivity of BrCl over chlorine and the delicate problem of BrCl dissociation have determined the necessity of fundamental design changes.

On-site generation of bromine from bromide salts using chlorine as the oxidizer has interesting possibilities. Injecting a bromide salt into a chlorinator solution discharge line (similar to the generation of chlorine dioxide) produces the formation of elemental bromine which hydrolyzes immediately to hypobromous acid. This concept can be used to great advantage at existing chlorination installations for the following reasons: the existing chlorination system can be arranged to produce a 2-5 min. controlled chlorine residual of 0.3-0.5 mg/liter (instead of the usual 5-7 mg/liter). This sequesters the halogen demand. Following this is the formation of elemental bromine from the combination of the chlorine solution discharge with the bromide salt injection from the second chlorinator application. This becomes the second point of "chlorine" application downstream from the original point of chlorination. This system has many desirable features: it can utilize existing chlorination facilities; it eliminates the hazardous problems of handling bromine; and it can utilize the superior germicidal properties of bromine without sacrificing other design considerations. Moreover, the bromamine residual die-away phenomenon may be so rapid that debromination by SO_2 may not be required, and the contact time for these "hot" bromine residuals can be as low as 5 min. and not longer than 15 min. This would bring contact chamber requirements to the lowest recommended for ozone which are the lowest for all disinfectants.

Iodine has been used as a temporary expedient for the disinfection of small water supplies and as an alternative to chlorine in the treatment of swimming pools.

It has never been tried even on a pilot-plant experimental basis for the disinfection of wastewater. The cost and reliability of iodine supply makes it an impossible choice as a disinfectant for wastewaters or waters for reuse.

Ultraviolet radiation has significant disinfection qualities. These qualities are best exemplified in waters of high quality—those which have been polished by various degrees of pretreatment.

The equipment available for the practitioner of UV disinfection does not have to conform to any rigid performance standards. Therefore, the reliability of commercial equipment is not necessarily trustworthy.

The UV process does not provide a persisting residual which could or would define a situation that "disinfection" had been achieved.

There are a great many interferences limiting the germicidal efficiency of UV radiation. These have to do with the quality of the wastewater being treated. Therefore, UV radiation as a disinfectant is hampered by too many variables, particularly those usually found in sewage plant effluents. Experience shows that UV radiation as a disinfectant is only effective when treating highly polished waters. It has no place in the disinfection of wastewaters.

REFERENCES

1. Mills, J. F. "Interhalogens and Halogen Mixtures as Disinfectants." Chapter 6 *Disinfection: Water and Wastewater*, J. D. Johnson, (editor). Ann Arbor Science, Ann Arbor, Mich. (1975).

2. White, G. C. *Handbook of Chlorination* Van Nostrand Reinhold, Co., N.Y. 1972.

3. "Bromine, Its Properties and Uses." Michigan Chem. Corp., Chicago, Ill. 1958).

4. "Bromine Outlook Tied to Clean Air Rules." Chem. and Engr. News, p. 11, (Feb. 25, 1974).

5. Galal-Gorchev, Hend and Morris, J. C. "Formation and Stability of Bromamide and Nitrogen Tribromide in Aqueous Solution." *Inorganic Chem.* 4:899 (1965).

6. Johnson, D. J. and Overby, R. "Bromine and Bromamine Disinfection Chemistry." *ASCE J. San. Eng. Div.* p. 617 (Oct. 1971).

7. La Pointe, T. F. Inman, G., and Johnson, J. D. "Kinetics of Tribromamine Decomposition," in Chapter 15 *Disinfection: Water and Wastewater.* J. D. Johnson (editor), Ann Arbor Science, Ann Arbor, Mich. (1975).

8. Brown, J. R., McLean, D. M., and Nixon, M. C. "Bromine Disinfection of a Large Swimming Pool." *Can. J. Public Health* **55**: 251 (June 1964).

9. Mills, J. F. assignor to Dow Chemical Co. "Control of Microorganisms With Polyhalide Resins." U.S. Patent 3,462,363 (Aug. 19, 1969).

10. Mills, J. F., Goodenough, R. D., and Nekervis, W. F. assignors to Dow Chemical Co. "Process for Treating Water With Bromine." U.S. Patent No. 3,316,173 (April 25, 1967).

11. Goodenough, R. D., Mills, J. F., and Place, J. "Anion Exchange Resin (Polybromide Form) as a Source of Active Bromine for Water Disinfection." *Env. Sci. Tech.* Vol. 3 p. 854 (Sept. 1969).

12. Regunathan, P. and Brejcha, R. J. "A New Technology in Potable Water Disinfection for Offshore Rigs," A paper presented Annual Mtg. Soc. of Petroleum Engrs. of AIME, Dallas, Tex. (Sept. 28–Oct. 1, 1975).

13. U.S. Patent 2,443,429, Procedure for Disinfecting Aqueous Liquid; Marks and Strandkov, and Wallace and Tiernan Co., Inc. Belleville, N.J. (June 15, 1948).

14. Derreumaux, A. "Process and Apparatus for Disinfecting or Sterilizing by the Combination of Chlorine and Other Halotens," French patent No. 2,171, 890 (1973).

15. Johnson, J. D., and Sun, W. "Bromine Disinfection of Wastewater," Chapter 9 in *Disinfection: Water and Wastewater*; J. D. Johnson (editor), Ann Arbor Science, Ann Arbor, Mich. (1975).

16. Sollo, F. W., Mueller, H. F., and Larson, T. E. "Prechlorination Enhances Disinfection by Bromine." Paper presented at the 7th Int. Conf. on Water Poll. Res., Paris, France (Sept. 9–12, 1974).

17. Venosa, A. D. "Disinfection: State of the Art, Alternatives to Chlorination for Wastewater." Presented at a Disinfection Seminar, Dept. of Ecology, Univ. of Wash., Seattle, Wash. (May 26–27, 1976).

18. Mills, J. F. and Oakes, B. D. "Bromine Chloride Less Corrosive than Bromine." *Chem. Eng.* pp. 102–106 (Aug. 1973).

19. Mills, J. F. "Disinfection of Sewage by Chlorobromination." Paper presented at Ann. Chem. Soc. Meeting, Dallas, Texas (April 1973).

20. Cole, L. G. and Elverum, G. W. "Thermodynamic Properties of the Diatomic Interhalogens from Spectroscopic Data," *J. Chem. Phys.* 20:1543 (1952).

21. Jolles, Z. E. "Bromine and Its Compounds." Ernest Benn Ltd., London, England, Chapter 1 (1966).

22. Ward, R. W., Giffin, R. D., De Graeve, G. M., and Stone, R. A., "Disinfection Efficiency and Residual Toxicity of Several Wastewater Disinfectants, Vol. I, Grandville, Mich." EPA Pre-Publication Report, Project No. S-802292 (1976).

23. Collins, H. F., Selleck, R. E., and White, G. C. "Problems in Obtaining Adequate Sewage Disinfection." *ASCE J. Env. Eng. Div.* 97:549 (Oct. 1971).

24. Cole, S. A. "Experimental Equipment for Feeding Bromine Chloride." Paper presented at the International Water Conference, Pittsburgh, Pa. (Oct. 31, 1974).

25. Mills, J. F., Dow Chemical Co., Midland, Mich. Private communication. (1976).

26. Wyss, O. and Stockton, R. J. "The Germicidal Action of Bromine." *Arch. Biochem.* **12**:267 (1947).

27. Marks, H. C. and Strandkov, F. B. "Halogens and Their Mode of Action." *Ann. N.Y. Acad. Sci.* **53**:163 (1950).

28. Johannesson, J. K. "Anomalous Bactericidal Action of Bromamine." *Nature* **181**:1799 (1958).

29. McKee, J. E., Brokaw, C. J., and McLaughlin, R. T. "Chemical and Colicidal Effects of Halogen in Sewage," *J. WPCF* **32**:795 (1960).

30. Koski, T. A., Stuart, L. S., and Ortenzio, L. F. "Comparison of Chlorine, Bromine, and Iodine as Disinfectants for Swimming Pool Water." *Appl. Micro.* p. 276 (March, 1966).

31. Schaffer, R. B. and Mills, J. F. "Proceedings of the National Symposium on Quality Standards for Natural Waters." Univ. of Mich. Press, Ann Arbor, Mich. p. 158 (1966).

32. Sollo, F. W., Mueller, H. F., Larson, T. E., and Johnson, J. D. "Bromine Disinfection of Wastewater Effluent." Chapter 8, *Disinfection: Water and Wastewater*, J. D. Johnson (editor), Ann Arbor Press, Ann Arbor, Mich. (1975).

33. Johnson, J. D. Univ. of North Carolina. Private communication (1975).

34. Kott, Y. "Effect of Halogens on Algae—III Field Experiment." Water Research, Pergamon Press, p. 265 (1969).

35. Derreumaux, A. CIFEC, Neuilly-sur-Seine, France, Private communication. (1976).

36. Mills, J. F. "A Spectrophotometric Method for Determining Microquantities of Various Halogen Species." Dow Chemical Co., Midland, Mich. (1971).

37. Larson, T. and Sollo, F. W. "Determination of Free Bromine in Water." Annual Progress Report, U.S. Army Medical R & D Command, Contract No. DA-49-193-MD-2909 (1967).

38. Palin, A. T. "Analytical Control of Water Disinfection With Special Reference to Differential DPD Methods for Chlorine, Chlorine Dioxide, Bromine, Iodine and Ozone." *J. Inst. Water Eng.* **28**:139 (1974).

39. "Standard Methods for the Examination of Water and Wastewater." 14th Ed., *Am. Pub. Health Assoc. WPCF AWWA*, N.Y. (1975).

40. "Mineral Industry Surveys." U.S. Dept. of Interior Bureau of Mines, Washington, D.C. (1975).

41. Lawrence, C. A. and Block, S. S. "Disinfection, Sterilization and Preservation." Lea and Febiger, Philadelphia, Pa. (1968).

42. Chang, S. L. and Morris, J. C. "Elemental Iodine as a Disinfectant for Drinking Water," *Ind. Eng. Chem.* **45**:1009 (1953).

43. Morris, J. C., Chang, S. L., Fair, G. M., and Conant, G. H., Jr. "Disinfection of Drinking Water under Field Conditions," *Ind. Eng. Chem.* 45:1013 (1953).

44. Chang, S. L. "Iodination of Water." U.S. Dept. H.E.W., Taft San. Engr. Center, Cincinnati, Ohio (1966).

45. Huff, C. B., Smith, B. S., Boring, W. D., and Clarke, N. A. "Study of Ultraviolet Disinfection of Water and Factors in Treatment Efficiency." Public Health Reports 80:695 (Aug. 1965).

46. Witherell, L. E., Solomon, R., and Stone, K. M. "A Demonstration Project to Determine the Feasibility of Using Ozone and Ultraviolet Radiation Disinfection for Small Community Water Systems, EPA 68-03-2182." a paper presented at the Disinfection Seminar AWWA Ann. Conf. Anaheim, Calif. (May 8, 1977).

47. Oliver, B. G. and Cosgrove, E. G. "The Disinfection of Sewage Treatment Plant Effluents Using Ultraviolet Light," *Can. J. Chem. Eng.* 53:170 (April 1975).

48. Oliver, B. G. and Carey, J. H. "Ultraviolet Disinfection: An Alternative to Chlorination for Sewage Effluents." a paper presented at 48th Annual Conf. WPCF, Miami Beach, Fla. (Oct. 7, 1975).

49. Roeber, J. A. and Hoot, F. M. "Ultraviolet Disinfection of Activated Sludge Effluent Discharging to Shellfish Waters, EPA 600/2-75-060." Municipal Environmental Research Lab., Cincinnati, Ohio (Dec. 1975).

50. Petrasek, A. C., Jr., Andrews, D. C., and Wolf, H. W. "Ultraviolet Disinfection of Wastewater Effluents." a paper presented for the Disinfection Seminar, AWWA Ann. Conference, Anaheim, Calif. (May 8, 1977).

51. Luckiesh, M. and Holladay, L. L. "Disinfection of Water by Means of Germicidal Lamps." General Electric Review, p. 45 (April 1944).

52. Loofbourow, J. R. "The Effects of Ultraviolet Radiation on Cells." *Growth* 12:77 (1948).

53. Kawabata, T. and Harada, T. "The Disinfection of Water by the Germicidal Lamp." *J. Illumination Soc.* 36:89 (1959).

54. Vilagines, R., Inst. Pasteur de Paris, Private communication. (1977).

55. "Policy Statement on the use of the Ultra-Violet Process for Disinfection of Water." Dept. of H.E.W. Div. of Env. Engr. and Food Prot., Washington, D.C. (April 1, 1967).

56. Task Force Report "Disinfection of Wastewater, EPA-430/9-75-012." U.S. Env. Prot. Agency, Cincinnati, Ohio (March 1976).

57. Mills, J. F. Private communication. Dow Chemical Company, Midland, Mich. (Nov. 1977).

8

Ozone

INTRODUCTION

Historical Background

Ozone (O_3) has been used for more than sixty yr. for water treatment on the European continent. More than a thousand municipal water treatment plants use ozone as part of their chemical treatment.[1] Most of these are in Western Europe, particularly France and Switzerland, but usage is spreading to other countries. Most of these installations are primarily for taste and odor control, and color removal; they are almost without exception backed up by chlorination in spite of the fact that ozone is an admirable disinfectant. The first experiments with ozone were performed in France in 1886 by de Merintence.[2] Ozone was first used on a full-scale basis at Nice, France in 1906 for the treatment of the municipal water supply. This water comes from the river Var which originates close by in the Alps. The quality of this water is usually excellent and only deteriorates during flood periods following heavy rainfalls. The resulting pollution consists mainly of suspended materials which are easily removed by sand filtration. Ozone is used to destroy coliform organisms and the grassy or earthy tastes which accompany periods of high runoff.[3]

The Nice plant is still in use. Enlargements made in 1922 and 1951 increased its capacity to about 25 mgd. Today (1976) the largest ozone installations are used for water treatment in the Paris area. Three separate plants treat water from the Oise, Marne, and Seine Rivers. The combined capacity of these treatment plants is 360 mgd and the ozone production (from air) is approximately 5 tons/day.

In Switzerland, ozone has been used during the past 25 yr. for treating spring waters, ground waters, and surface waters. Following a massive phenol discharge in 1957 at St. Gall, disinfection has been by ozone instead of chlorine. As of 1977, there are about 150 installations using ozone for treatment of potable water and industrial processes. The largest has a capacity of 1750 lb/day. This facility is at the Lengg Lake plant in Zurich. The next largest systems are 650 lb/day in Geneva and St. Gall.[4]

Montreal, Canada is currently building a water-treatment plant with ozone as the only disinfecting process. This facility will have a capacity on the order of 7.5 tons of ozone/day.

The first large Russian filtration-ozonation plant was built in 1911 in St. Petersburg (Leningrad) having a total capacity of 12 mgd. Owing to difficulties in operation and maintenance this unit was shut down in 1922.[2] Subsequently, large ozonation installations have been included in water treatment plants for Moscow (317 mgd at 4 mg/liter O_3), Kiev (106 mgd at 5 mg/liter O_3), and Gorski (82 mgd at 2.0 mg/liter O_3). Other plants (under construction in 1976) using ozone as a disinfectant are those of Singapore, Malaya; Lodz, Poland; and Chiba, Japan. None of these designs are for ozone usage capacities in excess of 5 mg/liter with the median design at 3.4 mg/liter.[5]

The United States currently has installations at Whiting, Indiana, dating from 1940, used primarily for the destruction of phenolic tastes and odors; the Belmont plant, City of Philadelphia commissioned in 1949 and operated successfully until 1957 when the ozone process was discontinued in favor of chlorination. This resulted primarily from the increase in the plant capacity which provided contact basins exceeding 24 hr, making free residual chlorination more economical than ozone.

The use of ozone in North America has not been popular compared to chlorine, mostly because the water available for potable use in North America is generally far superior in quality to its counterparts in Western Europe.

Ozone can be preferably more attractive for the treatment of low quality potable water supplies where taste, odors, and color are a dominant factor for removal. These situations are far more abundant in Western Europe than in North America. Ozone has long been considered the superpolishing agent for potable waters (i.e., color, taste, and odor removal plus disinfection).

Now, in the 1970's however, there is a more compelling situation in North America which may provide the needed impetus to establish ozone as a possible alternate disinfectant to chlorine, particularly in wastewater treatment and water reuse situations. There are several reasons for this supposition: 1) new pollution control requirements demand a higher degree of wastewater treatment, and ozone is known as an excellent water polishing agent; 2) a trend toward wastewater reuse which requires virus inactivation and polishing; 3) the installation of oxygen (Unox) activated sludge secondary treatment plants; 4) toxicity problems associated with chlorine treatment; 5) the additional benefits of ozone which produces a high level of D.O. in the effluent along with certain types of color removal and possible reductions in TOC and COD. A final note: as more ozone is used, there are sure to be technological advances which will improve the status of ozone as a wastewater disinfectant.

Owing to these factors many studies of the practical value of ozone have been initiated (ca. 1975). These include pilot-plant studies of secondary effluents at Fairfax, Virginia; St. Paul, Minnesota; Louisville, Kentucky; Hendersonville, Tennessee; Chicago, Illinois; Redbridge, London, England; and Dallas, Texas (Dallas Water Reclamation Research Center).

These and other small scale pilot-plant studies have resulted in the incorporation of ozone into tertiary plants now under design (1975). These include the 50 mgd Westerly Wastewater Treatment plant, Cleveland, Ohio; the 5 mgd Chino Basin plant, San Bernardino County, California; a 100 gpm plant in Suffolk County, New York to provide design information for several large tertiary plants on Long Island, New York; and in Springfield, Missouri construction is underway on a 32 mgd two-stage activated sludge plant which will utilize ozone for disinfection.

Considerable work has been done by the Sanitation Districts of Los Angeles County at their Pomona Research Center evaluating the efficiency of ozone as a disinfectant for both secondary and tertiary effluents. However, as of 1977, information on the practical value of ozone as a wastewater disinfectant is meager, except that we know it is a good virucide. These efforts and others, which are being encouraged by the EPA, will develop sufficient data so that by 1980 the designers will be able to resolve the position of ozone as a wastewater treatment tool.

Physical and Chemical Properties

Ozone (molecular weight 48), an allotropic form of oxygen, is an unstable blue gas with a characteristic pungent odor to which it owes its name. It is derived from the Greek word "ozein," meaning "to smell." Discovered by van Marum in 1785, it is produced commercially from dry air or oxygen formed by the corona discharge of high-voltage (4000–30,000 V) electricity. Ozone is also formed photochemically in the earth's atmosphere. It is one of the hazardous elements of smog, and its concentration in the atmosphere is an indicator of the smog intensity. Ozone content in the atmosphere in excess of 0.25 ppm is generally considered injurious to the health of man. Ozone levels of 1.0 ppm in the atmosphere are extremely hazardous to health. Ozone weighs approximately 0.135 lb/ft^3 at one atm. It is a powerful oxidizing agent. Its oxidation potential is -2.07 V referred to the hydrogen electrode at 25°C and a unit H-ion activity. Only Fluorine has a more electronegative oxidation potential.[6] Ozone is extremely corrosive, so that materials of construction must be very carefully chosen. Porcelain and glass do not react with ozone. PVC is used, but it is suspected that a reaction takes place, resulting in the loss of ozone.[7] *It completely disintegrates rubber and attacks all plant life.*

The solubility of ozone in water is a limiting factor that greatly affects the process of ozonation. At 20°C the solubility of ozone is only 570 mg/liter.[8] While ozone is more soluble than oxygen, chlorine is twelve times more soluble. In pure aqueous solution, ozone is thought to decompose as follows.[1]

$$O_3 + H_2O \longrightarrow HO_3^+ + OH^- \tag{8-1}$$

$$HO_3^+ + OH^- \longleftarrow 2HO_2 \tag{8-2}$$

$$O_3 + HO_2 \longrightarrow HO + 2O_2 \tag{8-3}$$

$$HO + HO_2 \longrightarrow H_2O + O_2 \tag{8-4}$$

The free radicals (HO_2 and HO) that form when ozone decomposes in aqueous solutions have great oxidizing power, and in addition to disappearing rapidly (Eq. 4), may react with impurities; e.g., metal salts, organic matter, hydrogen and hydroxide ions present in solution. These free radicals formed by the decomposition of ozone in water are apparently the principal reacting species. Ozone, while it exists, does not lose its oxidizing capacity in an aqueous solution.

Inorganic Reactions

Ozone reacts rapidly to oxidize ferrous and manganous ions into their insoluble ions resulting in either a floc that precipitates, or a scum that clings to the water surface. Sulfides and sulfites are readily oxidized to sulfates, and nitrites to nitrate. The oxidation of iodides to iodine is the basis of the usual analytical determination of ozone. Bromides and chlorides are similarly oxidized to bromine (Br_2) and chlorine (Cl_2) respectively, and these reactions are slow and dependent upon the concentration of reactants.

The ammonium ion (NH_4^+) is apparently not attacked under the conditions normally found in wastewater treatment, so there is no waste of ozone oxidizing capacity or side reactions with the ammonia nitrogen in wastewater.*

*Ammonia is oxidized completely to nitrate by ozone if the wastewater pH remains alkaline. The molar ratio of ozone consumed per ammonia oxidized is about 12 to 1.[64]

Organic Reactions

Ozone reacts readily with unsaturated organic compounds, adding all three oxygen atoms at a double or triple bond. The resulting compounds are called ozonides. Decomposition of ozonides results in a rupture at the position of the double bond, causing the formation of aldehydes, ketones, and acids. Ozone readily destroys phenolic compounds and is capable of bleaching organic color found in some waters. These last two characteristics are responsible for the popularity of ozone in the treatment of low quality surface waters in Western Europe.

Glaze et al.[9] has reported on the ability of ozone to destroy humic acid which is the precursor of THM (trihalomethane) formation.* Guirguis et al.[10] reported that ozone makes organic compounds more adsorbable by carbon. Prengle et al.[11] reports that, with time and proper dosage, ozone plus UV light can reduce malathion to carbon dioxide and water after forming three or four intermediate compounds such as alcohol, aldehydes, and oxalic acid after a one hr contact time. Not only was the pesticide (malathion) destroyed, as indicated by gas chromatographic analysis, but also the total organic content of the water. Likewise Richard[12] revealed from his studies that ozone can degrade the pesticides parathion and marathion to phosphoric acid.

Toxicity of Ozone

During the International Ozone Institute meeting in Cincinnati, Ohio (Nov. 1976), several speakers presented information on some of the known toxic properties of ozone. Falk and Moyer[13] reported on a review of scientific literature dealing with reactions of ozone with organic materials as they might occur under treatment conditions. All organic substances considered were taken from a list of compounds that had been found in drinking water by the EPA. Ozonolysis of some of these compounds forms a variety of hydroperoxides which are known to be mutagenic. Ozonolysis of pesticides can produce epoxides, some of which have been shown to be carcinogenic.

Hartemann, Block and Maugras from France[14] reported on the preliminary results of their studies that the toxicity induced by the ozonation of organic materials may be lower than those induced by chlorination which was attributable to the chlorine residual.

Kinman et al.[15] found that ozonated wastewater was more toxic than unozonated wastewater.

Spanggord and McClurg[16] of Stanford Research Institute (SRI) found that a very high concentration of ozone produces mutagenic compounds when reacted with ethanol.

*This illustrates the value of preozonation in an ozone-chlorination combination.

Simmon and Eckford[17] also of SRI found an increase in mutagenesis after ozonation of ethanol, benzidine, and nitrilotriacetic acid, but 26 other compounds studied were not found to be mutagenic after ozonation.

The public health significance of all these findings is not known as of this writing. The French who have been ozonating for at least 50 yr. do not express any concern, even though they are continuing their toxicity investigations.

CHEMISTRY OF OZONATION

Introduction

The role of ozone in wastewater treatment may be classified as both an oxidant and a germicidal compound. These are the same properties exhibited by aqueous chlorine; therefore, there is tendency to view the two substances as competitors. However, the competition is not exact nor universal, for the ways in which the functions are accomplished are somewhat different.[18] It should be emphasized that ozone and an aqueous chlorine solution can act in complementary fashion, each performing some tasks more usefully than the other. *Therefore, there are two distinct faces of ozone: 1) as a powerful chemical oxidant; and 2) as a germicidal agent.*

Ozone: The Disinfecting Agent

In the past several years (1965–1975) there have been a number of laboratory and pilot-scale studies to evaluate the effectiveness of ozone in disinfecting wastewater. This work has also provided information on the mechanisms by which ozonation occurs.

The potent germicidal properties of ozone have been attributed to its high oxidation potential. Research studies indicate that disinfection by ozone is a direct result of bacterial cell wall disintegration. This is known as the "lysis phenomenon." This mechanism of disinfection by ozone is indeed different from that by chlorine. Although the exact chemical action of chlorine is uncertain, it is generally agreed that the chlorine residual in an aqueous solution diffuses through the cell wall of the microorganism and attacks the enzyme group, the destruction of which results in the death of the microorganism (see p. 215, Ref 19).

Ozone has long been recognized as an excellent disinfecting agent, but reliable quantitative studies of the fundamental germicidal activity of ozone are so few that our knowledge of its real potency is meager compared to what we know about chlorine and other disinfectants. This is partly because of the superior oxidizing power of ozone. It is most difficult to experimentally obtain the necessary time-dependent relations with extremely small ozone concentrations.[18]

Venosa[21] pointed out in his comprehensive review of the literature dealing with the germicidal efficiency of ozone, that there exists much controversy, contradiction, confusion, and nonfactual subjective judgment on the use of ozone. One of the most serious failures by the various investigators has been their inability to distinguish between the concentration of ozone applied and the residual ozone necessary for effective disinfection. It must be recognized that the same principle for controlling chlorination, which is by residual, should also be applied to the control of ozone.

In 1976, Morris[18] presented an excellent summary of what is currently known about the germicidal efficiency of ozone. First he laid to rest the fallacy that ozone displays an "all-or-nothing" effect on bacterial kill. This was the result of the interpretation of the work done by Fetner and Ingols in 1956[22] and has been often quoted to substantiate this effect. Morris has emphasized that this so-called all-or-nothing effect is neither real nor significant.[18] The effect appears simply because of the inability or failure of investigators to space the dosage concentrations of ozone reagent close enough. Ozone is so strong a germicide that concentrations of only a few micrograms per liter are needed to measure germicidal action. The spacing of the concentrations used by Fetner and Ingols was about 0.1 mg/liter or 100 μg/liter, a large enough gap to go from zero kill at a dosage just equal to demand, all the way to a very rapid kill at a concentration equal to demand plus 0.1 mg/liter. This example is just one of many which confronts the researcher in attempting to evaluate the potency of disinfectants. To overcome this situation Morris[23] developed the concept of the lethality coefficient for a given disinfectant. He treated all of the significant developed data on ozone, beginning with the work by Kessel et al. in 1943 to develop this lethality coefficient:

$$\Lambda = 4.6/Ct_{99} \tag{8-5}$$

Where

C = residual concentration in mg/liter.

t_{99} = time in min. for 99 percent microorganism destruction (2-log destruction).

The weighted mean results of these evaluations are shown in Table 8-1. The values are considered by Morris to be valid only within a *factor of two*. The ranges of values do not warrant any greater confidence than this. Comparison of the values in this table with similar values obtained for chlorine (HOCl) are shown in Table 8-2. These values of the lethality coefficient are computed from the 1967 tabulation by Morris.[23] These tabulations by Morris clearly illustrate that ozone is a more powerful germicidal against all classes of organisms listed, by factors of 10–100. The relative sensitivities of the various types of organisms are the same as for HOCl. As would be expected, bacteria are the most sensitive and

TABLE 8-1 Parameters For Disinfection By Ozone[18]
(pH 7; 10-15°C)

Organism	Λ[a]	$C_{99:10}$[b]
Escherichia coli	500	0.001
Streptococcus faecalis	300	.0015
Polio virus	50	.01
Endamoeba histolytica	5	.1
Bacillus megatherium (spores)	15	.03
Mycobacterium tuberculosam	100	.005

[a]Λ = specific lethality coefficient = $\ln 100 \div Ct_{99}$
[b]$C_{99:10}$ = concentration in mg/liter for 99 percent destruction or inactivation in 10 min.

the usual types of vegetative bacteria examined all exhibit about the same sensitivity. Viruses are more resistant by about a factor of ten, but not enough forms have been tested with ozone to determine the range of resistances of the difficult viruses. Cysts and spores, as with aqueous chlorine, are about a factor of ten times more resistant than viruses.

Not much experimentation has been done with the activity of ozone at various pH levels. The germicidal efficiency of ozone does not seem to be affected significantly within the pH range of 6-8.5. Even less is known about the effect of treated wastewater temperature on germicidal efficiency. One thing is certain however, the higher the water temperature the lower the efficiency of ozone mass transfer, which translates to lower germicidal efficiency.

For oxidizing compounds used as disinfectants, their superior oxidizing characteristics might provide them with high-lethality coefficients; however, these same characteristics might also cause them to have higher consumption rates if placed in an environment such as wastewater abounding in compounds which react rapidly with oxidizing agents. Therefore, the more reactive the compound the fewer the "miles per gallon."

TABLE 8-2 Values of Λ At 5°C
$[(mg/liter)^{-1} (min.)^{-1}]$

Agent	Enteric Bacteria	Amoebic Cysts	Viruses	Spores
O_3	500	0.5	5	2
HOCl as Cl_2	20	0.05	1.0 up	0.05
OCl^- as Cl_2	0.2	0.0005	<0.02	<0.0005
NH_2Cl as Cl_2	0.1	0.02	0.005	0.001

Practical Standards for Ozone

There is a great need for establishing standards for the practical application of ozone as a disinfectant. In France, the parameter is a detectable residual at a variable contact time depending upon the water temperature. This is for high quality mountain runoff waters.[24] However, with polluted sources, such as the rivers adjacent to Paris, the current criterion for satisfactory disinfection is 0.4 mg/liter residual ozone for a minimum contact period of 4-5 min. Such a practice is termed "full ozonation." This term translates directly the definition of the 4-5 min. "ozone demand." Application of ozone in wastewater is an entirely different chemical situation than the application to low-quality surface water supplies. In order for the ozone to be germicidally effective it first must overcome the initial ozone demand. Owing to the high reactivity of ozone, the ozone demand can be extremely high and unpredictable in secondary effluents. Therefore, disinfection by ozone must ultimately be controlled either by a predetermined variable dosage rate based on laboratory determinations of the ozone demand or by a given ozone residual for a specified contact time. The latter may be difficult to achieve because of the wide diurnal variations of wastewater ozone demand in any given 24-hr period. By way of comparison, the application of chlorine has been controlled by the residual contact time concept established as long ago as 1912.[19]

Ozone as an Oxidant

Ozone has a wide array of attributes attractive to its use in potable water treatment such as taste and odor control, color removal, and iron and manganese removal. These oxidizing powers are quite valuable in the polishing of low-quality supplies including water reuse situations.

Ozone oxidizes inorganic substances completely and rapidly, e.g., sulfides to sulfates, ferrous iron to ferric, manganous ion to manganese dioxide or permanganate, and nitrites to nitrate. Of even greater importance is ozone's capability of breaking down organic complexes of both iron and manganese which usually defy the usual procedures of iron and manganese removal from potable waters.

Oxidation of organic materials by ozone is more selective and incomplete at the concentrations and pH values of aqueous ozonation. Unsaturated and aromatic compounds are oxidized and split at the classical double bonds, producing carboxylic acids and ketones as products.[25] Because of the high reactivity of ozone, oxidation of organic matter in the aqueous environment whether it be potable water or wastewater will consume ozone in varying amounts. Therefore, one of the most significant parameters for evaluating ozone is the determination of the immediate ozone demand. Oxidation of the (organic) material is usually incomplete. It is estimated that the reduction in TOC may be only 10-20 percent, although decreases in COD and BOD are generally greater, ranging up to 50 percent COD reduction as with Montreal water.[26] There are also instances where COD appeared to increase, resulting from conversion to more readily oxidized compounds.

Ozone exerts a powerful and effective bleaching action on the organic compounds which contribute to the color in the wastewater and potable water. The ability of ozone to attack these compounds which contribute to the color, some of which are the humates and fulvates, makes ozone a fine wastewater polishing agent.

The ability of ozone to destroy taste forming phenolic compounds is probably its most important contribution to the field of potable water treatment. Moreover, this ability appears to be capable of destroying other taste forming compounds of unknown origin. There are two major mechanisms by which ozone may react with organic material.[27] The first of these is a direct additive attack in which ozonides and ultimately peroxides are formed together with a splitting of the organic molecule. The other mechanism is an accompaniment to the decomposition of ozone. This decomposition proceeds by way of the formation of the free radicals OH, HO_2 and HO_3 as described above. These free radicals, especially OH, are highly reactive against all sorts of organic material and may lead to autooxidation of a wide variety of organic matter, particularly those present in wastewater effluents. The free radical autooxidation mechanism may well be involved in the disappearance of residual ozone after the initial rapid demand has been satisfied.[18]

EFFICIENCY OF OZONE IN WASTEWATER

The most difficult question to answer is: "How much Ozone will be required for disinfection?" In order to answer this question two factors must be known: 1) disinfection must be defined, and 2) the 5 min. ozone demand must be determined.

Disinfection Parameters

The Environmental Protection Agency relates disinfection to the MPN/100 ml of *fecal coliforms* whereas California and some other state agencies (as do water producers) rely on the identification of *total coliforms*. In Chapter 1 the differences between these two concepts of microbial indicators are discussed. For purposes of evaluating data presented in either way, it is safe to say that a *fecal coliform MPN requirement of 200-400/100 ml is comparable to a range of 2000-4000/100 ml MPN of total coliform.*

There are two popular ways of reporting disinfection efficiency. One is the log reduction of organisms and the other is the percent destruction of the organisms.

For example, a 99.0 percent kill is a 2-log reduction; a 99.9 percent kill is a 3-log reduction; a 99.99 percent kill is a 4-log reduction, and a 99.999 percent kill is a 5-log reduction. Disinfection of a high quality activated sludge effluent will require a 4- to 5-log reduction depending upon the receiving water requirements.

When comparing fecal to total coliforms it is generous to say that 200/100 ml MPN fecal coliforms is comparable to 2300/100 ml MPN total coliform. This is the lower limit where infectious diseases are likely to occur (see Chapter 1).

Ozone Demand

Choosing a number such as the "3-min. O_3 demand" is based largely upon the consensus that ozone demonstrates its maximum kill at 3 min.[28] Elsewhere in the literature, this figure varies from 2 to 10 min. It may be more convenient to choose 5 min. since it is easier on the laboratory personnel who have to perform these studies. At any rate, it is imperative to know the ozone demand of a wastewater in order to size the equipment for a particular project. Furthermore, it is necessary to know the diurnal variation of this demand. Since ozone is much more highly reactive than most other disinfectants such as chlorine or chlorine dioxide, it is likely this range will be as great as 6 to 1, compared to chlorine at 3 to 1. This estimate is based upon a well oxidized domestic sewage effluent as will be described below. Sudden wide variations in chemical demand presents a serious control problem. Both ozone and bromine are extremely sensitive to demand changes.

Recent Studies

Redbridge England. Boucher et al.[29] reported on the use of ozone in a pilot plant at the Eastern Sewage Works, Redbridge, England. The pilot plant received the equivalent of a secondary effluent and provided for microstraining (35 μ), prechlorination, ozonation, coagulation, and rapid sand filtration for a flow of about 35 gal./min. The ozone dose was kept between 20 and 25 mg/liter to keep the color at or below 10° Hazen. When chlorine was used, the maximum dose used was 20 mg/liter. With these dosages and on a microstrained secondary effluent containing 1.7×10^6 MPN total coliforms/100 ml; ozone alone produced a 4.5-log reduction down to 90/100 ml while chlorine alone achieved a 6-log reduction down to 1 coliform/100 ml. (No data was available on contact time). However, ozone performed well as a polishing agent. It achieved good color removal and was observed to break down the detergents.

Los Angeles County California. The Sanitation Districts of Los Angeles County have made a comprehensive study on the use of ozone at their Pomona Water

Renovation Plant. This work has been reported by Ghan et al.[28,30] They used two different effluent disinfection requirements and a variety of effluents. Using a well oxidized secondary effluent it required 50 mg/liter ozone to meet the California requirement which specified a seven-day median total coliform of 2.2 or less per 100 ml. A tertiary effluent filtered through activated carbon before ozonation required only 5 mg/liter ozone dose. They observed the best kill at 3 min. contact time in the carbon column effluent. To achieve the former EPA standard of 200/100 ml fecal coliforms required only 10 mg/liter ozone for the secondary effluent and 1 mg/liter in tertiary carbon filtered effluent. They also found that ozone can achieve consistently only a 2-log reduction in a well oxidized effluent of total coliforms at 10 mg/liter (dose) or less. A greater removal requires more ozone or a better quality effluent. While the rate of kill was found to be best at 3 min. in the carbon effluent, the secondary effluent required more time to absorb the higher dosages of ozone. This study further revealed that the best disinfection performance occurs when the DCOD (dissolved chemical oxygen demand) is less than 12 mg/liter and nitrite does not exceed 0.15 mg/liter. It was also discovered that the removal of suspended solids from *7 to 1 mg/liter in the Pomona effluent was a major advantage of the ozone process.*

Fort Southworth, Kentucky. In 1973 researchers at the Fort Southworth Sewage Treatment Plant in Louisville,[31] reported on ozone disinfection of secondary effluents. These studies confirmed that it was necessary to dose the pilot-plant effluent (4.0 GPM maximum) at 15 mg/liter with 10 min. contact time to achieve the ORSANCO requirement: not to exceed 200 fecal coliform/100 ml as a monthly geometric mean, or 400/100 ml in more than 10 percent of the samples during the months of May through October. *This is not a stringent disinfection requirement.*

Grandville, Michigan. Ward et al.[32] reported that a plant-scale investigation at Grandville, Michigan revealed an inability to meet an effluent disinfection requirement of 1000/100 ml MPN total coliforms with ozone dosages of 2.5–8.5 mg/liter. The effluent treated was a well oxidized activated sludge effluent with one stage of tertiary treatment. The contact time was about 10 min. and the total suspended solids averaged 20 mg/liter.

Paris, France. Gomella[33] reported that to achieve a total coliform MPN of 200/100 ml in the effluent of the Colombes, Paris, France wastewater plant, required as high as 17 mg/liter ozone at a contact time of about 4 min. Filtration of this effluent reduced the ozone required by about 25 percent.

Cleveland, Ohio. Guirguis et al.[34] reported only 40 percent compliance to achieve a *fecal* coliform MPN of 200/100 ml with ozone dosages ranging from 3 to 11 mg/liter in the effluent of a 25 GPM physical-chemical pilot plant which was used for the purpose of establishing design criteria for Cleveland's Westerly AWT plant. This plant is located on the shore of Lake Erie west of the mouth of the Cuyahoga River.

New York City. Naimie[35] reported that a 10 GPM ozone pilot plant at New York City's Wards Island Sewage Plant indicated a 2.5 mg/liter ozone dosage can achieve the fecal coliform requirement of 200/100 ml. Using this information, Naimie claimed the cost estimates for the Cleveland Westerly plant were conservative as it was assumed the effluent ozone demand for the finished plant would only be 2 mg/liter in excess of the disinfection requirements. Naimie claimed that their studies of ozone demand revealed suspended solids to be a critical factor. From these studies the following expression was developed:

$$\text{Ozone Disinfection Demand} = 1.5 \times 0.38 \text{ TSS.} \tag{8-6}$$

Accordingly, an effluent with 20 mg/liter TSS should have an ozone disinfection requirement of about 9 mg/liter. (Contact time is presumably 10 min.) However, Ghan et al.[28] believe that the disinfection efficiency of ozone is not significantly affected by suspended solids (15 mg/liter or less) but that it is greatly affected by DCOD and Nitrites.

EPA Observations. Venosa[21] of the EPA made some important observations resulting from the Grandville, Michigan project. He points out that the most serious practical problem for both ozone and bromine chloride is inadequate process control. Chlorine produced the most consistent results because the dosage required to achieve consistent disinfection was easily controlled by continuous residual control equipment. In order to achieve results comparable to chlorine, both ozone and bromine chloride application had to be by manual dosage control due to the high intensity of the chemical reactivity displayed by ozone and bromine chloride with wastewater constituents.

Venosa suggests the practical limits of ozone for wastewater treatment, particularly its use as a disinfectant, may require polishing an activated sludge effluent with filtration preceded by coagulation and sedimentation. Such a polished effluent would be comparable to an acceptable but rather poor raw water supply for potable purposes.

Virus Destruction. Selna, Miele, and Baird[36] of the Sanitation Districts, Los Angeles County, made a comprehensive study of water reuse disinfection for unrestricted recreational purposes. According to the California Department of Public Health guidelines, in order to qualify for such use a well oxidized secondary effluent must be coagulated, settled, filtered, and disinfected to achieve a median total coliform MPN of 2.2/100 ml, or less. The required treatment is expensive both from a capital and operational standpoint; therefore, the Sanitation Districts investigated less costly tertiary treatment alternatives to the required system during a two year study at the Pomona research facility.

One of the objectives of this study was to provide an effluent which would protect swimmers against viral illnesses. The pilot systems employed were from 25 to 100 GPM. Four treatment systems were evaluated as follows:

1. Coagulation, sedimentation, filtration, and disinfection.
2. Coagulation, filtration, and disinfection.
3. Two-stage carbon adsorption and disinfection.
4. Nitrification, filtration, and disinfection.

Ozonation was tested as an alternative to chlorination in systems 1, 2, and 3. System 4 was an investigation of free residual chlorination with a 2-hr contact time. All ozone contact times were designed for 18 min. Ozone dosages varied from 10 mg/liter for System 1; both 10 and 50 mg/liter for System 2, and 6 mg/liter for System 3 (Carbon adsorption). Ammonia nitrogen in effluents 1, 2, and 3 were approximately 20 mg/liter and suspended solids (predisinfection) about 1.5 mg/liter. The total COD was on the order of 20-25 mg/liter. Cumulative virus removal was best in System 1 (i.e., 5.5-log removal with an ozone dose of 10 mg/liter.) System 2 with an ozone dose of 50 mg/liter was only about 5.4 logs removal. System 2 with a lowered ozone dose of 10 mg/liter did almost as well in virus destruction as with 50 mg/liter. System 3 using carbon adsorption and a 6 mg/liter dose provided a 5.25-log removal. These results compared generally with chlorine (free or combined) of 4.6-5.25 logs removal depending upon the system (see Chapter 2).

This investigation defines with confidence the virucidal capabilities of ozone under full-scale treatment plant conditions. It also reveals that the final ozonated effluent coliform concentrations did not routinely meet the required MPN standard of 2.2 MPN/100 ml. This continues to confirm the fact that while ozone is a superior virucide, it is not a reliable bactericide. The public health significance of this may be impossible to determine.

Preozonation as a Process. Trussell et al.[37] reported in 1975 on a comprehensive study of an alternative tertiary treatment process which would produce an effluent to meet the *strictest requirements of the California State Department of Health*. The conventional process usually proposed by the State for similar situations requires coagulation-flocculation, followed by sedimentation, filtration, and disinfection by chlorine. The "alternate" process reported on here is the one suggested by the consultants directing this project.* This alternate process consists of preozonation of a secondary effluent followed by flocculation, coagula-

---

*James M. Montgomery Consulting Engineers, Pasadena, California.

tion, sedimentation, filtration, and disinfection with either ozone or chlorine. This investigation demonstrated the many benefits of preozonation followed by chlorine for final disinfection. Moreover, it proved to be less costly than the conventional method.

This alternate process was chosen because it was believed it would produce an effluent to meet the following objectives.

1. Reduced incremental salt increase.
2. Greater virus removal efficiency.
3. Lower tertiary treatment costs.
4. Reduced organics.
5. Increased color and turbidity removal.
6. Disinfection to 2.2/100 ml coliforms.

The performance of ozone in the role of preozonation was exemplary. It was most dramatic in the case of color removal. The reductions were approximately linear with applied dosage up to 10 mg/liter, after which the reductions taper off.

As for turbidity and suspended solids removal, similar patterns were observed (i.e., dosages greater than 10 mg/liter do not result in any significant reductions). As to coliform destruction, preozonation appears limited to a consistent 2-log reduction at 10 mg/liter dosage. Higher dosages did not produce a significant improvement in disinfection.

Therefore, on the basis of this investigation it appears that a point of diminishing returns for the removal of turbidity, color, suspended solids, and viruses occurs at about 10 mg/liter ozone dosage. At this dosage virus removal of 3–4 logs appears typical although somewhat inconsistent. Contact time seemed to have but little influence on the effectiveness of ozonation provided the applied dosage was kept constant and that there was sufficient time for the transference of the ozone to the wastewater to satisfy the ozone demand. The optimum contact time for this secondary effluent was about 5 min.; however, the system was operated at a contact time of 10 min. owing to the original contactor design. This contact time proved just as effective as longer contact times, and undoubtedly provided an optimum ozone transference efficiency, exceeding that at 5 min.

One other significant achievement not directly related to disinfection was the effect of the preozonation dose on the coagulant dose required for turbidity removal in the secondary effluent.

The lowest effluent turbidities could be achieved with an alum dose of 90–160 mg/liter in the secondary effluent. By ozonating that same effluent, the alum dosage producing the lowest turbidities was reduced to 50–90 mg/liter; a 50 percent reduction with 10 mg/liter of ozone.

Final Disinfection. Both chlorine and ozone were compared as final disinfectants for bacterial destruction. Figure 8-1 shows this comparison. As shown, reozonation of the final effluent with ozone dosages as high as 10 mg/liter failed to meet the 2.2/100 ml coliform objective. The chlorine residual in the contact basin was all "combined" because the NH_3-N content of this effluent is about 35 mg/liter.

The recommended chlorine dose for the full-scale plant should be 5 mg/liter with a 30-min. "detention" time. The latter proved satisfactory in the pilot-plant work. The dispersion coefficient for the 2- and 4-in. pipe used for the pilot-plant contact chamber and the existing contact chamber was 0.016, 0.042, and 0.023 respectively. The 2-in. pipe was selected for the chlorine contact chamber because its Peclet Number of 62 indicates it has good "plug flow" characteristics. While the calculated "detention" time was 31 min., the first appearance of dye (T_i) in this chlorine contact chamber configuration was 22 min. This can be considered the "effective" contact time comparable to laboratory studies of chlorine demand versus contact time.

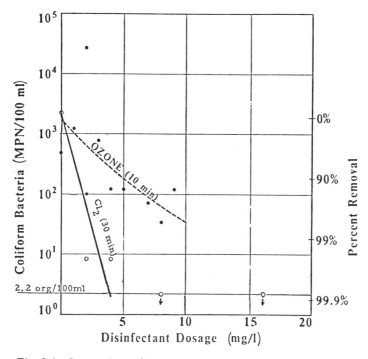

Fig. 8-1 Comparison of bacteria destruction: ozone vs. chlorine.

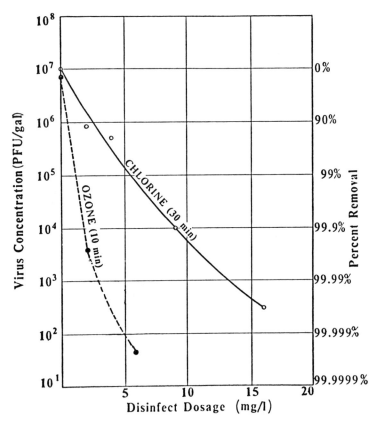

Fig. 8-2 Comparison of virus destruction: ozone vs. chlorine.

TABLE 8-3 Projected Quality Parameters of Full Scale Plant Effluent[37]

Parameter	Units	Secondary Effluent	Tertiary Effluent	Percent Removal
Turbidity	TU	3–5	1.0	80
Color	SCU	20–40	5–10	75
SS	mg/liter	10–15	1–3	75
COD	mg/liter	40–60	10–20	60
Coliform[a]	MPN/100ml	10^4–10^5	≤2.2	99.998
Viruses[b]	PFU/liter	100	0.02	98.98

[a]The secondary coliform MPN is usually much higher than is shown in this table. The MPN for this investigation varied from 4.9×10^4 to 3.5×10^6 total coliforms.
[b]Seeded viruses.

Virus destruction. As described above, preozonation gave a 3-4-log removal of seeded viruses. Since ozone failed in the final disinfection process, reozonation for virus destruction was omitted from the investigation.[38] However, Fig. 8-2 illustrates the shape of the curves comparing combined chlorine with ozone for seeded virus destruction with a one- or two-run experiment. Therefore, the 5-log removal shown for ozone is not to be considered reliable.[38]

Projected Full-Scale Plant. Expectations: Table 8-3 illustrates the probable effectiveness of ozone as a pretreatment process for wastewater reuse or tertiary effluents.

COMMERCIAL OZONE GENERATORS

Introduction

Ozone must be generated at the point of use because it is subject to rapid decomposition, reverting back to oxygen. Ozone can be generated from atmospheric air or from pure commercially produced oxygen. The process may be by electrical discharge or by photochemical action using ultraviolet light. The latter produces the ozone layer in the upper atmosphere. The most practical method for producing ozone is by electrical discharge.

When ozone is produced by electrical discharge in the corona, the gas stream feeding the generator must be dried to at least -40°F dewpoint, as water vapor lowers ozone production efficiency significantly.

Theory of Generation

Ozone is produced from the oxygen in air or pure oxygen when a high voltage alternating current is imposed across a discharge gap in the presence of either of these gases. Ozone generation by electric discharge is as follows.[39] The key to the process is in the field between the electrodes of stray electrons left over from previous discharge or background radiation. These electrons become excited and accelerated within the high energy field between the electrodes. The alternating current causes changing polarity which in turn causes the negatively charged electrons to be attracted first to one electrode and then to the other, like bouncing balls. As the velocity of the electrons increases, they attain enough energy to split some O_2 molecules in two, which in turn combine with other O_2 molecules to form O_3. The visible effect of the incomplete oxygen molecule breakdown in the air gap between the highly charged electrodes is known as the "*corona glow.*" Factors which must be considered in the design of electric discharge generators are: voltage, frequency, dielectric material property and thickness, discharge gap, and absolute pressure within the discharge gap.

Under optimum conditions, ozone production depends upon the following relationships.[40]

$$V \sim pg \tag{8-7}$$

and

$$(Y/A) \sim f \in V^2/d \tag{8-8}$$

Where

Y/A = ozone yield per unit area of electrode surface under optimum conditions.
V = Voltage across the discharge gap (peak V).
p = gas pressure in the discharge gap (psia).
g = width of discharge gap (in.).
f = frequency of applied voltage (Hz).
$\in$ = dielectric constant.
d = dielectric thickness (in.).

When studying the above relationships it is obvious that the ozone generator manufacturers are confronted with formidable problems. First of all, the basic method is inherently inefficient. Commercially available corona discharge generators produce ozone in the exiting gas in concentrations varying from 0.5 to 4.0 percent by weight of the carrier gas. Therefore, when making ozone from oxygen most of the oxygen passing through the generator is unchanged. So for economic reasons this oxygen must be used elsewhere in the overall scheme or be dried out and recycled through the generator.

The yield of a generator is related to the square of the voltage, yet high voltages increase the possibility of electrode failure caused by dielectric or electrode puncture. Higher voltages also result in higher pressures (the discharge gap being set), and this means higher operating temperatures. High operating temperatures increase the rate of ozone decomposition.

Dielectric thickness appears in the denominator of the yield relationship which indicates that a thin dielectric is desirable. Thin dielectrics, however, are more susceptible to failure by puncture. It is evident that the problem is to attain maximum yield while simultaneously economizing on maintenance and replacement costs. The solutions to these problems are difficult to achieve. Some of the methods to increase ozonator efficiencies are as follows: high frequency is less damaging to dielectric surfaces than high voltage. Frequencies as high as 2000 Hz are now in use, emphasis is placed upon this parameter. Glass has been found to be a practical dielectric material. It is cheap and readily available. Ingenious generator designs optimize heat removal. In addition, solid-state circuitry, and acid-resistant materials reduce power requirement cost and increase reliability.

So in a modern ozone generator with the latest manifestations of the ingenuity of the designers, only about 10 percent of the energy applied results in the production of ozone.

The largest portion of this energy loss is by the heat generated. The remainder of the loss is through light and sound. The decomposition of ozone back to oxygen is greatly accelerated with increasing temperature, so all high concentration ozonators must employ a method of heat removal.

Assuming a clean dry oxygen-rich gas is being fed to an ozone generator and an efficient method of heat removal is available, production of ozone per unit area of electrode surface under optimum conditions is a function of:

1. Peak voltage across discharge gap.
2. Absolute gas pressure in gap.
3. Width of discharge gap.
4. Frequency of applied voltage.
5. Dielectric constant.
6. Thickness of the dielectric.

Therefore, to optimize the ozone yield, the following conditions should exist.

1. The combination of gas pressure and gap width should be arranged so the voltage may be kept relatively low for reasonable operating pressures.*
2. A thin material with a high dielectric constant should be used.
3. High frequency current (300–2000 Hz) should be used because high frequency increases the ozone yield and is less damaging to the dielectric than high voltage and prolongs the life of the equipment.
4. An efficient heat removal system is essential.

The most significant factor is the efficiency of heat removal. While the voltage and/or frequency can be continually increased to produce more and more ozone, the additional heat produced must be removed as efficiently as possible. Otherwise if the temperature rises, the additional ozone produced will only decompose back to oxygen.

Commercial Generators

Currently there are three types of ozone generators in use (1977). These are: the Lowther Plate; the Otto Plate; and the Tube. All three are illustrated in Fig. 8-3.

Otto Plate. This is the first of the ozone generator designs developed by Otto in 1905. It is known as the "Otto partial injection system." (See Fig. 13-4, p. 685 Ref. 19.) This design is still being used extensively in Western Europe. This ozonator is made up of a series of sections arranged in the following sequence: 1) a water-cooled cast aluminum block which acts as the grounded electrode;

*Lower voltages protects the dielectric and/or the electrode surfaces from high voltage failure.

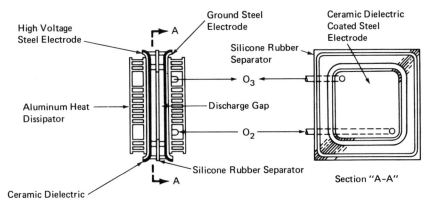

Lowther Plate Generator Unit

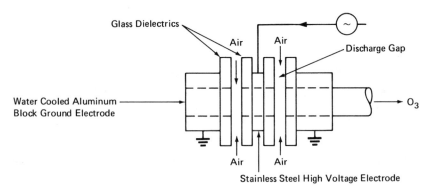

Otto Plate Type Generator Unit

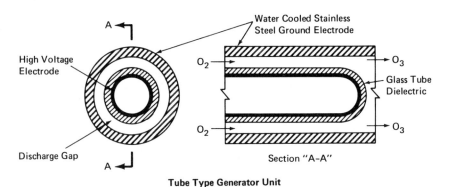

Tube Type Generator Unit

Fig. 8-3   Schematic illustration of commercial ozone generators.

2) a glass plate dielectric; 3) followed by an air gap and another glass dielectric; 4) and then a high voltage stainless steel electrode. A complete unit includes the mirror image of the dielectrics and their corresponding air gaps together with a water cooled, grounded electrode. One of the inherent disadvantages of the Otto-type generator is that the air blown into the ozonator discharge gap area is limited to low pressure.

Tube Type. This generator is composed of a number of tubular units. The outer electrodes are stainless steel tubes fastened into stainless steel tube spacers and surrounded by cooling water. Centered inside the stainless steel tubes are tubular glass dielectrics whose inner surfaces are coated with a conductor which acts as the second electrode. The outer electrodes are arranged in parallel and are sealed into a cooling water system. This group of water-cooled tubular units is then enclosed in a gas-tight vessel so that either air or oxygen may be fed to the unit at one end and ozone collected at the other. The glass tube is sealed so that the feed gas passes only through the discharge gap. Figure 8-4 illustrates a large tube-type ozonation system.

Lowther Plate Unit. This is the most recent species of ozone generating equipment. It is air cooled and utilizes either atmospheric air or pure oxygen. This generator is made up of a gas-tight "sandwich" consisting of an aluminum heat dissipator; a steel electrode coated with a ceramic dielectric; a silicone rubber spacer which provides the precise amount of discharge gap and; a second ceramic-coated steel electrode with an air (or oxygen) inlet and ozone outlet. The ozone exits through a second aluminum heat dissipator.

These sandwich-type units are pressed together in a frame and are manifolded for either air or oxygen inlet flow. Cooling is accomplished by a fan which moves ambient air across the heat dissipators. The composite arrangement of a 30-cell module is shown in Fig. 8-5.

The manufacturers using this type of a generator cite the following superior characteristics.[40]

1. Uses air cooling.
2. Thin dielectrics provide greater ozone yield which in turn enhances heat removal.
3. The small discharge gap allows higher operating pressures.
4. Low operating voltages provide longer dielectric life.
5. High frequency operation allows higher ozone production without resorting to higher voltages.
6. Heat removal is more efficient.

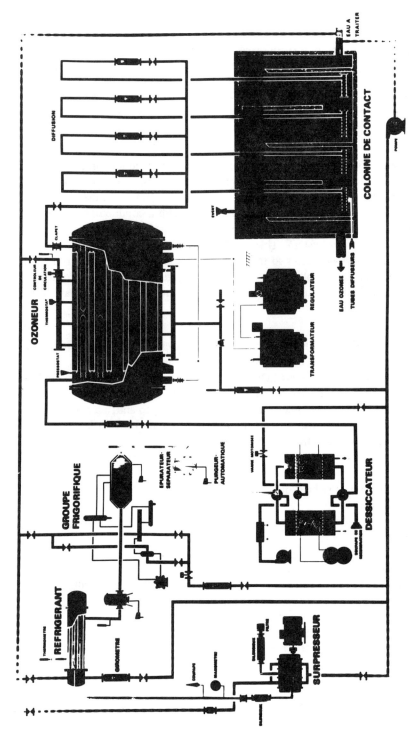

Fig. 8-4 Ozonation system schematic for the Choisy-Le-Roi treatment plant, city of Paris water supply (*courtesy* Compagnie Générale Des Eaux).

Fig. 8-4a Typical tube type ozone generator (*courtesy* Compagnie Générale Des Eaux, Paris, France).

Fig. 8-5   Composite arrangement of 30 cell Lowther plate unit (*courtesy* Union Carbide Co.).

TABLE 8-4 Typical Ozonator Operating Characteristics

Type	Feed	Dew Point of Feed (°F)	Cooling	Pressure (psig)	Discharge gap (in.)	Voltage (kv, peak)	Frequency (Hz)	Dielectric Thickness(in.)
Otto	air	-40	water	~0	0.125	7.5-20	50-500	0.12-0.19
Tube	air, oxygen	-60	water	3-15	0.10	15-19	60	0.10
Lowther	air, oxygen	-40	air	1-12	0.05	8-10	2000	0.02

7. Provides a high-yield efficiency which results in smaller space requirements for the equipment.
8. Power requirements are less than for generators based on older technology.

The operating characteristics of the different types of generator are shown in Table 8-4. The power requirements for the different type of generators are shown in the following table.

Ozone Generating Power Requirements[a]

	Power Required (kWh/lb)	
Type	Air	Oxygen
Otto	10.2	—
Tube	7.5-10.0	3.75-5.0
Lowther	6.8-8.8	2.5-3.5

[a]For 1.0 percent by wt ozone.

Current Practice of Manufactures

United States Manufacturers. The producers of the largest ozone generating units in the United States are Union Carbide Co., Tonawanda, New York, and Emery Industries, Cincinnati, Ohio. The latter uses ozone mainly for the commercial production of organic chemicals. The generating units made by Union Carbide use the Lowther Plate design. Emery Industries are still using the water-cooled tube type.

All other generator manufacturers in the United States use the water cooled tubes. These include Crane, Welsbach, and Orec. P.C.I. also uses water cooled tubes, with the addition of a fluid cooled dielectric. The low-voltage side is cooled with water and the high-voltage side by a transformer oil cooling type of compound.

Foreign Manufacturers. Western Europe and Japan are among the foreign manufacturers of ozone generating units. The European manufacturers: Trailagaz, Degremont, Gebruder-Hermann and Kerag-Suisse, are all tube cooled systems. However, in 1976 Sauter introduced a new Otto plate-type unit. In Japan the Suni-sonic system uses the Lowther plate; Mitsubishi uses the water cooled tube system.

Comparison of Systems. The older Otto plate design developed at the turn of the century is the least efficient. The water cooled tube design and the Lowther plate design incorporate modern materials and design technology and are therefore more efficient than the Otto plate system. The most recent development is the Lowther Plate, hence, this is the most efficient. All of these generators produce an exiting gas mixture containing 0.5–3 percent by weight of ozone. When oxygen is used as the supply gas, it will produce about twice as much ozone as compared to air.

OZONE FACILITY DESIGN

General Considerations

The key elements of an ozone system are: pretreatment of the air supply, unless oxygen is used; generation of ozone; measurement and metering of the ozone produced; mass transfer of ozone to the process flow (called variously contacting, mixing and or dispersion); and an ozone decomposer.

The ozone supply system must be equipped with the necessary pretreatment facilities to provide operating personnel with an instantaneous and continuous quantitative readout of the ozone feed-rate.

The mixing (contacting) system should be capable of dispersing the ozone into the process stream so that the bubbles of ozone will stand "shoulder-to-shoulder" within the regular distribution of microorganisms. Each shoulder should contact a microorganism so as to produce *lysis* which is the lethal effect of ozone.[41]

Three Ozone Systems

Once Through Oxygen System. According to Rosen[41] this is the simplest and most cost-effective system for wastewater disinfection, and by far, the one for greatest commercial application. The process flow is G-D-E-F as shown in Fig. 8-6. Oxygen is produced on-site by a cryogenic or pressure swing adsorption process. Dry oxygen is fed to the ozone generator and the off-gas from the

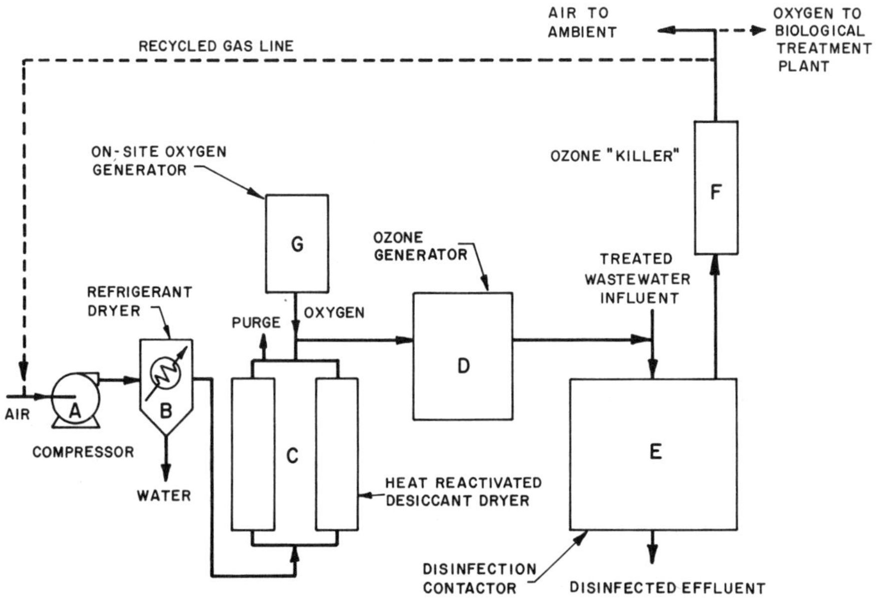

THE OZONE DISINFECTION PROCESS

Fig. 8-6 Flow diagram of the various ozone disinfecting systems (*courtesy* Union Carbide Co.).

contacting unit exits through the ozone decomposer and the resulting oxygen is used to feed an oxygen activated sludge process.

Oxygen Recycle. In this process (G-D-E-F-A-B-C), Fig. 8-6, the oxygen-rich off-gas is recycled to the compressor "A" and the refrigerant dryer "B" for cleaning and drying before returning to the ozone generator. Oxygen lost from this recycle loop by ozone conversion and reaction with the wastewater plus the dissolved oxygen added to the wastewater is made up from the oxygen source. Additionally, recycled gas is purged, and makeup oxygen added to maintain oxygen concentration in the recycle loop. This purging procedure is necessary because of the oxygen dilution by nitrogen gas, which is stripped from the ozone mixing chamber, and the carbon dioxide produced by ozone oxidation of organic matter.

Once Through Air.   This process is (A-B-C-D-E-F- discharge to ambient) as shown in Fig. 8-6. The choice favoring this system is purely economic: power costs; ozone generator efficiency; air versus oxygen costs; and plant size.

If an oxygen activated sludge system is not part of the treatment process, the best method is either recycled oxygen or once-through air.

Metering System

Until recently the measurement of the ozone output of the generator has been based upon the electrical energy input to the generator. This concept is so vague that it cannot be classified as an inferential measurement. Modern generators measure the flow of ozone based upon the computation of an assumed ozone concentration in the gas flow produced by the generator. This again, is an inferential measurement, crude in terms of modern technology.

Equipment is currently available which can continuously measure both the ozone concentration and volumetric flow of the gas exiting the generator. This technology is provided by the Dasibi ozone photometer.[42] It is a fully automatic instrument designed for continuous monitoring of ozone generation systems. This instrument utilizes the ultraviolet absorption technique—a direct measurement of ozone concentration. This technique is absolute, based on the Beer-Lambert law. The response of this photometer to ozone is a direct function of: optical path length; ozone-UV absorption coefficient; and ozone concentration. No explosive gas reagents or wet chemicals are required.

Union Carbide recently demonstrated the Dasibi unit. It has a digital display panel which reads out the instantaneous ozone flow from the generator (g/hr). This is accomplished by the use of a minicomputer which calculates the product of the gas flow times the ozone concentration. This reading is comparable to the gas flow reading on the rotameter of a chlorinator or sulfonator. The Dasibi metering unit is connected by a feedback loop which controls the gas feed compressor. This provides a constant ozone concentration for any given ozone generator power setting.

Control Systems

Current Practices. For wastewater ozonation, whether for disinfection or color removal, ozone is usually applied at a constant dose in proportion to the flow being treated. This is because at economical dosages the ozone demand of the wastewater consumes the residual before it can be measured.

Most ozone installations in operation today (circa 1977) whether for potable water or wastewater are based upon manual control of the dosage. Some installations make an inferential attempt to vary the ozone dosage in accordance with the variation in the process flow. Such a system translates the variation in flow signal to a linear change in the applied voltage (or frequency) to the ozone generator. This results in a corresponding change in the ozone output of the generator.

Ths most difficult practical problem of current ozone practice is determining the optimum ozone dosage. This can only be accomplished, with any degree of reliability, by a laboratory ozone demand. The better the quality of effluent, the more stable and consistent will be the ozone demand.

The 10-min. ozone demand for potable water and/or highly polished tertiary effluents for reuse are on the order of 5 mg/liter and are fairly constant over a 24-hr period. On these types of waters, residual control of ozone dose becomes practical; however, there is insufficient actual experience to make any further comments on ozonation residual control.

Flow Proportional Control. This type of control is currently available from all reputable manufacturers of ozonation equipment. The output signal from the wastewater flow meter is transmitted via a ratio station to the generator voltage input or the generator current frequency. This reflects in a linear change in the generator output of ozone. Current technique is to change ozonator output as a function of current frequency (Hz). This is less cumbersome than the previous technique of input voltage variation.

Residual Control. The rapid die-away of ozone residual, whether it be in potable water, highly polished wastewaters, or distilled water, presents a technical problem for residual control. However, with current expertise and available instrumentation, it is reasonable to assume that if the diurnal ozone demand change is on the order of 1.5 mg/liter/hr, an ozone facility designed for compound-loop control should be practical.

A prototype installation would require a Dasibi[42] ozone analyzer-computer system, to provide an accurate measurement of the ozone generator system output flow rate. Next, it would require the provision for a reliable signal based upon the concentration of the ozone residual in the treated water x min. after ozonation. The ozone residual analyzer for this purpose should be similar to the Fischer and Porter Series 17L-2000.[43] In summary, such a facility would include but not be limited to:

1. Wastewater effluent flow meter signal to be sent via a 0.4:4.0 ratio station to an intermediate computer station. This is input No. 1.

2. The continuous ozone analyzer output signal to be sent to the (above) computer station via a similar ratio control station. This is input No. 2.

3. The output signals of No. 1 and No. 2 described above are arranged through appropriate instrumentation to adjust both the gas flow from the supply source (ambient air or oxygen compressor) and the power or current frequency through another ratio control station to the ozone generator. This provides a compound loop effect.

4. In the meantime, the Dasibi photometer analyzing the ozone content discharging from the ozone generator will, through its computer system, change as required the gas-feed compressor output so as to maintain the desired ozone concentration in the gas leaving the generators. This provides the operator with instantaneous readout of the lb/day of ozone being generated.

The practicality of residual control for wastewater disinfection is yet to be determined. As of 1977 there are no known operating installations using residual control.

Mixing (Contacting)

The solubility of ozone in water is a limiting factor that greatly affects the ozonation process. At a water temperature of 20°C it is only soluble to the extent of 0.57 g/liter.[6] Therefore, special attention must be given to applying the ozone. This is generally referred to as the "contacting system." Successful mixing involves transferring a maximum of ozone from the gas phase to the liquid phase.

An ideal mixing device will produce large bubble-area to gas-volume ratios. This allows maximum diffusion area by producing many small bubbles of ozone. The mixer should also provide enough turbulence to reduce film thickness which offers resistance to ozone transfer. It should also provide for large ozone concentration differences between the gas and liquid phases. Devices currently in use include the Otto partial-injection system, porous diffusers in baffled chambers, and turbine impellers in a recirculated water flow.

Masschelein et al.[44] have suggested that a pilot-plant study should determine the contactor system performance for injecting ozone into a particular water. They have reported on various types of mixers, all showing the ozone loss in the mixing chamber vent to range from about 6 to 36 percent using ozone dosages of 1.5 mg/liter. The two systems with the best overall performance of those studied appear to be the liquid-circulating aeration turbine and the water recirculation system. These gave good dispersion of ozone with ozone losses of about 6–8.5 percent at dosages of 1.5 mg/liter.

Union Carbide prefers the diffuser system in a baffled contact chamber. Their current design is illustrated in Fig. 8-7. The diffusers are sintered alumina discs. Rosen[45] advises that this method of mixing will have an ozone transfer efficiency of 90 percent.

Ozone Decomposer. Each ozone mixing chamber or contacting device must be enclosed so that the ozone that is not put into solution can be withdrawn and destroyed in an ozone "killer." The ozone is converted to oxygen by allowing the ozone to pass through a metallic oxide catalytic decomposer. The ozone in the decomposer exit gas should be less than 0.1 percent by volume.

In once-through air systems the decomposer exit is discharged to atmosphere. In the oxygen recycle system it is returned to the system itself.

Safety Considerations. The air space in all mixing chambers and contacting systems should be monitored for hydrocarbons in order to alert operating personnel to explosive conditions.

Ambient air in the ozone generator room should be continuously monitored for ozone leaks. If the ozone generator is air cooled the ambient conditions surrounding the generator can be purged into the ventilation system. This air can be used to heat the building, as there is normally a 15°F temperature rise in the cooling air.

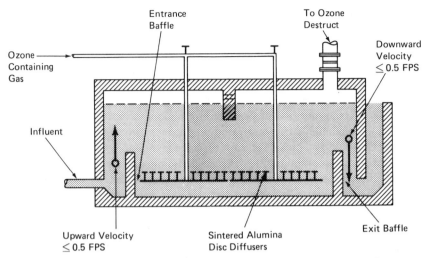

Fig. 8-7 Ozone contacting system (*courtesy* Union Carbide Co.).

COMPARATIVE COSTS

General Considerations

To evaluate ozone on cost alone is an unfair evaluation of its capabilities. The cost-effectiveness of ozone as a bactericide for secondary effluent as compared to chlorine will always demonstrate the superiority of chlorine. The only chance ozone has as a secondary effluent bactericide is when it is part of the UNOX process. This is the system recommended by Rosen.[41] However, ozone for potable-water treatment and wastewater reuse is a different matter. There are situations where cost alone is not the dominant factor. Ozone must be evaluated on its capabilities as a powerful oxidant in addition to its ability to destroy viruses and bacteria. Ozonation is an excellent process for polishing low-grade potable waters and wastewater for reuse.[37]

Cost Studies

Several ozone feasibility studies have been made recently.[46-49]

EPA Summary. Some of these studies and others have been summarized in the EPA Task Force Report, March 1976.[50] These findings are shown in Table 8-5 for the ozonation process as compared with the chlorination/dechlorination process shown in Table 8-6.

The costs outlined above should be considered as tentative, because the parameters of "disinfection" are not specific. The following cost comparisons are specific and therefore are of more practical value for the designer.

TABLE 8-5   Ozone Disinfection Cost

A. Ozone Generated From Air			
Plant size (mgd)	1	10	100
Capital cost	190,000	1,070,000	6,880,000
Disinfection cost ϕ/1000 gal.	7.31	4.02	2.84
B. Ozone Generated From Oxygen			
Capital cost[a]	160,000	700,000	4,210.000
Disinfection cost ϕ/1000 gal.	7.15	3.49	2.36

[a]The reduced cost here represents the elimination of drying atmospheric air for the process.

Sacramento, California. In 1975, the Sacramento Regional County Sanitation District of California made an ozone feasibility study.[46] The disinfection requirement for the plant discharge is a total coliform MPN of 23/100 ml. The PWWF is 240 mgd. The effluent from this plant is assumed to be a well-oxidized secondary effluent resulting from the activated sludge process. There is some cannery waste during the dry weather flow season of July to September. For this situation ozone equipment manufacturers recommended design capacity of 20 mg/liter. However, the design engineers thought this would be much too low in view of recent operating experiences by the County Sanitation Districts of Los Angeles at their Pomona water reclamation plant. These studies indicated that a 40 mg/liter dosage would be more appropriate to achieve the disinfection requirement of 23/100 ml MPN total coliforms. The results of this report are shown in Table 8-7.

Power cost is based on 0.9 ϕ/kWh.

This analysis shows the capital cost of an ozonation system is between seven and eleven times as expensive as the chlorination/dechlorination system, and the annual operating cost for ozone is somewhere between 1.2 and 2.3 times as expensive as the chlorination/dechlorination method. On an average annual cost basis, the ozonation system is between three and five times as expensive as the chlorination/dechlorination system.[46]

TABLE 8-6   Chlorination Disinfection Cost

Plant size	1	10	100
Capital cost	60,000	190,000	840,000
Disinfection cost ϕ/1000 gal.	3.49	1.42	0.70

Dechlorination With Sulfur Dioxide			
Plant size	1	10	1000
Capital cost	11,000	29,000	94,000
Disinfection cost ϕ/1000 gal.	0.88	0.35	0.19

TABLE 8-7 Cost Comparison of Ozonation and Chlorination/Dechlorination 240 mgd Design Capacity Current Cost Basis—ENR 2300 ($)

Type of Cost	Chlorine 10 mg/liter Dosage	Ozone 20 mg/liter Dosage	Ozone 40 mg/liter Dosage
Total capital cost	2,150,000	14,250,000	23,450,000
Average annual cost			
capital[a]	173,000	1,148,000	1,890,000
operating[b]	373,000	439,000	858,000
total	546,000	1,587,000	2,747,000
Average unit cost[c]			
(¢/1000 gal.)	1.2	3.5	6.0
Average annual local cost[d]	395,000	583,000	1.094,000

[a] Capital recovery at 7% for 30 years.
[b] Based on 125 mgd average annual flow.
[c] Based on total average annual cost.
[d] Sum of operating cost and local amortized capital cost share based on 87.5% Federal grant for capital facilities.

Richmond, California. Another analysis for a less stringent disinfection requirement was made for the City of Richmond, California, discharging a well oxidized activated sludge effluent into San Francisco Bay.[48] The effluent standards to be met are total coliform MPN of 70/100 ml with no more than 10 percent of these samples exceeding 230/100 ml.

Little reliable data are available to predict the ozone dosage required to achieve this particular disinfection objective. However, the following design parameters were selected based upon proposals from the ozone equipment suppliers:

- Ozone dosage ADWF 15 mg/liter.
- Ozone dosage PWWF 6 mg/liter.
- Contact time at PWWF 10 min.

The system chosen, as recommended by the manufacturers, was based upon the use of high-purity oxygen recycled to the generators. Oxygen was to be supplied locally by the Airco plant at $50-$60/ton. Owing to certain cost differentials between the two ozone equipment suppliers, separate estimates are shown in Table 8-8. A separate calculation was made for a chlorination/dechlorination system which is included in the tabulation. The analysis shown in Table 8-8 is based upon the present worth concept.

From the above two analyses using the most optimistic dosage and contact times proposed by the ozone equipment manufacturers, disinfection by ozone is substantially more costly than by chlorination/dechlorination.

TABLE 8-8 Present Worth Analysis: Chlorination/Dechlorination Versus Ozone

Cost Element	Chlor./Dechlor.	PCI Ozone	Union Carbide Ozone
Total Capital	$990,000	1,380,000	1,570,000
Total O and M	2,600,000	4,620,000	4,870,000
Total Project	3,590,000	6,000,000	6,440,000
Annual O and M Cost	93,000	130,000	149,000

Cleveland, Ohio. However, a report prepared in 1974 for the Cleveland Regional Sewer District[49] for the Westerly Advanced Wastewater Treatment Facility resulted in a recommendation for ozone rather than chlorination. This was based upon an ozone disinfection dosage of 6 mg/liter. Disinfection was not defined for MPN total coliforms. The comparative cost of the two systems is summarized in the following tables.

It is clear from Tables 8-9 and 8-10 that the cost of disinfection by ozone is considerably greater than by chlorination. However, in this case, ozonation was selected on the basis of longer equipment life (20–40 yr); the relatively inexpensive cost of deodorizing the screening and degritting facilities; and the control of the biological slimes in the carbon columns. That these additional factors might influence the choice of ozonation are at the best tenuous. More operating data concerning these parameters must be developed.

Other long range economic evaluations have been made attempting to justify the use of an ozonation disinfection system instead of an existing chlorination process requiring the addition of a dechlorination facility. One such system involved the conversion of a primary treatment plant to provide secondary treatment. The UNOX process was chosen for this conversion. Moreover, the existing chlorination system needed to be supplemented with dechlorination capability. In spite of all these factors the long range economic factors as presented by the

TABLE 8-9 Annual Operating Cost

Operation of Maintenance	Chlorination	Ozonation
Disinfection 50 mgd flow	77,000[a]	91,000[b]
Deodorizing at headworks	16,000	1,000
Control of carbon column biofouling	16,500	5,000
Total O&M cost	109,500	97,000

[a] Includes cost of chemical, electrical power and maintenance.
[b] Includes cost of electric power and maintenance.

TABLE 8-10 Total Annual Cost of Disinfection

Itemized Costs	Chlorination	Ozonation
Capital cost	1,593,000	2,457,000
Amortized annual cost	136,000	205,000
Oper. and maintenance cost	109,500	97,000
Total annual cost	245,000	302,000

engineers demonstrated that the chlorination/dechlorination process was the one of choice.[51]

Houston, Texas. Consultants for the city of Houston made a different type of disinfectant evaluation. They compared ozone, liquid-gas chlorine and liquid hypochorite for their new 400 mgd (peak flow) wastewater-treatment plant.[47] Then evaluations were prepared to determine capital cost and annual operation costs for each disinfection system. The present worth method was used to relate capital cost to annual cost over a 20-yr. design period. Parenthetically, it should be mentioned that the high-purity oxygen activated sludge treatment process (UNOX) was found to be the most cost effective process.

The design dosage of 5 mg/liter chlorine was assumed to provide a 1.0 mg/liter residual at the end of 20 min. contact time at peak flows. It was further assumed that a 5 mg/liter ozone dose would be sufficient to meet or exceed the state standards for disinfection. This was also expected of chlorine, and no claim was made that either one would be more effective in eliminating total coliform organisms. In developing the costs of chlorine (in tank cars) versus ozone, the capital cost assessed chlorine at $4,640,000 for the contact chamber, and $1,000,000 for ozone. This calculates to an ozone contact time of 4.3 min. at peak flow.

Table 8-11 tabulates the Houston analysis; the ozone figures are based upon an arithmetical average of three equipment suppliers.

TABLE 8-11 Present Worth Analysis of Chlorine Disinfection Versus Ozone[a]

Cost Items	Chlorine	Ozone
Capital cost	4,840,000	7,137,000
Present worth of interest	2,411,000	3,552,000
Present worth of annual costs	1,226,000	919,000
Net present worth	8,477,000	11,608,000

[a]A planning period of 20 yr. and an interest rate of 6.125 percent was used.

MEASUREMENT OF OZONE RESIDUALS

Introduction

Analytical procedures for the quantitative measurement of ozone residuals are burdened with many problems. Ozone cannot be measured successfully if the data are collected in a slovenly manner. Ozone is such a dynamic oxidizing agent that the life of a measurable residual is short. The longer the time between collection and analysis the greater the error in measurement will be.

When ozone is applied to water or wastewater the following occurs: 1) ozone reacts with organic and inorganic compounds and other material present in the substrate; 2) some ozone is decomposed directly to oxygen; and 3) some is lost to the atmosphere through incomplete gas transfer. The analytical procedure must quantify the ozone molecules remaining as a residual.

The nature of inorganic chemistry is such that any analytical procedure is subject to chemical interferences. In the case of ozone this dilemma is aggravated by the wide range of compounds normally found in wastewaters which will interfere and produce significant errors in the quantitative analysis of ozone residuals. These are the ferric ion, manganese dioxide, nitrites, and peroxides. As if this were not enough, the ozonation process itself produces oxidants which interfere with the quantitative selectivity of analysis of a true ozone residual. These interfering oxidants are nongermicidal—if they were germicidal they could not be considered as interferences. They are broadly classified as ozonides and hydroperoxides. There is a parallel case with the chlorination process which involves the dichloramine fraction of a total chlorine residual. It has been found that a combined chlorine residual in wastewater usually contains a small fraction of organic chloramines which have little or no germicidal effect. In the forward titration procedure these compounds seem to appear in the dichloramine fraction and are attributed to the organic nitrogen always present in wastewater effluents. Therefore, it is reasonable to assume that whatever the analytical procedure may be for measuring ozone residuals in aqueous solutions it will probably not be specific for ozone.

The following methods are considered to be the best of available techniques.

Iodometric

Standard Methods.[52] This is probably the only method which could be considred free of interferences as it is based upon washing the ozone residual out of the sample and measuring the ozone in the gas phase. It is also capable of good precision. The procedure (see p. 455 Ref. 52) consists of passing a pure air or nitrogen stream through the sample and then through an absorber containing a potassium iodide solution. This solution is then treated with sulfuric acid to reach a pH of 2.0 and titrated with 0.005 N sodium thiosulfate titrant until the yellow color of the liberated iodine is almost discharged. This is followed by the addition of a starch indicator solution which imparts a blue color. The titration is then continued carefully but rapidly until the blue color just disappears. This is the classical iodometric endpoint. The endpoint can also be determined amperometrically described below using phenylarseneoxide as the titrant.

The chemical relationship which describes the ozone-iodometric analysis is as follows

$$O_3 + 2I^- + H_2O \longrightarrow O_2 + I_2 + 2(OH)^- \tag{8-9}$$

The iodine released in Eq. 8-9 is titrated against either thiosulphate or phenylarseneoxide to provide a quantitative answer.

The precision given for the iodometric method is ±1.0 percent. This is probably close for high concentration of ozone in solution, but it is not correct for low levels of ozone concentration.[53] The starch-iodide procedure begins to show significant error when the released iodine is less than about 2.0 mg/liter. This concentration of iodine in a neutral solution is equivalent to an ozone concentration of about 0.4 mg/liter. *Therefore, the starch-iodide procedure has a large error when the ozone concentration is less than about 1.0 mg/liter.*

Amperometric. Both Dailey and Morrow[54] and Kinman[53] describe how the amperometric titration technique can be applied to measuring ozone residuals. The procedure involves the addition of an excess of potassium iodide so that all of the ozone is reduced immediately, therefore minimizing loss of ozone residual due to its various routes of decomposition. Moreover, the iodine released by the reaction between KI and O_3 is more stable in aqueous solution than is ozone. Another feature which improves the sensitivity of this procedure is that for each 0.2 mg/liter ozone, a full 1.0 mg/liter of iodine is released. One molecule of ozone gives one molecule of iodine. The ratio of molecular weights would be:

$$I_2/O_3 = 253.82/48 = 5.29/1.0 \tag{8-10}$$

So for a titrator with a sensitivity of 0.01 mg/liter iodine, the sensitivity for ozone would be 0.002 mg/liter. Titrators with this order of sensitivity are available from either Wallace and Tiernan, Div. of Pennwalt Corp. or Fischer and Porter Co.

Amperometric titration is not free of analytical problems, so the analyst should be aware that certain ions in wastewater may poison the electrodes. In all of the copper-noble metal electrode combinations, 0.3 mg/liter of copper, and 0.1 mg/liter aluminum are likely to cause analytical errors.[53]

One of the severest criticisms of the amperometric titration procedure is the excessive agitation inherent in the titrating device. This in itself would cause dissipation of any ozone residual by air stripping caused by the motor driven agitator. Any loss of ozone due to this agitation is avoided by carefully adding the sample with ozone residual to the sample jar containing the necessary amount of KI. Therefore, the ozone reacts to liberate iodine before any aeration can take place.

Except that total agreement is lacking on the arithmetical results of amperometric titration of ozone residuals, it appears to be a valid method. Dailey and Morrow[54] claim the following:*

$$O_3 \text{ mg/l} = \text{ml PAO} \times 0.675 \qquad (8\text{-}11)$$

Leuco Crystal Violet Method. This method was first suggested in 1970 by Layton and Kinman.[55] They discovered that Leuco crystal violet (LCV), a redox indicator for both iodine and chlorine was found to be very sensitive to low concentrations of ozone in aqueous solutions. The use of this colorimetric indicator is prepared so as to produce a final sample pH of 2.5. This is one of the most serious objections to analytical procedures which quantify residuals in a pH environment totally different from the environment of the reactants. After the addition of LCV the absorbance is measured at 592 mμ on a spectrophotometer. The ozone residual is observed from a curve prepared by plotting absorbance versus ozone concentration. The precision of this method is good and the reliability is dependable when working with solutions which do not have appreciable turbidity or contain other constituents which could absorb light energy. According to Kinman, most of the ions usually found in wastewater do not interfere with the color development of this reagent. The color development is rapid and the color is stable. While the method is still under investigation, it is probably faster than procedures using the potassium iodide-starch indicator.

Acid Chrome Violet K. The acid chrome violet K (ACVK) method has been suggested by Masschelein and Fransolet[56] as a procedure specific for ozone. The authors claim it does not suffer any significant interference caused by the formation of ozone reaction products such as ozonides and peroxides. Nor is it affected by the presence of chlorine residual species (below 10 mg/liter) nor chlorinated oxidants.

Ozone reacts with the dye ACVK. The change in reagent color caused by the presence of an ozone residual must then be measured by a spectrophotometer. The sensitivity of this method when measuring the decrease of absorbance at 550 mμ with optical path cells of 5 cm is about 25 μg/liter ozone. The method

*This is using the usual 0.00564N phenylarseneoxide reagent (PAO). This relationship is reported to show good agreement (±1 percent) in multiple titrations over the range pH 4-10.

is applicable in the range of 0-1 mg/liter ozone residual. The measurements involve comparing the absorbance of a blank sample (ACVK without ozone) against the sample with an ozone residual. First the spectrophotometer has to be adjusted for 100 percent transmittance by using the sample without any reagent. The values of ΔA (difference in absorbance between the blank sample and the ozonated sample) vary linearly as a function of the concentration in ozone.

DPD (Palin Method).[57] In the absence of iodide, residual ozone gives a color with DPD which corresponds to only a fraction of its total equivalent "available chlorine" concentration. Subsequent addition of iodide has little effect upon the color. However, if the titration is performed with the iodide added before or jointly with the DPD, the full ozone content is obtained. The latter is therefore the preferred procedure and is accomplished as follows.

To a 100 ml sample add approximately 0.5 g of potassium iodide followed by approximately 0.5 g of DPD powder No. 1.* Mix and titrate with standard FAS solution.

Should the need arise for the separate determination of ozone and residual chlorine in the same sample, this can be executed by introducing this step: Add about 0.2 g of glycine to a 100 ml sample. This destroys the ozone instantaneously whereas there is no loss of total chlorine residual.** If by chance there might be any free chlorine present it reacts with the glycine to form chloraminoacetic acid, which together with any combined chlorine originally present responds fully to DPD in the presence of excess iodide. The result is a complete response by the DPD to the ozone-chlorine residual mixture. One disadvantage that could prove to be a nuisance in measuring ozone residuals in wastewaters is that owing to color fading in the presence of glycine the two-min. standing period after the iodide addition should be omitted. However, this could lead to significant error in the presence of high dichloramine residuals. This can be the case in wastewater owing to the so-called dichloramine species which might titrate at pH levels of 7-9. These species are attributed to the presence of organic nitrogen, and are probably organic chloramines.[19] If these residuals are present it is advisable to retain the normal standing period.[57]

The procedure described above gives the total ozone plus the total chlorine residual. To separate these two, take a separate 100 ml sample and add 2 ml of 10 percent wt/v glycine solution and mix. Then add approximately 0.5 g potassium iodide and approximately 0.5 g DPD powder No. 1. Then mix and titrate immediately with standard FAS solution to obtain total available chlorine only. The reading obtained in the first procedure minus the reading obtained in the

---

*This powder reagent can be substituted by using DPD solution in combination with a phosphate buffer solution.
**Glycine converts any free chlorine present to combined chlorine.

second procedure is total residual ozone. For 100 ml samples 1 ml FAS solution is equivalent to 1 mg/liter available chlorine. To express the readings in terms of ozone residual multiply the results by 0.7 or by 48/71.

This method is similar to the iodometric method as the DPD procedure consists of determining the amount of free iodine liberated by ozone from neutral potassium iodide. There will, therefore, be general agreement between the two methods in potable water. In view of this similarity both the DPD and Iodometric methods will be subject to the same interferences listed in Ref.52 p. 455, except that nitrite does not interfere with the DPD Method.[58] The extent to which ozonides and peroxides may interfere with the DPD ozone determination has not been fully explored. It is presumed that this method will measure these compounds as ozone residual.

Continuous Residual Recording

Introduction. Development of dependable instrumentation to provide reliable and continuous analysis of the true ozone residual would be a singular achievement and at the same time enhance the case for ozone.

Significant progress has been made by Fischer and Porter Co. to provide an on-line analyzer for this task.[43,54] Additionally, Johnson et al. are pursuing the development of a membrane cell analyzer specific for ozone.[59,60] The use of this type equipment is mandatory if ozonation equipment is to have residual control capability.

Amperometric Cell. Fischer and Porter Co. have investigated two types of amperometric cells; the galvanic analyzer and the Ozotrol analyzer.[54] The series 17L 2000 ozone analyzer is a flow through continuous measuring cell. It uses a gold measuring electrode and a copper reference electrode without any bias voltage across the electrodes and does not require any chemicals. This is known as a *galvanic analyzer.* This unit will respond only to free ozone residuals when it is used without chemicals.[61] This unit can be modified by adding a 5 percent potassium iodide solution which will respond to total ozone residuals (i.e., free ozone plus ozonides and hydroperoxides). This analyzer, the 17L 2000, standard or modified version is the one recommended for tertiary effluents. Either the standard or modified unit is subject to interferences from nitrites, ferric ion, and manganese dioxide to the same extent as residual chlorine analyzers. The standard free ozone unit is also subject to interference from aluminum ion (>0.1 mg/liter) and copper ion (>0.3 mg/liter). These interferences, if significant, can be dealt with by the addition of a sample conditioning chemical. Additionally, residual chlorine directly interferes with this analyzer, either the standard or modified.

The other unit investigated by Fischer and Porter[54] utilizes dual copper electrodes with a bias voltage impressed across the electrodes. This is referred to as the Ozotrol analyzer which is a modification of their 17K 1000 Chlor-trol unit which measures free chlorine residual without any chemicals. The use of two copper electrodes with a biased voltage produces current-voltage curves similar to those obtained with the gold-copper combination. Current output is directly proportional to ozone concentration. It responds only to free ozone and cannot be modified to measure total ozone residuals. This unit is not suitable for tertiary effluents. It is applicable only to special situations of clean waters such as is found in the water-bottling industry. If there is any iron in the water the ozone residual present would cause the electrodes to become fouled by the ozone precipitation of the iron.

Field installations of the 17L 2000 galvanic analyzer have shown the unit to be accurate and reliable under the conditions selected. It is too soon to make any statements about the application of sampling tertiary effluents. As with chlorine residual analyzers care will have to be taken to provide a continuous sample which will not foul the unit due to debris, suspended solids and/or grease.

Membrane Cell. An amperometric membrane cell has been investigated by Johnson and Dunn[60] as a sensitive and selective analytical technique for the determination of free ozone residuals. This work is an outgrowth of the investigation by Johnson et al.[62] in the development of a similar type cell for the measurement of undissociated HOCl.

The advantage claimed for the membrane cell is greater selectivity of oxidants to be measured and elimination of electrode fouling by the use of a microporous flourocarbon membrane. Interfering substances are supposedly unable to pass through the membrane; therefore, they are neither measured as a false concentration nor able to foul the noble metal electrode surface. This cell is specific for free ozone residual. It will not measure the combined ozone residuals containing ozonides and hydroperoxides. Moreover, chlorine residuals can be excluded from the ozone residual by changing the biased voltage across the electrodes.[63]

The configuration of the membrane cell is such that it is better described as a probe. It is similar to the dissolved oxygen probe but with a new and improved membrane material and a positive applied voltage. It is available from Delta Scientific Corp. A brief description of the cell follows.

A 1.6 molar solution of potassium chloride which is saturated with silver chloride fills a cavity in the base of the probe and acts as an electrolyte. This solution is in contact with a solid silver reference electrode. As voltage is applied across this silver-silver chloride interface the silver electrode is oxidized; losing one electron and combines with the chloride ion to form a silver chloride coating on the electrode surface. This reaction supplies the electrons necessary to reduce the ozone and maintains a constant reference potential. The released electrons flow through the gold (positive) measuring electrode to the ozone in spite of the opposing biased voltage force supplied by the battery. At the gold electrode, ozone residual in the sample passes through the membrane and is reduced to oxygen.

$$O_3 + 2H^+ + 2e^- \longrightarrow O_2 + H_2O \qquad (8\text{-}12)$$

This reaction "consumes" two electrons per molecule of ozone reduced. This flow of electrons through the gold electrode is measured and recorded as an electric current. The selectivity of this method is dependent upon the microporous membrane to allow only small diameter molecules to reach the gold electrode and by the high positive opposing voltage which prevents the weaker oxidants from reaching the gold measuring electrode.

This measuring system has not had enough field application to determine whether or not it is a reliable method for measuring ozone residuals in tertiary effluents.

Chapter 8—Summary

Ozone is an allotropic form of oxygen which is an unstable compound. It is a severe irritant to the respiratory system, highly corrosive to metals, completely disintegrates rubber products, and attacks all plant life. The low solubility of ozone in water (570 mg/liter at 20°C) is a limiting factor that adversely affects the ozonation process.

However, there are two important facets of ozone: 1) It is not only a powerful oxidant (viz. destroys cynadies whether organic or inorganic; destroys phenols, and oxidizes iron and manganese, both the organic and inorganic complexes), but it is also an effective disinfectant. 2) It is a better virucide than bactericide. There are many questions to be answered regarding the effectiveness of ozone as a bactericide.

The enactment of ever increasingly stringent requirements placed upon sewage and industrial waste discharges may develop the use of ozone as a most important polishing agent for these situations. Up until now ozone has been used primarily for the elimination of tastes, odors, and organic color in the low quality surface water supplies of Western Europe. In practically all of these uses ozone is considered primarily a polishing agent because it is nearly always supplemented by chlorine and/or chlorine dioxide as the disinfectant. This is the direct result of one of ozones most defeating properties: it is so reactive and unstable that residuals will not persist more than a few minutes, even in highly polished potable waters. This in itself adds to another ozone dilemma, that of feed rate control. The use of residual control of ozone is not an accomplished fact as of 1978, not even in potable water. The situation becomes more exaggerated in wastewater treatment.

Ozone can be effectively used as a polishing agent for tertiary effluents, but cannot be considered an alternative to chlorination-dechlorination as a wastewater disinfectant. However, as a tertiary treatment unit process, preozonation can eliminate color, reduce the COD and chlorine demand, and render the effluent highly susceptible to the chlorination disinfection process.

Ozone is an efficient virucide. It has been amply demonstrated that the combination of preozonation and final chlorination of a tertiary effluent can produce a virus free effluent if properly engineered.

The use of ozone as an alternative to chlorine for disinfection of a well oxidized secondary effluent is not competitive with the chlorination-dechlorination method. Ozone dosage requirements for these situations to produce a 1 or 2 min. residual can be as high as 50 mg/liter compared to 15 mg/liter for chlorine.

The overall potentials of ozone must be considered. One of these is the increased amount of dissolved oxygen put into the effluent as a result of the ozonation. The others have been mentioned, namely, color removal, COD removal, oxidation of cyanides, iron and manganese, and reduction of chlorine demand.

Finally, ozone should be considered primarily as an integral unit treatment process for polishing tertiary effluents the same as that accorded to activated carbon in similar situations.

Commercial ozone generators have been around for a long time. These include the tube type, the Otto plate, and the Lowther plate unit. Production of ozone occurs under the severest conditions for the elements employed in each of the individual pieces of equipment. These severe conditions cause many trying and frequent maintenance problems but those operators who have belief in the ozone process have learned to live with the maintenance problems. Manufacturers in the United States have continually improved their designs to reduce the common maintenance complaint. There are three types of systems available: 1) once through oxygen, 2) oxygen recycle, and 3) once through air. The best method is either recycled oxygen or once through air unless there is an oxygen activated sludge treatment system for the wastewater. In the latter the choice would be once through oxygen.

The development of ozone metering and control equipment has been advanced to a point where flow proportional control is reliable. Heretofore, the actual feed rate of an ozonator has been inferential. Now at considerable additional cost an ozone stream analyzer, which computes the percentage of ozone in the gas flow rate through the ozonator to lb/day ozone, is available.

The ozone contacting system, which is as much a part of the ozonation facility as the injector water supply and diffuser system is for a chlorinator installation, determines the efficiency of the ozone transfer to the wastewater treated. The design considerations are critical because of the low solubility of ozone in water.

The cost studies to date demonstrate the significant increase in cost required for any ozonation process compared to the chlorination-dechlorination method, regardless of its ultimate worth as a wastewater treatment process. *However, the selection of ozone should not be based on cost, but on its merits.*

Information available on the continuous measurement of ozone residuals is nil. There is some disagreement among the experts as to which residual species is being measured. The various analytical techniques include but are not limited to: 1) standard methods iodometric procedure, 2) amperometric, 3) Leuco crystal violet, 4) acid chrome violet K and 5) Palin's DPD method.

A continuous residual recording analyzer has been developed by Fischer and Porter. This unit has been field demonstrated to indicate that it is a practical method. Professor J. Don Johnson has proposed the membrane cell as a specific ozone residual analyzer which, if operationally practical, might be superior to the amperometric cell since the membrane cell is specific for O_3.

REFERENCES

1. McCarthy, J. J. and Smith, C. H. "A Review of Ozone and Its Application to Domestic Wastewater Treatment." *J. AWWA* **66**:718 (Dec. 1974).

2. Dyachov, A. V. "Recent Advances in Water Disinfection." A Paper Presented at the Annual Conf. of the Int. Water Poll. Research, Amsterdam, Neth. (Sept. 1976).

3. Richards, W. N. and Shaw, B. "Developments in the Microbiology and Disinfection of Water Supplies." *J. Inst. Water Eng. Sci.* **30**:191 (June 1976).

4. Schalekamp, M. "Experience in Switzerland with Ozone, Particularly the Neutralization of Hygienically Undesirable Elements Present in Water." A Paper Presented at The Annual AWWA Conf., Anaheim, Calif. (May 1977).

5. Stone, B. G. and Trussell, R. "Application of Ozone for Viral Disinfection." A Paper Presented at The Annual Calif. Sect. Mtg. AWWA, San Diego, Calif. (Oct. 30, 1975).

6. Hann, V. A. and Manley, T. C. "Ozone." *Encyclopedia of Chemical Technology* pp 735–753 Interscience, N.Y. (1952).

7. O'Donovan, D. C. "Treatment With Ozone." *J. AWWA* **57**:1167 (1965).

8. Kinman, R. N. "Ozone in Water Disinfection." p. 123 in "Ozone in Water and Wastewater Treatment." by Evans, F. L., Ann Arbor Science, Ann Arbor, Mich. (1972).

9. Glaze, W. H., Rawley, R., and Lin, S. "By-Products of Organic Compounds in the Presence of Ozone and UV Light." A Paper Presented at the IOI Meeting, Cincinnati, Ohio (Nov. 17–19, 1976).

10. Guirguis, W. A., Srivastava, P., Meister, T., Prober, R., and Hanna, Y. "Ozone Reactions with Organic Material in Sewage Non-Sorbable by Acti-

vated Carbon." A Paper Presented at the IOI Meeting, Cincinnati, Ohio (Nov. 17–19, 1976).

11. Prengle, H. W. Jr., Mauk, C. E., and Payne, J. E. "Ozone-UV Oxidation of Pesticides in Aqueous Solution." A Paper Presented at the IOI Meeting Cincinnati, Ohio, (Nov. 17–19, 1976).

12. Richard, Y. "Organic Materials Produced Upon Ozonation of Water." A Paper Presented at the IOI Meeting, Cincinnati, Ohio (Nov. 17–19, 1976).

13. Falk, H. L. and Moyer, J. E. "Ozone as a Disinfectant of Water." A Paper Presented at the IOI Meeting, Cincinnati, Ohio (Nov. 17–19, 1976).

14. Hartemann, P., Block, J. C., and Maugras, M. "Biochemical Aspects of the Toxicity Involved by the Ozone Oxidation Products in Water." A Paper Presented at the IOI Meeting, Cincinnati, Ohio (Nov. 17–19, 1976).

15. Kinman, R., Rickabaugh, J., Elia, V., McGinnis, K., Cody, T., Clark, S., and Christian, R. "Effect of Ozone on Hospital Wastewater Cytotoxicity." A Paper Presented at the IOI Meeting, Cincinnati, Ohio (Nov. 17–19, 1976).

16. Spanggord, R. J. and McClurg, B. J. "Ozonation Methods and Ozone Chemistry for Selected Organic Compounds in Water." A Paper Presented at the IOI Meeting in Cincinnati, Ohio (Nov. 17–19, 1976).

17. Simmon, V. F. and Eckford, S. L. "Methods for Evaluating the Mutagenic Activity of Ozonated Chemicals." A Paper Presented at the IOI Meeting, Cincinnati, Ohio (Nov. 17–19, 1976).

18. Morris, J. C. "The Role of Ozone in Water Treatment." A Paper Presented at the Annual Conference AWWA, New Orleans, La. (June 24, 1976).

19. White, G. C. *Handbook of Chlorination.* Van Nostrand Reinhold Co., N.Y. (1972).

20. Evans, F. L., III "Ozone in Water & Wastewater Treatment." Ann Arbor Science, Ann Arbor, Mich. (1972).

21. Venosa, A. D. "Comparative Disinfection of Wastewater Effluent with Chlorine, Bromine Chloride, and Ozone," A Paper Presented at the Forum on Disinfection with Ozone, Chicago, Ill. (June 2–4, 1976).

22. Fetner, R. H. and Ingols, R. S. "A Comparison of the Bactericidal Activity of Ozone and Chlorine against *Esch. coli* at $1°C$." *J. Gen. Mircrobiol.* **15**:381 (1956).

23. Morris, J. C. "Aspects of the Quantitative Assessment of Germicidal Efficiency." In Chapter 1 *"Disinfection, Water and Wastewater."* J. D. Johnson (editor), Ann Arbor Science (1975).

24. Gomella, C. "Ozone Practices in France." *J. AWWA* **64**: 39 (June 1972).

25. Bailey, P. S. "Reactivity of Ozone with Various Organic Functional Groups Important to Water Purification." First Int. Symposium on Ozone for Water & Wastewater Treatment, Proc. IOI Waterbury, Conn. (1975).

26. Dellah, A. "Study of Ozone Reactions Involved in Water Treatment and the Present Chlorination Controversy." Proc. 2nd Int'l. Symposium on Ozone Technology, Montreal, Canada (May 11–14, 1975).

27. Hoigne, J. and Bader, H. "Identification and Kinetic Properties of the Oxidizing Decomposition Products of Ozone in Water and its Impact on Water Purification." Proc. 2nd. Int'l. Symposium on Ozone Technology, Montreal (May 11–14, 1975).

28. Ghan, H. B., Chen, C. L. and Miele, R. P. "The Significance of Water Quality on Wastewater Disinfection with Ozone." A Paper Presented at the Forum on Disinfection with Ozone, Chicago, Ill. (June 2–4, 1976).

29. Boucher, P. L., Lowndes, M. R., Truesdale, G. A., Taylor, E. W., Burman, N. P., and Poynter, S. B. "Use of Ozone in the Reclamation of Water from Sewage Effluent." A Joint Paper Presented at a Meeting of the Inst. of Pub. Hlth. Engrs., Caxton Hall, Westminster, London Eng. (Dec. 11, 1967).

30. Ghan, H. B., Chen, C. L., Miele, R. P., and Kugelman, I. J. "Wastewater Disinfection with Ozone." A Paper Presented at the CWPCA Annual Conference, Los Angeles, Calif. (April 1975).

31. Nebel, C., Gottschling, R. D., Hutchison, R. L., McBride, T. J., Taylor, D. M., Pavoni, J. L., Tittlebaum, M. E., Spencer, H. E., and Fleischman, M. "Ozone Disinfection of Industrial-Municipal Secondary Effluents." *J. WPCF* **45**: 2493 (Dec. 1973).

32. Ward, R. W., Giffin, R. D., DeGraeve, G. M., Stone, R. A. "Disinfection Efficiency and Residual Toxicity of Several Wastewater Disinfectants." EPA Report, Grant No. S-802292 Cincinnati, Ohio (1976).

33. Gomella, C. "Contribution to the Study of Treated Sewage Disinfection by Ozone." A Paper Presented at the Forum on Disinfection with Ozone, Chicago, Ill. (June 2–4, 1976).

34. Guirguis, W. A. Jain, J. S., Hanna, Y. A., and Srivastava, P. K. "Ozone Application for Disinfection in the Westerly Advanced Wastewater Treatment Facility." A Paper Presented at the Forum on Disinfection with Ozone, Chicago, Ill. (June 2–4, 1976).

35. Naimie, H. "Ozone, an Alternative to Chlorine Disinfection and Comparative Cost Estimation." A Paper Presented at the Forum on Disinfection with Ozone, Chicago, Ill. (June 2–4, 1976).

36. Selna, M. W., Miele, R. P., and Baird, R. B. "Disinfection for Water Reuse." A Paper Presented at the Disinfection Seminar at the Annual AWWA Conf. Anaheim, Calif. (May 8, 1977).

37. Trussell, R., Nowak, T., Ismail, F., Jopling, W., and Cooper, R. "Ozone as a Pretreatment for Coagulation, Filtration and Disinfection." A Paper Presented at the CWPCA Annual Conf. Los Angeles, Calif. (April 1975).

38. Trussell, R., J. M. Montgomery Engrs., Pasadena, Calif. Private communication. (1977).

39. Ogden, M. "Ozonation Today." *Ind. Water Eng.* 7: 36 (June 1970).

40. Rosen, H. M. "Use of Ozone and Oxygen in Advanced Wastewater Treatment." *J. WPCF* 45: 521 (Dec. 1973).

41. Rosen, H. M. "Wastewater Ozonation. A Process Whose Time Has Come." Civ. Engineering–ASCE, p. 65 (March 1976).

42. Dasibi Model 1003-AH "Ambient Air Quality Ozone Photometer." Glendale, Calif. (1975).

43. "Dissolved Ozone Analyzer." Specification 17L 2000, Fischer and Porter Co., Warminster, Pa. (1973).

44. Masschelein, W., Fransolet, G., and Genot, J. "Techniques for Dispersing and Dissolving Ozone in Water Parts I and II, *Water Sewage Works* 122: 57 (Dec. 1975) and 123: 34 (Jan., 1976).

45. Rosen, H. M., Union Carbide Co., Tonawonda, N.Y., Private communication (Nov. 1976).

46. Hoag, L. N., and Salo, J. E. "Ozonation Feasibility Study." Sacramento Regional Wastewater Management Program, Sacramento, Calif. (June 1975).

47. Matson, J. V., and Coneway, C. R. "Economics of Disinfection." A Paper Presented at the IOI Forum on Ozone Disinfection, Chicago, Ill. (June 2–4, 1976).

48. Calmer, J. and Adams, R. M. "Evaluation of Disinfection by Ozone." In-House Report, Kennedy Engineers, San Francisco, Calif. (1977).

49. "Feasibility Study of Ozone Disinfection of Wastewater Effluent for Westerly Advanced Wastewater Treatment Facility, Cleveland Sewer District." Engineering Science Ltd., Cleveland, Ohio (March 1974).

50. "Disinfection of Wastewater; EPA Task Force Report No. 430/9-75-012." Cincinnati, Ohio (March 1976).

51. Consoer, Townsend and Assoc. Cons. Engrs. Chicago, Ill., East Bay Mun. Util. Report.; "Disinfection and Chlorine Residual Removal." (1974).

52. Rand, M. C., Greenberg, A. E., and Taras, M. J. "Standard Methods for the Examination of Water and Wastewater." 14th Ed. Wash., D.C. (1976).

53. Kinman, R. N. "Analysis of Ozone: Fundamental Principles." First Int. Symposium on Ozone for Wastewater Treatment Wash., D.C., p. 56 (Dec. 1973).

54. Dailey, L. and Morrow, J. J. "On Stream Analysis of Ozone Residual." A Paper Presented at the First IOI Conf., Wash., D.C. (1970).

55. Layton, R. F. and Kinman, R. N. "A New Analytical Method for Ozone." National Speciality Conference on Disinfection, Univ. of Mass. Amherst, Mass. (July 8–10, 1970).

56. Masschelein, W. J. and Fransolet, G. "Spectrophotometric Determination of Residual Ozone in Water with Acid Chrome Violet K." A Paper Presented for Publication in the *J. AWWA* (Dec. 1976).

57. Palin, A. T. "Analytical Control of Water Disinfection with Special Reference to Differential DPD Methods for Chlorine, Chlorine Dioxide, Bromine, Iodine and Ozone." *J. Inst. Water Eng.* 28: 139 (1974).

58. Palin, A. T. Private communication. (July, 1977).

59. Singer, P. C. and Johnson, J. D. "Reactions of Ozone in Aqueous Systems: Water Treatment and Analytical Implications." A Paper Presented at the IOI Meeting, Cincinnati, Ohio (Nov. 17–19, 1976).

60. Johnson, J. D. and Dunn, J. F. "Ozone Amperometric Membrane Electrode." Publication 415, Dept. Env. Sciences & Eng., Sch. of Pub. Hlth., Univ. of North Carolina, Chapel Hill, N.C. (1976).

61. Hayes, T. J. Private communication. Fischer and Porter Co., Warminster, Pa. (July 22, 1977).

62. Johnson, J. D., Edwards, J. W., and Keeslar, F. "Real Free Chlorine Residual Probe: HOCl Amperometric Membrane Electrode." A Paper Presented at the AWWA Annual Conf. Minneapolis, Minn. (June 1975).

63. Johnson, J. D. Private communication. IOI Meeting (Nov. 18, 1976).

64. Singer, P. C. and Zilli, W. B. "Ozonation of Ammonia in Wastewater." *Water Research* 9: 127 (1975).

Book summary

The best and most widely used index of pollution considered detrimental to public health is the most probable number (MPN) of coliform organisms present in a 100 ml sample. Health authorities believe that when the factors of dilution of sewage discharges are considered they can be reasonably certain that there will not be any outbreak of disease if the MPN coliform concentration is maintained below an arbitrary level. These arbitrary levels are chosen from long experience with drinking water standards. These levels will vary from 1000/100 ml in open surf waters down to 2.2/100 ml in waters for reuse or those discharging into ephemeral streams.

The coliform concentration level concept of pollution carries with it not only the need for disinfection as a unit process but also an optimum degree of treatment. This is why it is virtually impossible to rely solely on the disinfection process to treat stormwater overflows. Similarly, primary effluents are often impractical to disinfect particularly if the coliform MPN requirement is less than 1000/100 ml.

Well oxidized secondary effluents are amenable to the disinfection unit process when the coliform MPN requirements are between 23.2 and 240/100 ml. However, some form of tertiary treatment is required to achieve the drinking-water standard of 2.2/100 ml.

The most popular and therefore the most widely used disinfection process is the chlorination-dechlorination method. It is the least expensive of all the alternatives; there is a wide variety of equipment available from reliable and reputable manufacturers; the equipment for dechlorination is the same as for chlorination; it is reliable and flexible under widely variable conditions because it responds quickly and accurately to the various control systems; and finally the operation and maintenance of the chlorination-dechlorination systems is better understood by designers, operators, and health agencies, than any other method.

In the 1970's, this method has undergone intensive field evaluation which has revealed the elements of a system necessary to provide optimum and predictable results. These include: a good metering and control system; rapid mixing at the point of application; adequate contact time in a chamber approaching plug flow conditions; and competent operating personnel.

The most widely used dechlorinating agent is sulfur dioxide. Activated carbon is effective, but it is a considerably more expensive dechlorination agent. When used for dechlorination it is usually incorporated in the overall wastewater-treatment scheme for other purposes.

There still exist some differences in opinion about the techniques to be used to achieve maximum germicidal effeciency of chlorine. Owing to the presence of substantial quantities of ammonia nitrogen in wastewater, the predominating chlorine species for disinfection is monochloramine followed by the less germicidal organic chloramines formed by the presence of organic nitrogen. It was once believed that to achieve virus destruction a free chlorine residual was required. If that were true then nitrification (an expensive process) would be required. It was also believed that a coliform MPN of 2.2/100 ml could not be consistently met without a free chlorine residual. Both of these beliefs have been recently (1977) *disproved.* It has been found that if properly designed; a chlorination system producing all combined chlorine residuals cannot only meet a 2.2/100 ml coliform MPN easily, but that virus destruction can also be accomplished. These particular studies also revealed a surprise that the virucidal efficiency of free chlorine residuals compared to combined residuals was only slightly greater. Other investigations have shown that the presence of organic nitrogen in sewage has a significant effect on the germicidal efficiency of chlorine residuals.

The discovery of objectionable organo-chlorine compounds in the nation's drinking water raised some serious doubts as to the long-range desirability of chlorinated sewage effluents. Of primary concern was the formation of the trihalomethanes, most notably chloroform, a known carcinogen which is usually the result of the practice of free residual chlorination involving waters of low quality. As it turns out the practice of chlorination which forms chloramines (as in wastewaters) *does not promote the formation of these feared trihalomethanes.*

Safety is the watchword around any chlorine liquid-gas installation. Designers are warned to observe certain precautions and operators are advised how to prevent accidents and what to do in the event of a bad leak.

The practice of hypochlorination in those instances normally using the liquid-gas equipment has found a certain degree of popularity in congested metropolitan areas where the prime consideration is the potential hazard associated with liquid-gas storage. These systems are practical but expensive to operate owing to the comparatively high chemical cost for imported hypochlorite.

On-site generation of hypochlorite is also available for both the large and small installations alike. The designer can select either a brine system as the chlorine source or saline waters ranging from ocean estuaries to surf waters.

Sometimes on-site manufacture of hypochlorite (instead of imported hypochlorite) is attractive. A unique system for this approach is now available to the designer.

Chlorine dioxide is a potent and intriguing disinfectant. Its value in wastewater and water reuse is only speculative but should be investigated. Chlorine dioxide has properties similar to ozone. It does not react with water to form any compounds with water or its constituents (i.e., it does not react with ammonia nitrogen); it is easily destroyed by sunlight and it aerates out of water on the slightest provocation; its germicidal efficiency is not affected by pH; and it shows off its maximum efficiency when it destroys phenols; and, it must be generated on-site. It differs from ozone in that it provides a persisting residual.

The major difficulties encountered with the use of chlorine dioxide results from the inability to produce a pure solution. The techniques for differentiating chlorine dioxide residuals from other chlorine species is an arduous task; therefore, a system which can turn out a pure ClO_2 solution will be paramount to its acceptance as a competitive disinfectant. Only then will the current methods of chlorine residual determinations be equally applicable to chlorine dioxide. This will allow the same flexibility of control for a ClO_2 system as now exists with a chlorine system.

Chlorine dioxide as a disinfectant has a great advantage over alternative methods since it can be adapted to any existing chlorination system. Investigative work is directed now at how to get the greatest yield from current on-site methods of generation. These methods consist of the addition of a sodium chlorite solution to an aqueous chlorine solution at a controlled pH. The dilemma here is: what should the controlled pH be? Some say 1.0 others say 3.0 or 4.0.

Chlorine dioxide has some other favorable characterisics. Even though it does not react with ammonia nitrogen as does chlorine to form chloramines which in turn does not promote the formation of objectionable trihalomethanes, neither does chlorine dioxide promote these reactions. Similarly, it can be expected that since ClO_2 does not enter into reactions with either inorganic or organic nitrogen normally present in wastewater it will therefore not be subject to the inhibiting effects of these constituents which plague the chlorination process.

Other halogens such as bromine, bromine chloride, and iodine have been considered as alternative disinfectants. Iodine is impractical for wastewater treatment for a variety of reasons, one is its cost; another is its limited availability. The only possible practical use of iodine as a disinfectant is for temporary or emergency treatment of small remote water supplies.

Bromine compounds have some validity as wastewater disinfectants for these reasons: 1) bromine, whether it is applied as an aqueous solution of bromine or bromine chloride, it hydrolyzes to form hypobromous acid, this forms bromamines in wastewaters and the germicidal efficiency of this compound exceeds greatly similar chloramine compounds and is recognized as nearly equal in potential activity as free chlorine; 2) bromine residuals are so active their decay is rapid. This phenomenon could conceivably eliminate the necessity of the "dechlorination" step; and 3) an on-site generation process utilizing existing chlorination facilities could make the use of bromine in sequence with chlorination attractive in some wastewater treatment situations. This process is being used to limited degree in France.

The major disadvantages of the liquid-gas bromine and bromine chloride systems have to do with the difficulty of metering and control. An entirely new species of equipment has had to be developed for the metering of bromine. Moreover, this achievement was not possible until the development of commercially available bromine chloride. The vapor pressure of bromine chloride is sufficient (35 psi at $20°C$) to be handled as a liquid-gas similarly to that of sulfur dioxide, whereas molecular bromine which is a liquid at room temperature is practically impossible to handle either as a vapor, liquid, or solution. Bromine is highly reactive and therefore the residuals die-away within minutes. This makes dosage by residual control difficult. More experience is needed in the application of bromine.

Ultraviolet radiation is highly regarded as a disinfection process for the treatment of polished waters, particularly in situations when chlorine residuals cannot be tolerated. It has been demonstrated that the use of UV radiation as a wastewater treatment process will require at least a tertiary effluent, and, must be backed up by chlorination. There are many engineering problems to be solved in order to achieve an optimum UV design. The reliability of the UV process is low, it is very costly, and the ability to control it is marginal. There is also the potential danger of mutagenesis of surviving viruses caused by the radiation. This process suffers from the inherent problem of no residual disinfectant.

Ozone is the most powerful oxidant among all of the disinfecting compounds used in water treating processes. It is an excellent virucide, destroys phenols, removes color and reduces COD. It is not as good a bactericide as it is a virucide particularly in wastewaters where the ozone demand is high.

Ozone has been used for at least 50 yr in Western Europe as a special polishing agent for potable water. It is highly reactive with constituents in wastewater, so there is little chance to maintain a residual. Therefore, dosage control must be feed-forward based on a laboratory determination of ozone demand.

Ozone treatment of wastewater effluents as an alternative disinfection process is too expensive to be competitive with chlorination-dechlorination. When all of the attributes of ozone are evaluated it appears that in wastewater treatment, ozone is most cost-effective when used as a pretreatment process for tertiary effluents. The use of ozone in tandem with chlorine will probably emerge as the best way to disinfect and polish most waters reclaimed for reuse. This is the same approach now being used in Western Europe for low quality waters being processed for potable use.

Finally it should be recognized that the disinfection process efficiency is a direct reflection upon the quality of the effluent being treated. Therefore the better the quality of the effluent, the greater the efficiency of the disinfection process. This means that the disinfection process efficiency is dependent upon the other unit processes in the overall treatment scheme. *It follows then that a disinfection facility provides the operator with a continuous monitoring system of the effluent quality.* This is an important function of any disinfection process which is often overlooked.

Appendix

Dispersion Index "d"

An example of the computation of the Chemical Engineering Dispersion Index Number d where C = fluorometer units (dye concentration), and t = residence time (min.).[*]

t	C	$t \times C$	$t^2 \times C$
40	0	—	—
45	10	450	20,250
50	49	2450	122,500
55	78	4290	235,950
60	72	4320	259,200
65	61	3965	257,725
70	50	3500	245,000
75	37	2775	208,125
80	21	1680	134,400
85	15	1275	108,375
90	9	810	72,900
95	5	475	45,125
100	4	400	40,000
105	2	210	22,050
Sums	413	26,600	1,771,600

[*]*Courtesy* Endel Sepp California State Dept. of Health.

$$T = V/Q = 72.3 \text{ min.}$$

$$t_g^* = \frac{\Sigma tC}{\Sigma C} = \frac{26{,}600}{413} = 64.4 \text{ min.}$$

$$t_g^2 = 4148.2$$

$$\sigma_t^2 = \frac{\Sigma t^2 C}{\Sigma C} - t_g^2 = \frac{1{,}771{,}600}{413} - 4148.2$$

$$= 4289.6 - 4148.2 = 141.4$$

$$\sigma^2 = \sigma_t^2 / t_g^2 = \frac{141.4}{4148.2} = 0.034$$

$$d = \sigma^2/2 = \frac{0.034}{2} = 0.017$$

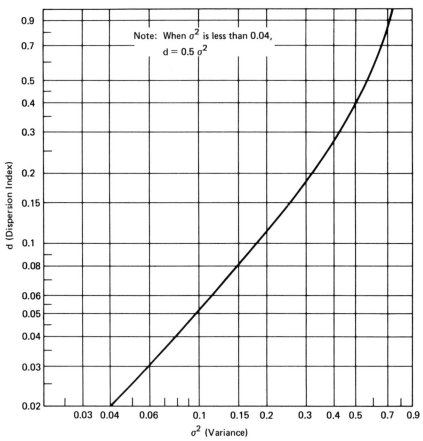

A-1 σ^2 (variance) versus "d" the chemical engineering index.

*t_g = ave. contact time = time to centroid of curve.

Calculator Solution

The following is the keystroke sequence for a HP 21 calculator to determine the value of d for any given value of σ^2 based on the formula:

$$\sigma^2 = 2d - 2d^2(1 - e^{-1/d})$$

First assume $d = \frac{1}{2}\sigma^2$, then by trial and error find d by selecting values of d on either side of $\frac{1}{2}\sigma^2$, as follows:

$$DSP, 3; d = \text{approximately } 0.5\sigma^2$$

ENTER	ENTER	ENTER
$1/x$	CHS	e^x
1	$x \leftrightarrows y$	$-$
$x \rightleftarrows y$	ENTER	2
///////	y^x	2
$\times$	$\times$	$x \rightleftarrows y$
2	$\times$	$x \rightleftarrows y$
$-$		

Read σ^2 value on the register

Repeat this process until the value of d is within 0.005 for a given value of σ^2.

Morrill Index

An example of the Morrill Index computation, where

$$t = \text{residence time (min.)}$$
$$C = \text{fluorometer units (dye concentration)}$$

t	C	Cumulative C	$\dfrac{C}{\Sigma C}$ (percent)
0	0	—	—
20	10	10	6
25	15	25	14
30	25	50	28
35	30	80	44
40	60	140	78
45	30	170	94
50	10	180[a]	100

[a] $\Sigma C = 180$

The values of $C/\Sigma C$ are plotted against time on the probability scale.

The *Morrill Index* is the ratio of the time required for the passage of 90 percent of the dye to the time of passage of 10 percent of the dye.

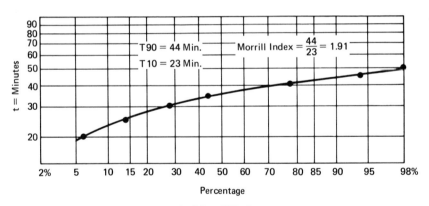

A-2 Morrill Index.

CONVERSION FACTORS

U.S. (English System)

1 GPM	= 1440 GPD
1 GPM	= 0.0022 cfs
1 gal.	= 0.832 Imperial gal.
1 gal.	= 0.1337 ft^3
1 ft^3	= 7.48 gal.
1 mgd	= 1.547 cfs
1 mgd	= 694.44 GPM
1 cfs	= 0.646 mgd
1 cfs	= 448.8 GPM
1 day	= 1440 min
1 day	= 86,400 sec
1 hp	= 550 ft-lb per sec
1 psi	= 2.04 in. Hg at 60°F
1 psi	= 2.31 ft H$_2$O

U.S. to Metric

1 in.	= 2.54 cm
1 ft	= 0.3048 meters
1 in.^{3}	= 16.387 cm^3
1 ft^3	= 0.0283 cu meters
1 oz	= 0.0295 liters
1 gal.	= 3.8754 liters
1 gal.	= 0.003785 cu meters
1 lb (mass)	= 0.45359 kilograms
1 oz	= 28.349 grams
1 watt hour	= 3600 joules
1 hp	= 746 W
1 lb (force)	= 4.448 Newtons
1 GPM	= 0.63 liters per sec
1 cfs	= 2446.6 cu meters per day
1 cfs	= 0.02832 cu meters per sec
1 mgd	= 3785.4 cu meters per day
1 psi	= 6.893 Kilo-Pascals
1 ppm (by wt.) $\times$ S.G.	= milligrams per liter

Metric to U.S. (English System)

To convert from metric to U.S. take the reciprocal of the metric values.

For Example:
To convert from cubic meters to gal. multiply cubic meters by 1/0.003785.

$$\therefore \ 1 \text{ cu meter} \times \frac{1}{0.003785} = 264.2 \text{ gal.}$$

Temperature
Fahrenheit = $1.8°C + 32$
Centigrade = $\frac{5}{9} \, (°F + 32)$
Kelvin = $°C + 273.15$
Kelvin = $\frac{5}{9} \, (°F + 459.67)$

Other Calculator Solutions

The following is the keystroke sequence for a HP 21 calculator for two often used equations in this text.

$$\Delta P \text{ (in. Hg)} = \frac{L \times 5.83 \times f \times W^2 \times 2.04}{10^9 \times \rho \times d^5} \tag{3-1}*$$

Solving for ΔP:

and

$$y/y_0 = [1 + 0.23 \ ct]^{-3} \tag{2-11}†$$

Solving for ct:

*Equation 3-1 is shown on page 110 this text.
†Equation 2-11 is in this text.

Chlorine Gas Density
(for use with Eq. 3-1 p 111)

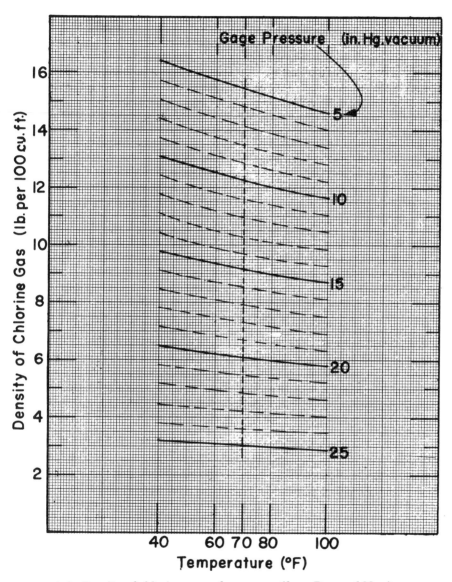

A-3   Density of chlorine gas under vacuum (from Ross and Mass).

Pipe Specifications

Nominal Diameter	Inside Diameter	A (in.^{2})	Volume (ft^3/ft) Area (ft^2)
Schedule 80 Steel for Gas and Liquid Chlorine Supply Lines			
¾	0.742	0.4324	0.00300
1	0.957	0.7193	0.00499
1¼	1.278	1.2813	0.00891
1½	1.500	1.767	0.01225
2	1.939	2.953	0.02050
Schedule 80 PVC for Chlorine Solution Piping and Injector Vacuum Lines			
½	0.546	0.2341	0.00163
¾	0.742	0.4324	0.00300
1	0.957	0.7193	0.00499
1¼	1.278	1.2813	0.00891
1½	1.500	1.767	0.01225
2	1.939	2.953	0.02050
2½	2.323	4.298	0.02942
3	2.900	6.605	0.04587
4	3.826	11.50	0.07986
6	5.761	26.07	0.1810
8	7.625	45.66	0.3171

Cubic Equation

The following cubic equation was used to generate the data shown in Tables 2-1 and 2-2 (Chapter 2). These data illustrate the molecular chlorine, hypochlorous acid, and hypochlorite ion equilibria under various conditions of concentration, temperature, and pH.

$$\left\{ -\left(\frac{1}{K_D}\right)^2 (H^+)^3 - K_A \left(\frac{1}{K_D}\right)^2 (H^+)^2 + 4\left(\frac{1}{K_D}\right) K_3 ([H^+] + K_A)^2 \right\} [HOCL]^3$$

$$\left\{ -4\left(\frac{1}{K_D}\right) K_3 T [H^+] ([H^+] + K_A) \right\} [HOCL]^2$$

$$\left\{ +\left(\frac{1}{K_D}\right) T [H^+]^2 + K_A + [H^+] + \left(\frac{1}{K_D}\right) K_3 T^2 [H^+]^2 \right\} [HOCL] - T[H^+] = 0$$

Where

T = total halogen added
K_D = hydrolysis of Cl_2 at 25°C = 3.944×10^{-4}
K_A = acid dissociation of HOCL = 2.904×10^{-8}
K_3 = formation of trichloride ion = 0.191

Index